Palgrave Studies in Educational Futures

Series Editor
jan jagodzinski
Department of Secondary Education
University of Alberta
Edmonton, AB, Canada

The series Educational Futures would be a call on all aspects of education, not only specific subject specialists, but policy makers, religious education leaders, curriculum theorists, and those involved in shaping the educational imagination through its foundations and both psychoanalytical and psychological investments with youth to address this extraordinary precarity and anxiety that is continually rising as things do not get better but worsen. A global de-territorialization is taking place, and new voices and visions need to be seen and heard. The series would address the following questions and concerns. The three key signifiers of the book series title address this state of risk and emergency:

1. **The Anthropocene**: The 'human world,' the world-for-us is drifting toward a global situation where human extinction is not out of the question due to economic industrialization and overdevelopment, as well as the exponential growth of global population. How to we address this ecologically and educationally to still make a difference?
2. **Ecology**: What might be ways of re-thinking our relationships with the non-human forms of existence and in-human forms of artificial intelligence that have emerged? Are there possibilities to rework the ecological imagination educationally from its over-romanticized view of Nature, as many have argued: Nature and culture are no longer tenable separate signifiers. Can teachers and professors address the ideas that surround differentiated subjectivity where agency is no long attributed to the 'human' alone?
3. **Aesthetic Imaginaries**: What are the creative responses that can fabulate aesthetic imaginaries that are viable in specific contexts where the emergent ideas, which are able to gather heterogeneous elements together to present projects that address the two former descriptors: the Anthropocene and the every changing modulating ecologies. Can educators drawn on these aesthetic imaginaries to offer exploratory hope for what is a changing globe that is in constant crisis?

The series Educational Futures: Anthropocene, Ecology, and Aesthetic Imaginaries attempts to secure manuscripts that are aware of the precarity that reverberates throughout all life, and attempts to explore and experiment to develop an educational imagination which, at the very least, makes conscious what is a dire situation.

More information about this series at
https://link.springer.com/bookseries/15418

Michael Paulsen • jan jagodzinski
Shé M. Hawke
Editors

Pedagogy in the Anthropocene

Re-Wilding Education for a New Earth

palgrave
macmillan

Editors
Michael Paulsen
Department of the Studies of Culture
University of Southern Denmark
Odense, Denmark

jan jagodzinski
University of Alberta
Edmonton, AB, Canada

Shé M. Hawke
Mediterranean Institute for
Environmental Studies
Science and Research Centre of Koper
Koper, Slovenia

Palgrave Studies in Educational Futures
ISBN 978-3-030-90979-6 ISBN 978-3-030-90980-2 (eBook)
https://doi.org/10.1007/978-3-030-90980-2

This Palgrave Macmillan imprint is published by the registered company Springer Nature Switzerland AG.
The registered company address is: Gewerbestrasse 11, 6330 Cham, Switzerland

To our children and grandchildren

Preface

The idea for this book arose in the spring of 2020. It began when one of us, Michael Paulsen, in December 2019 wanted to join the conference "Anthropocene—Reworking the Wound" in Katowice, Poland on June 17 to June 20, 2020, and offered to chair a panel session on "Pedagogies of the Anthropocene (s)". The conference organizers, Ania Malinowska and Karolina Lebek, then wrote back affirmatively, after which Michael began to look for researchers to join the panel. The interest was indeed great, and already on January 10, 2020, we registered four panel sessions with a total of 20 researchers from around the world. Everyone was very excited to meet in Katowice. However, then COVID-19 pandemic came—the physical version of the conference was canceled, although some panels continued virtually. However, one of the panelists, jan jagodzinski, suggested that we "instead" transmute the panels into a book on Pedagogy and the Anthropocene. There seemed to be enough interest to reach the group that was to appear at the conference sessions. 'I edit a series for Palgrave press called Educational Futures', jan told Michael. Michael agreed to co-edit and wrote to the panelists to garner interest for such a book project. Most responded with a yes! One panelist, Shé Hawke, had been considering editing a Special Journal Issue on the Anthropocene and Inter-disciplinarity, with Reingard Spannring, and offered to consolidate editing energy into one robust and diverse book. As Reingard was now involved in another book, our editing team of three got to work. We drafted a proposal for Palgrave, which led to the publishing contract for us, for which we are extremely grateful. A huge thank you goes to Milana Vernikova, Antony Sami and everyone at Palgrave who helped with the execution of the book. Many thanks also to both the original panelists and the authors of the 16 chapters of this book. It has been a fantastic journey, with valued commitment from so many good people full of energy and passion for the project. Many of us have yet to meet in person, but we are not strangers to one another's work, or our shared commitment to a 'new earth'.

Odense, Denmark — Michael Paulsen
Edmonton, AB, Canada — jan jagodzinski
Koper, Slovenia — Shé M. Hawke
August 2021

Praise for *Pedagogy in the Anthropocene*

"Much has been written about the Anthropocene but surprising little about its implications for education. This book tackles that fundamental issue head-on. The definitions and interpretations of the Anthropocene are vast, but they all point towards the same formidable challenge—we need to examine who we are and what relationship we should have with the rest of the planet. The next generations will feel the full force of the Anthropocene, so there is nothing more important than preparing them for the uncertain future of the human epoch."

—Will Steffen, Emeritus Professor, *The Australian National University, Canberra*

"This book, written by authors of passion and conviction, charts pedagogical pathways for an unknown future. The offerings range from those who believe that the future is dark yet hold a flickering torch of hope, others who believe that our hope for the future lies in our ability to re-member, re-new, re-gain our deep relationality to all species, to re-wild and re-pair the extraordinary damage that we have done to the beautiful planet we call home, and those whose call is to re-vitalize our spiritual consciousness and connection to the nonhuman and more-than-human world. Wherever we are on this continuum, this book points to critical pedagogies for a liveable future—for our children, for our grandchildren and for our more-than-human relations. It is a much-needed and magnificent book. I am honored to write its endorsement."

—Makere Stewart-Harawira, Professor, *Faculty of Education, University of Alberta, Canada*. Dr. Stewart-Harawira is of Māori and Scots descent, her primary iwi affiliation being Waitaha ki Te Waipounamu. She is the author of *The New Imperial Order: Indigenous Responses to Globalization.*

Contents

Notes on Contributors

Stefan Bengtsson holds an associate professorship of didactics at Uppsala University and in his research focusing on issues of sustainability in education. He is the author of several acclaimed titles focusing on renegotiating concepts of educational approaches to environmental, sustainability and climate change issues though education drawing on continental philosophy mixed with new materialist approaches. He has been national coordinator for the implementation of Education for Sustainable Development (ESD) (2015–2019) in Sweden and link convenor for the Environmental and Sustainability Education network of the European Education Research Association (EERA). Parallel to his academic work, he has been providing technical assistance to international efforts to address sustainability and gender issues in education, in particular in the East and South-East Asian context. As a national coordinator and as a researcher, he has authored and contributed to a number of reports and research projects focusing on ESD policy formulation and implementation.

Sean Blenkinsop is a professor in the faculty of education at Simon Fraser University, Vancouver, Canada. His research explores teacher education, school change, and the challenges of justice and the environmental crisis in a rapidly changing world. Blenkinsop has always been involved in creating and researching three innovative public elementary schools in British Columbia that are focused on being much more community, place, and nature-based in both pedagogy and curriculum. Important strands in this work include ideas related to nature as co-teacher, questions of equity, teacher as activist, cultural change, and eco-social justice. His most recent book was *Wild Pedagogies: Touchstones for Re-Negotiating Education and the Environment in the Anthropocene*. His next book, with Peter Lang, due in the summer of 2021 is called *Ecoportraiture: The art of research when nature matters*. It explores, complexifies, and offers examples of what research becomes if human researchers actively seek to engage with the more-than-human community as co-researchers.

Thomas Burø is a musician, cultural organizer and researcher based in Copenhagen. Burø has specialized in using mapping as a way to document the complexities of ecologies of culture. He is interested in promoting cultures of sustainability and is generally drawn to the idea of 'livable futures.' His latest work to be published is his PhD thesis: *Land of light. Assembling the Ecology of Culture in Odsherred 2000–2018*. https://www.uia.no/en/kk/profile/heleneil

Joe Gray lives on the North Atlantic island of Great Britain. As an amateur field naturalist, he has run courses for people of all ages on natural history and reconnecting with nature. Joe is also a co-founder of *The Ecological Citizen*, a knowledge advisor on ecological ethics for the United Nations' Harmony with Nature programme, and author of *Thirteen Paces by Four: Backyard Biophilia and the Emerging Earth Ethic* (2021). More information on Joe's work can be found on his website, https://deep-green.earth/.

Shé M. Hawke is an inter-disciplinary scholar and Head of the Mediterranean Institute for Environmental Studies at the Science and Research Centre of Koper, Koper, Slovenia (2019–2021). She is also an Honorary Associate in the Department of Gender and Cultural Studies, University of Sydney, Australia where she taught from 2005 to 2013. Water is her primary research area as it intersects with sociology, environmental pedagogy, environmental philosophy and heritage studies. Her contribution to this anthology is part of the joint project "Surviving the Anthropocene" together with Reingard Spannring (FWF, I 4342-G, Austria and ARRS J7-1824, Slovenia).

Marianne Presthus Heggen is a professor in the faculty of education and sports at Western Norway University of Applied Sciences, Bergen, Norway. With a background in ecology and climate related research, she teaches early childhood teacher education on topics in natural sciences, outdoor learning and environmental education. Her research foci are on education for sustainability and environmental education in early childhood, with particular interest on children's reflections and contributions within these. She is occupied with the project "Being and becoming eco-citizens" at KINDknow—Kindergarten Knowledge Centre for Systemic Research on Diversity and Sustainable Futures. She tries to hold her research close to the children.

Helene Illeris is Professor of Art Education at the University of Agder (UiA) in Norway. Her research interests include art education in schools and galleries, with a special focus on aesthetic learning processes, contemporary art forms, and ecological awareness. Illeris is a leader of the *Art and Social Relations* research group and a coordinator of the *Arts in Context* research platform, both at UiA. She has written more than 60 books, chapters. and articles in English, Danish, Swedish, and Italian. One of the latest is "Intimacy, solidarity, fragility. Everyday objects and ecological awareness in arts and crafts education", which can be retrieved from http://www.rixc.org/press/AS.Vol.18-OSLOFJORD-ECOLOGIES-download.pdf.

jan jagodzinski is Professor of Visual Art and Media Education, University of Alberta in Edmonton, Alberta, Canada. He is series editor for *Educational Futures* (Palgrave-Springer). He is the author of 18 books. His most recent titles include *The Deconstruction of the Oral Eye: Art and Its Education in an Era of Designer Capitalism* (Palgrave, 2010), *Arts Based Research: A Critique and Proposal* (2013). jagodzinski, j. ed. *The Precarious Future of Education* (Palgrave, 2017) and jagodzinski, j. ed. *What is Art Education? After Deleuze and Guattari* (Palgrave, 2017). jagodzinski, j. ed. *Interrogating the Anthropocene: Ecology, Art, Pedagogy, the Future in Question* (Springer-Palgrave, 2018), *Schizoanalytic Ventures at the End of the World: Film, Video, Art and Pedagogy* (Springer-Palgrave, 2019). *Pedagogical Explorations in a Posthuman Age: Essays on Designer Capitalism, Eco-Aestheticism, Visual and Popular Culture as West-East Meet* (Springer-Palgrave, 2020) and with Jessie Beier as editor, *Ahuman Pedagogy: A Manifesto from the Outside*, (Palgrave-Springer, forthcoming); and *Pedagogical Encounters and the Post-Future Anthropocene* (forthcoming).

Bob Jickling is Professor Emeritus at Lakehead University and has interests in environmental philosophy; environmental, experiential, and outdoor education; and philosophy of education. In his most recent book, *Wild Pedagogies: Touchstones for Re-Negotiating Education and the*

Environment in the Anthropocene he and others of the *Crex Crex Collective* attempt to find openings for radical re-visioning of education. As a long-time wilderness traveller, much of his inspiration is derived from the landscape of his home in Canada's Yukon.

Nick Kleese is a doctoral candidate in Literacy Education at the University of Minnesota, whose research focuses on representations of rurality and the environment in children's and young adult literature. Born on a farm in Southeast Iowa, he continues to share his love of talking about the weather with anyone within earshot.

Oleg Koefoed is Director of Growing Pathways and Co-founder of the Action University. Exploring possibilities of participation in urban planning and greening, based on the local action and shared reflection, he works locally in Copenhagen as well as on a translocal level in European and virtual places. He is the author of a series of publications related to participatory urban nature making, placemaking, and sustainable cities, learning to live in a time where lines of flight cross chaotic movements. His most recently published work is "Organising Spirit" (Burø & Koefoed, 2021).

Estella Carolye Kuchta is an English instructor at Langara College in Vancouver. Her SSHRC-funded ecocritical research of Canadian literary love stories resulted in the novel *Finding the Daydreamer* (2020). Her research focuses on relational ontologies, love of land, ecocriticism, and theories of ecologizing education. She is a member of the International Love Research Network and holds an MA in English Literature and BFA in Creative Writing from the University of British Columbia. She is working towards a PhD in Philosophy of Education at Simon Fraser University.

Ole Andreas Kvamme is Associate Professor in the Department of Teacher Education and School Research, University of Oslo, Norway. He is a member of the research groups COSER (Challenges of Sustainability in Educational Research) and HumStud (Humanities Studies in Pedagogy). His PhD Thesis from 2020 is entitled *Recontextualizing Environmental Ethical Values in a Globalized World: Studies in Moral Education*. In an interdisciplinary approach involving philosophy of education, curriculum studies, and global education policy studies, Kvamme explores how environmental ethical values formulated within the context of United Nations are recontextualized in various formal and non-formal educational settings.

Jonas Andreasen Lysgaard is associate professor in education and pedagogy at the Danish School of Education, Aarhus University, Denmark. His research relates to overlapping challenges of education, environmental and sustainability issues in practice and policy. He is the former link-convenor (Chair) for the Environmental and Sustainability Education network of the European Education Research Association (EERA). He has participated in European, Asian and North American research projects related to development and implementation of environmental and sustainability education. Lysgaard is board member of several key scientific journals in the field and is involved in both national and regional policy development in relation to linking environmental, sustainability and educational agendas. Together with Stefan Bengtsson and Martin Hauberg-Lund Laugesen, he has authored the 2019 book *Dark Pedagogy*.

Marcus Morse is Senior Lecturer and Program Convenor for Outdoor Environmental Education, in the School of Education, at La Trobe University, Australia. He grew up in Tasmania, where he spent extended time on the island's rivers, coastlines and mountains and developed a love for being in wild places. Morse has a passion for extended journeys and his research interests are in the areas of outdoor environmental education, community engagement projects, place-based education and wild pedagogies.

Lennart Nørreklit is Professor emeritus. Danish philosopher, PhD in philosophy 1971. Dr. Phil. in philosophy 1987. Lecturer in theory and methodology, Aalborg University 1973. Professor in Philosophy and Leadership Aalborg University 2008. Head of Center for Philosophy and Theory of Science at Aalborg University and initiating the philosophy studies. Active member of research networks on pragmatic constructivism and the EU network on structural change (SUPI).

Marek Oziewicz is Professor of Literacy Education at the University of Minnesota, Twin Cities and holds the Sidney and Marguerite Henry Chair in Children's and Young Adult Literature. He teaches courses on speculative fiction, multiculturalism, and how literature empowers young people to develop climate change literacy.

Michael Paulsen is associate professor and head of Intercultural Pedagogy Studies, at the University of Southern Denmark. He was an associate professor in applied philosophy at Aalborg University, the department of Learning & Philosophy from 2012 to 2016, where he also was chair of the

Applied Philosophy Study Board. His research focuses on the ontology and axiology of education. He is working on developing a new understanding of education situated in the Anthropocene. See https://portal.findresearcher.sdu.dk/en/persons/mpaulsen.

Alexander Popov is Assistant Professor in the Department of English and American Studies at Sofia University "St. Kliment Ohridski". His research interests range from utopian and science fiction studies, posthumanism, ecocriticism, linguistics, to artificial intelligence and cognitive science. He has recently published articles on climate fiction, digital reading, and on possible worlds semantics as applied to fictional narrative. He is currently working on an interdisciplinary research project focused on artificial intelligence in the context of the humanities and social sciences.

Reingard Spannring is lecturer and researcher at the Institute for Educational Sciences, University of Innsbruck, Austria. Her main areas of interest are critical animal studies, environmental education research, philosophy of education, and youth research. Her contribution to this anthology is part of the joint project "Surviving the Anthropocene" together with Shé Hawke (FWF, I 4342-G, Austria and ARRS J7-1824 Slovenia).

Lars Petter Storm Torjussen is Associate Professor at The Department of Education, University of Bergen. His research areas are sustainable education, philosophy of education, philosophy and philosophy of science.

Jason J. Wallin is a Professor in the Faculty of Education at the University of Alberta, Canada. He is the author of "A Deleuzian Approach to Curriculum: Essays on a Pedagogical Life" (Palgrave Macmillan), co-author of "Arts-Based Research: A Critique and Proposal" (with jan jagodzinski, Sense Publishers), and co-producer of the 2016 extreme music documentary entitled "Blekkmetal" (with David Hall, Vivek Venkatesh and Owen Chapman). Jason was raised by wolves in the hinterlands of British Columbia.

Kerrie Willis was raised a big-city Texan, who grew up to become a small-town Iowan. She's a high school teacher and instructional coach driven to help beginning and veteran teachers alike find powerful ways to learn and grow in their practice, and she's dedicated to doing whatever it takes—coffee carts, pineapple buckets, and handwritten notes—to create a culture of learning in rural Washington, Iowa.

List of Figures

List of Boxes

CHAPTER 1

A Critical Introduction

Michael Paulsen, jan jagodzinski, and Shé M. Hawke

We now live in the geological epoch called the Anthropocene (Crutzen & Stoermer, 2000; Steffen et al., 2011; Zalasiewicz et al., 2008; Steffen et al., 2016; Morton, 2016; Sørlin, 2017; Ellis, 2018). In this age, Anthropos, through human activities, technologies and alterations of the global environment have begun to affect the whole life-critical zone of the Earth more than ever before, and more than anything else (Lin, 2010; Latour, 2017). The consequences are many: "the great acceleration" of technology, industry, agriculture, and an over-use of natural resources (McNeill & Engelke, 2016), mass extinctions (see for example Chap. 2), global warming (Oreskes & Conway, 2011), collapse of eco-systems

M. Paulsen (✉)
Department of the Studies of Culture, University of Southern Denmark, Odense, Denmark
e-mail: mpaulsen@sdu.dk

j. jagodzinski
University of Alberta, Edmonton, AB, Canada

S. M. Hawke
Mediterranean Institute for Environmental Studies, Science and Research Centre of Koper, Koper, Slovenia

M. Paulsen et al. (eds.), *Pedagogy in the Anthropocene*, Palgrave Studies in Educational Futures,
https://doi.org/10.1007/978-3-030-90980-2_1

(Steffen et al., 2006), and the spread of pandemics and multi-resistant bacteria. All this promises an ever more impoverished earth if we continue along the prevailing trajectory. Innovative, integrated, achievable, and inclusive pedagogical intervention into climate change and Anthropocene damage, form the two-pronged yet intertwined focus of this collection.

Integrated Pedagogy

Let us begin with the pedagogical stream of this collection. The overarching question for this aspect of the book is how transformations in pedagogy and education can change the face of the earth for the better.

We have intentionally invited scholars from pedagogy and beyond, to develop potent intersections with other scientific inquiries, and creative industries. For a more enlightened—or re-enlightened—sustainability praxis in the educational domain, in both the school and with the public, we engage inclusively with cultural and research fields, and ethico-political-critical pedagogy, as a significant focus of this collection of essays. Innovation and integration alongside disruption and challenge, form equally important parts of the book's direction and intention.

Pedagogy, and by extension language, ontology, and epistemology, is predicated by our social, cultural, and geographic situatedness and conjunctions. Even in the western world, the variance in teaching practice and curriculum is vast. But what has been observed in recent decades—partly because of the increase in digital cultures—is that learning and teaching is less-often enacted outside the classroom than in previous decades (Spannring & Hawke, 2021), and less-often cross pollinated with other cultural systems of knowledge. A return to First Nations' invitation to dialogue about cross-cultural sustainability learning practices is now increasing. This has perhaps been provoked by the planetary perils and catastrophes brought on by the Anthropocene Epoch and its hyper-productive agricultural practices, deforestation, and the residual effects of colonization and imperialism (Stewart-Harawira, 2012). While comprehensive cross-cultural engagement with pedagogical practices is beyond the scope of this collection, several chapters reference, and engage deeply with such ontology and pedagogy (see for example Chaps. 3 and 10) from the northern hemisphere to Australia in the deep south. Such critical pedagogies and cartographies of place, add valued dimensions to sustainable pathways in which culture and the environment converge.

The Anthropocene

'The 'Anthropocene' … is the current working term to describe anthropogenic changes since the late Holocene Geological Epoch. It is the language used by scientific and research organizations such as UNEP, and the European Commission, and refers to the current geological age, beginning with the Industrial Revolution' (Spannring & Hawke, 2021, 3). Yet not all fields agree on the term, its date of birth, or its applied application across the sciences. For purposes of delineation, we outline below some of the challenges and contestations about the term Anthropocene, and its understanding in both scholarly and everyday life, as they relate to the provocations presented in this book.

By manifold anthropogenic effects—including uncontrollable domino-effects, negative feedback loops, and further non-intended consequences (Tønder, 2020)—humankind has ended the geological epoch called "the Holocene", an unusual climatically stable period beginning about 11,500 years ago, after the last Ice age (Crutzen & Schwägerl, 2011). In the Holocene epoch (the forerunner of the Anthropocene), all human civilization—including cities, agriculture, democracy, science, and capitalism—developed, under climatically favorable conditions, which are now undermined by the very same development and its excessive consequences (see for example Chap. 11). DNA nanotechnology and other new inventions have made it possible to change and reshape the conditions of earthly life, to an incomprehensible and unpredictable extent (see for example Chap. 8). Thus, in all stratums of the life-critical zone, from nano to macro global planetary level, human activities now reshape the earthly conditions of life (Zalasiewicz et al., 2016, 2017). This has also led to the collapse of the old western distinction between cultivated land (the world of nomos and culture) and uncultivated wild areas (the world of physis and nature)—a distinction that has now deconstructed itself (Marris, 2011; Oppermann & Iovino, 2017; Emmett & Nye, 2017; Latour, 2018; Paulsen, 2021a). We now live in a world where the effects of humans can be found everywhere—result: the end of (untouched) nature (e.g., Mckibben, 2006; Morton, 2009) as we have known it.

The key ethical-political-pedagogical question of today is not so much how to change the world, but how to change it less, or in ways that are less harmful to it (see Purdy, 2015); to partner with nature rather than work against it (Spannring & Hawke, 2021)—towards a new earth. In a comparatively short time in Earth history, the zone slightly above and below the surface of the earth (Latour, 2017) has been reshaped to meet human needs, pointing towards the making of an artificial and human manipulated world; a 'man's world' one could say—the world of Anthropos, where a total ecological breakdown, threatens to remove the foundation of this world. Thus, we live in a time of transition and risk, but perhaps also in a world that has moved off the edge of history (Giddens). We are no longer living in a late modern society, in distinction to modern and premodern societies. This self-understanding has become obsolete, because it does not take account of what it set in the background, our interrelatedness and entanglement (for good and for worse) with our environment, i.e., the rest of life's critical zones, the multi-species and more-than-human world. While the concept of late modernity places us only in human history, seeing 'nature' as a mere background as a 'scene' for human development, the story of the Anthropocene inscribes our species in a larger earthly history, which opens a whole new view of ourselves, our past, present, and future, here on Earth (Paulsen, 2021b; Spannring & Hawke, 2021). This story tells us that we have become re-shapers of the earth, but with many overwhelmingly negative consequences for the life of the planet, including ourselves.

The Contested Anthropocene

Yet, the Anthropocene is a contested concept (see Paulsen, 2021c) and rejected outright by some (e.g., Malm & Hornborg, 2014), while others argue that it is inadequate and needs to be complemented by other stories (Haraway, 2016). Thus, two of the main objections to the Anthropocene story are: (1) that it gives humans/Man too big a role in earth history, by focusing on Anthropos (Haraway, 2016) and (2) that it blames humanity as such, while some humans and some human activities are much more guilty/damaging than others. For example, the industrialized world and owners of fossil-fuel driven activities (Moore, 2015) (see also Chap. 2). Our retort to the first objection is that it is precisely human industry and activities that have created the negative effects (Morton, 2017). Against the second objection we would extend the contestation to include more-than-humans and marginalized human communities, globally (Spannring & Hawke, 2021). This creates a double entanglement in its specific asymmetrical and contemporary forms, e.g., capitalistic, consumer-oriented, with inequality between both 'humans and humans' and 'humans and more-than-humans', that makes up the Anthropocene situation we now live in (Braidotti, 2013; Paulsen, 2021b).

The important thing here, however, is not whether some are right, and others are wrong, but rather that different understandings of the Anthropocene and different confirmations and rejections of the story—and so *eo ipso* of the time and situation we now live in—have political, ethical, practical, and pedagogical consequences. Or perhaps better stated: How one responds to the Anthropocene is framed by how one understands the world (and crisis) we live in. Depending on how we understand our current age, past and future, different solutions and paths might come to the fore, as many of the chapters in this collection demonstrate. Some chapters argue for the need of a radical change as a response to the Anthropocene (see for example Chap. 8), while others suggest more moderate rethinking and adjustments of our Holocene institutions (see for example Chap. 16).

There is also the further dimension of shifting away from *being* human-centred to *becoming* more life-centred (Chakrabarty, 2015). Numerous other options are available, such as: the hope we can solve the planetary crisis by advancing green technology, improving the environment-making state and regain our control of the spaceship/spaceshop Earth (Parenti, 2015) often found in political rhetoric, which others see as either human hubris or as an outright dystopic image of the world to come, a totally artificial and man-made *Technopolis* (to use a term coined by Postman, 2011). Some of the critics of such control-regimes want to roll back culture and thus "go back to nature", before we civilized everything (c.f. the re-wild movement); others (still!) deny that climate changes and other planetary calamities are serious problems, thus wanting to keep on with business as usual, in schools and elsewhere (see for example Chap. 15). There is also the backdrop of fossil fuel capitalism being the main driver, which creates the call for a needed solution that is a social revolution (Malm, 2014). Moreover, some of the chapters in this collection are optimistic and project utopias and active hope (see for example Chap. 12) or think that the 'anthropocene recognition' might open the space for cultivating and enjoying a wider range of life-possibilities and ways of life than in the late Holocene (see for example Chap. 11); while others are pessimistic and forecast dystopia and catastrophe, which might already have happened: we are 'in' it (see for example Chap. 6 and Morton, 2016). Such differences are reflected in the pedagogical responses to the Anthropocene, unfolding within a spectrum from light and optimistic pedagogies to dark and horror-acknowledging pedagogies (Lysgaard et al., 2019).

The Anthropocene term, despite coming out of geological science, has also spread to other disciplines and areas, including humanities and social science. It has also become a popular concept and attractor outside of academia, for example in creative industry such as art exhibitions (Davis & Turpin, 2015) and in pop music (for example the 2020 album *Miss Anthropocene* by Grimes, and Bjork's 2011 album and app *Biophilia* that turns the focus the other way, to name a few). This means that we cannot take for granted what we understand by the term 'Anthropocene'. One might even conclude that the term is so ambiguous that it is not useful. Or, that we should settle for the geological definition of the Anthropocene only. Yet we take the ambiguity of the concept as an advantage because of its great rhizomatic and transversal capacity. It has become a meeting point of so many different understandings, enterprises, controversies, discussions, and conflicts, both within and across different sciences, and attracted huge attention outside of academia as well (Latour, 2017). The aim of this critical introduction to pedagogy in the Anthropocene is therefore to explicate its immanent complexity. Further, we indicate that different pedagogical responses to the age we are now living in depend on what parts of the Anthropocene story are narrated, accounted for, and legitimated based on different worldviews and outlooks, theoretical and practical inclinations, and preferences. We fulfil this task in the following, by emphasizing four controversies of the Anthropocene.

Controversy # 1: When Did It All Start?

Although there have been forerunners in the late Holocene, it is only now that we really see and acknowledge that we are in an earth-historical border situation, with several phenomena that signify a transition from the Holocene that is ebbing out, to the Anthropocene, which is about to open. It is not "only" about the climate crisis and global warming, but about the entire life zone transformed throughout the Holocene, the great acceleration, the sixth mass extinction, the biodiversity crisis and much more, including its full acknowledgement. Only today, when life is more threatened than ever in the history of mankind, do we begin—more comprehensively—to understand ourselves as also living in an earth zone, and not just as citizens of society. However, a question now arises: When exactly does the Anthropocene era begin?

In one sense, the Anthropocene (understood as man mega-influencing overall life on earth) begins with the agrarian revolution that unfolds at the beginning of the Holocene era. However, the Anthropocene is reinforced with the Industrial Revolution (from 1750) and further with nuclear explosions and the great acceleration (of, among other things, resource consumption from 1945). In a slightly different sense (namely understood as recognition of ourselves as earthly and in the process of destroying all earthly life), one might highlight events such as lunar landings (1969), where we see the earth from the outside and the very naming of the Anthropocene era (2000). For us, however, it is not important to settle upon one exact start date. You can also not set an exact date on the modern or late modern.

More importantly, the belief that we are *now* living in the Anthropocene, is more unshakable than ever before. Most importantly, should we be aware of how different start dates frame different historical understandings, and therefore also understand what has given rise to our current situation? If, for instance, one thinks that the Anthropocene started with the agrarian revolution, the inventing of agrarian religions and what Morton (2018) calls "agricultural civilization" and "agrilogistics", it most likely implies that the adequate political, practical, ethical, and pedagogical response today will be to cancel and/or rethink monocultural agriculture, including all the societal institutions build around agrilogistical values and worldviews. This, for instance, could also imply a critique of the mainstream western (idea of the) school (including academia) as being understood itself mainly as a kind of agricultural treadmill, treating students as seeds to be cultivated and sorted in artificially made "greenhouses"—schools—as auotomated instruments, separated from everything else. Here, an understanding of education as taming of what is otherwise 'wild' (see for example Chap. 3) becomes apparent.

If, on the other hand, the starting date of the Anthropocene is identified with the industrial revolution and the rise of modern capitalism, from 1750 an onwards, it is more likely that one responds by criticizing capitalistic structures and the tight link between a capitalistic society, the school, and the damaging effects on the environment. This might also include a critique of the industrial view of the school, framing schools as factory halls and students as products produced by the society to meet societal needs, especially in relation to the labor market; transforming human beings into a workforce, human capital, a mere manipulable resource (see for example Chaps. 7 and 8) on a production line of life. Finally, if one sets the Anthropocene to start around 1945 or early 1950s, with nuclear proliferation and testing and many new technological inventions such as the modern computer (1936), the lunar landings (1969), and so-called modern agriculture, it is more likely to see modern science and technology as the great cause of the anthropogenic effects. This then enacts an option for political, practical, ethical and pedagogical responses that criticize and counter the increasing instrumentalization, technification and scientific management of both humans and the earth, as well as the school (see for example Chaps. 7, 11 and 13).

To agree on one specific start date therefore, is not necessarily the 'most' crucial factor. Rather, it can be an advantage to think and act on the basis of a multi-stranded understanding of history, so that, for example, both agrarian logic, capitalist logic and technical-scientific logic, all of which are important layers in the Holocene, are exposed to criticism and constructive alternatives. These can then be translated into (among other things), better pedagogical practice, supporting future generations to dissolve these Holocene logics, and to replace them with better approaches to life here on earth.

Controversy # 2: The Name of the Game

Two key objections have been raised to *designating* the time we live in as the Anthropocene.

Firstly, it has been objected that not all humans are equally guilty of its negative consequences (Moore, 2015, 2016; Malm, 2016). Global fossil companies and rich people are especially guilty, much like the unequal distribution as to who should take responsibility for the problems. However, objectionists have asserted that it is not fundamentally humanity as such (and thus the species homo sapiens) that are the cause of global warming, but a capitalist way of relating to the world that has spread from Europe to the entire globe from around the fifteenth century onwards. The story of the Anthropocene is too superficial, as it does not dive into the specific capitalist relations between humans and the more-than-human world that capitalism exploits. Instead of such a specific mosaic of relationships, Anthropocene history speaks only of an abstract homogeneous acting entity—humanity—facing the great forces of nature. The concept of the Anthropocene therefore implies, critics say, a reductionist narrative in which humanity as a collective actor faces Nature. It means that one conceptualizes the Anthropocene from a catalogue of geological and biospherical consequences, while what leads to these consequences appears as a black box consisting of categories such as industrialization, urbanization, population growth, and so forth.

These two features: (1) that humanity is abstractly seen as a cause, and (2) the Anthropocene is based on consequences for nature—implies a Cartesian dualism, where the world is divided into separate domains: human activities in one box, and nature in another. Such a narrative has political-pedagogical consequences: it draws attention to technical variables such as population growth and technologies for dealing with the challenges that have arisen. Part of this narrative is that the Anthropocene starts with coal and steam engines in mid-eighteenth century England. If, instead of periodizing based on external consequences, its emergence was placed in relation to capitalist structures and relations, then its decisive beginning is as early as the fifteenth century; politically, its these relations and structures that must be overcome. Having an eye for the origins of capitalism in early modernity and its extraordinary transformation of culture and nature before the steam engine, is therefore politically crucial, as our understanding of the origins of problems affects how we choose to respond pedagogically. Such a critique of capitalism questions the resource and technology determinism embedded in the concept of the Anthropocene. The concept of *the Capitalocene* focuses on the fact that today we live in a time shaped by structures that privilege an endless accumulation of capital. The hope in this critique is that the Anthropocene is not to be conceptualized by and through external consequences, but based on which structures produce these consequences.

We agree in principle—and to some extent—with this critique. *If* the Anthropocene is simply understood as a boxing match between Man and Nature, and viewed from the consequences, then one has both a weak and politically-pedagogically problematic concept of the Anthropocene. We have argued in this collection, one can well thematize that it is not Man abstractly theorized that is the cause, but specific assemblages of unequal and asymmetrical intricacies of 'humans and humans' and 'humans and non-humans' that result in negative effects. In the above critique, emphasis is placed on the fact that these intricacies have increasingly assumed a capitalist format since the fifteenth century. We do not necessarily disagree with this, but will nevertheless suggest that one should trace the cause a layer deeper, namely down to the underlying human-centred worldview developed especially in western culture, including and perhaps especially in the Renaissance (Paulsen, 2021b) (see further in Chap. 11). The advantage of such deeper explanations, among other things, is that it becomes possible to explain that structures other than specifically capitalist ones can also rest on the same worldview, and therefore have many similar negative consequences; for instance, the major environmental deficits in the planned economic arrangements in the so-called communist countries. This also makes it clear why a mere critique of capitalism is politically-pedagogically inadequate as it does not really penetrate to the underlying worldview and the need to change it. Yet it must be considered as part of the bigger picture (see for example Chaps. 7 and 8). Furthermore, the concept of the *Capitalocene*—as an alternative term preferred by some to the Anthropocene—has the added problem that it only focuses on intra-human relations. Secondarily, the focus on human capital does not open the gaze to the diversity of non-humans, including their potential and capacity for creating symbiotic life and being part of new solutions.

Secondly, it seems inappropriate to name an era of the earth's history as Man's new age—which is what the 'Anthropocene' literally means. Such a designation overemphasizes Man and his role (which is also the case with alternatives like the Capitalocene) (Haraway, 2016). By designating the current epoch of the Earth as the time of Man, the anthropocentrism that some proponents of the concept of the Anthropocene try to do away with is simply confirmed. This is sheer hubris—insofar as the term fortifies an attempt to take control of a geohistory that is being co-authored by cosmological earthly forces (see Chaps. 2 and 7).

It is apparent however, that anthropogenic activities have led to the plastic waste in the oceans, so much so that yet another designation has emerged: Plastocene. The 'Anthropocene' can signify specific arrangements (or entanglements) between humans and non-humans that have begun to play a major role in Earth's overall geobiohistory, which are new and therefore might justify the name. However, if an understanding of the Anthropocene is solely based on the fact that we live in a time where mega-influences of the planet are attributable to humans, then it leans towards a problematic anthropocentrism. Alternatively, incorporating understandings of specific entanglements between humans and non-humans, and developing new dialogical sensitivity to the latter, then this anthropocentrism is countered. Furthermore, it is precisely the ambiguous, ambivalent, and controversial nature of the name that enables this term to do what no other term can currently do; on the one hand, it emphasizes the crucial role of Man, but on the other hand it problematizes this very claim by bringing together the cultural history of Man and the geobiological history of the earth, so that Man (re)enters a geobiological time together with other beings. The Anthropocene invites thinking of the interweaving of the human species with other beings on a common fragile planet. Understood along such lines, the Anthropocene term does not involve a celebration of human impact on earth systems, but functions as a critical concept, which problematize our species' current earthly presence. The concept holds a critical potential—and now possesses a normative self-negating dimension; it points to a fictitious desirable post-Anthropocene future in which our species does not constitute a mega-influential centre, but has become one among other beings who stand in a dialogical relationship of care and co-existence. Alternative terms such as *Chthulucene*, proposed by Haraway (2016), only receive their meaning and role as supplementing the Anthropocene term. This is also the reason why Haraway (2016) tends to tell a three-fold story about the Anthropocene, Capitalocene and Chthulucene, arguing that all three narrative layers are important. We agree with such a multi-narrative starting point as it enables a way to counter too simple one-dimensional political-pedagogical solutions. The essays in this book can be seen as a contribution to such enabling, pointing to different dimensions of the Anthropocene and complementing solutions. Some chapters point out the need for education to address the current negative impact of human activities (see especially Chap. 15). Other chapters criticize the capitalistic structuring of the Earth (see for example Chaps. 2, 7, 8 and 16). But most chapters emphasize the need to pay more heed to entanglements and care for more-than-humans and interspecies relationships (see for example Chaps. 3, 4, 10, 11, 13 and 16), and take on the perspective of *Planetarianism* (see especially Chap. 12) and *Biophilia* (see for example Chap. 10).

The upshot is that the term Anthropocene is useful to gather and join a constellation of current forces and ideas that point towards new pedagogical practices suited for a better Earth and future.

Controversy # 3: A Narrow or Broad

Some argue that we today, due to the latest impact of human industrial manufacturing and urbanization on the global climate on Earth, live under Anthropocene *conditions*, characterized by global warming, extensive ecological destruction, biodiversity crisis/mass extinction and the loss of inhabitation. These conditions generate, among other things, a movement where urban families begin to flow out of the cities to become permacultural (see especially Chap. 17). By designating the Anthropocene as the condition for a new situation, one might avoid the question of just when the Anthropocene epoch began (or ended), but that ultimately many forms of regernation are necessary now.

Yet, another question, which cannot be avoided, is how narrow or broad the Anthropocene should be understood. For example: if the Anthropocene is understood only as an age where, through natural science, we can detect that human activities profoundly affect the environment, and that we now realize that there is a risk to destroying the environmental conditions of human life as we know it, the pedagogical response would probably only be about learning to manage resources—and spaceship Earth—better; that is, to try and solve planetary problems with the same Holocene instruments and logics, understanding the world in the same way that created the problems in the first place. If on the other hand, the Anthropocene is understood as opening up a new "more life-friendly" world understanding, i.e., a new way of being-in-the-world, the pedagogical response would be to support and encourage the emergence of alternative ways of being present in the world, and paying attention to new ways of understanding ourselves, and the world (see for example Chap. 11). While many politicians, policymakers and established regimes might be attentive to global warming and the need to respond to the global environmental problems of today more inclusively, in our view there is an over emphasis on political rhetoric (Kopnina et al., 2021) and basing responses on arguably redundant late Holocene logics. This is primarily a *technical* worldview, a narrow scientific-geological understanding of the Anthropocene that attempts to *control* the new situation through instrumentalized means (see for example Chap. 2).

On the other hand, many eco-critics, eco-artists, and eco-critical researchers respond on the basis of a broader epistemological concept of the Anthropocene. They argue for the need to change dominant world-views: how to understand ourselves as a species in the larger context of life and cosmos, situated together on a planet with other living beings (Chakrabarty, 2015, Paulsen, 2021a). It remains open as to how this new possibility to understand such inter-intra-relations should be encountered, actualized, and understood more precisely. On a theoretical-philosophical level, there are different implications for such a new life-centred world understanding.

A 'soft distinction' between two slightly different movements can be detected: the first movement, *new materialism* (Coole & Frost, 2010), mainly tries to map and understand how human and non-human bodies are entangled, and part of the same living earth (see for example Chaps. 6, 8, 9 and 17). One of the main theoretical references here is the differential 'cosmology' and rhizomatic conceptual framework and ideas put forward by Deleuze and Guattari, along with process-philosophical thinkers like Spinoza, Bergson, and Whitehead. It is important to notice that this kind of materialism is interested in creations, affections, relations, and what bodies can do together and to each other. The pedagogical consequence of this kind of thought, is perhaps most of all about creating new educational practices, in which students are encouraged to experiment with bodies (artistically), and develop ecological awareness of entanglements and affections (see for example Chap. 9). What is at stake is to see things as "vibrant matter" (Bennett, 2010), possessing virtual possible capacities for creating something new, possibly a 'new Earth' on a geopolitical scale (see especially Chap. 8) if these capacities are not suppressed and/or marginalized/restricted, but encountered with experimental openness and affirmative life affections. This might bring some hope. If the basic capacity for life and creation is not destroyed, and the actual possibilities of destruction are reduced, it might be possible to detect and further grasp the virtual life capacity of the Earth, and perhaps then have a direction to recreate a new and better Earth. Yet (as argued especially in Chaps. 7 and 8), the power of capitalism to reinvent itself and counter life and alternative movements is strong and should not be underestimated.

Aside from new materialism, there is a second movement, which makes it possible to talk of a *new idealism* (Paulsen, 2022). It mainly pays heed to the possibility and value of dialogical relationships between singular human beings and more-than-human creatures (see Chaps. 2, 3, 4, 5 and 10). This line of thought expands the realm of subjectivity and ethical beings to more-than-humans—animals and plants (see for example Chaps. 11 and 14), as well as other entities such as landscapes (see for example Chap. 3), waterscapes (see for example Chap. 9), and whole environments and collective entities (see for example Chap. 12). This way of thinking is often based on a reconceptualization of humanism and humanistic ethics, drawing on thinkers like Gadamer and Levinas as well as non-western thought. Ontologically speaking, it might involve a kind of pan-psychism (Skrbina, 2017), and overlaps with new materialism to the extent that both movements treat 'all beings' as 'actors' that are able to 'do' things in relation to their environments.

The pedagogical consequence seems to be slightly different between these two movements. While new materialism invites educational practices that contain possibilities for the creation of differences and building new ecological awareness of entanglements and possibilities of connecting things differently—it recreates how we connect to the world—new idealism invites (1) a change to our own (self)understanding and ways of being-in-the-world, and (2) more specifically, begin to develop life communities in which both human beings and more-than-human-beings can participate and enter dialogue with each other as to how to live well together. One of the main motives of new idealism is to (educationally) establish valuable relationship with more-than-humans. In this way, it becomes more likely that such an attunement will enhance more care and responsibility for others—including more-than-humans. When we recognize the intrinsic value of more-than-humans, that is, recognize them as singular unique irreplaceable beings, and thus not only as resources or background for human life, we become ethically committed to become responsible for more-than-humans, de-centring ourselves.

Difficult questions arise as we untangle a broad spectrum of interconnected issues. A number of the essays deal with these difficulties (see for example Chaps. 13 and 14), rather than completely resolve them. They are pedagogical-practical-ethical-political wicked problems, *fundamentally undecidable*. Also, there seems to be two strands in the new idealistic literature: on the one hand some seem to argue that we should pay attention to how many more-than-humans are *like* us. Given these similarities, we should treat them better, learn from them, and include them as near-equals (Fredriksen, 2020). On the other hand, others put forward a kind of Levinasian argument: it is precisely because the more-than-humans are *different* from ourselves than we are ethically called to take care of them as Others. It is precisely their radical Otherness as to why they cannot be totally understood or replaced by us. At the same time, it is also the reason that it might be profitable to be together with more-than-humans and learn with and from them (see for example Chap. 14 and Paulsen, 2021c). In our view, all these difficult questions presented by new materialisms and idealisms, generating exciting discussions, are part of responding to the current Anthropocene situation. These issues should be incorporated into educational discourses and practices, inviting new generations to participate in their further thinking and experimenting without arriving at final answers. The essays in this book demonstrate ways to engage in such an enterprise.

Controversy # 4: A Good or Bad Anthropocene—or, Does It Ever End?

The last controversy we want to highlight in this introduction is about what kind of future we can hope for, dream about, or aim for. How is this better state to be conceptualized and understood as a post-Anthropocene state or a different kind of Anthropocene than we live in now? This includes the question of whether we should "stay with the trouble" or try to find "a way out" of the Anthropocene, or indeed go further to imagine a spectrum between these binary positions. This controversy is linked to others previously mentioned. If for instance, one thinks that the Anthropocene is rightly called the Capitalocene, an age started with the growth and spread of capitalism, then there is the inclination to think that the needed future is a post-capitalistic one. If this is not possible, we are doomed in one way or another (see for example Chap. 7). Our future imaginaries are linked to our views on the past and present, and our pedagogical responses are also framed by this in relation to what kind of imaginations are encouraged in our schools.

Some argue that the Anthropocene is a time of catastrophe: we are in it, right now! Gloomy and disastrous, it will only become worse and worse over the next years, until the end of the world as we used to know it comes to a halt (Scranton, 2015; Morton, 2013, 2016). The only sane pedagogical response to this will be to empower new generations to cope and make the most out of it; that is, if human life is still possible. This might include learning to protect ourselves against the worst effects and minimize the negative effects as much as possible. Afterall, nothing lasts forever, so what we can do is only to act so as to live well and as long as possible. In this dystopic story, the Anthropocene is the end of history. Part of this narrative is to expect 'worst case scenarios'—ecological breakdowns, but also societal breakdowns, anarchy and new totalitarian regimes, mass extinction, wars and global crime. Thus, the Anthropocene is 'bad,' and remains so. We better learn fast to realize this, to "know our enemies," to counter, resist and modify the best we can. This seems to be the lesson of much current dystopic literature, film, gaming, and music that young people today grow up with, paving the way for apathy and depression (see for example Chap. 12). The world, when it comes to both nature as well as society, is bad, and cruel. When it strikes (back), it comes with death and terror. There is no way out. It's a strong narrative!

There is also a different kind of hope, proposed by a myriad of alternative storylines. Many agree that it all looks very hopeless, but that there is still "a good life" to strive for, even if there might not turn out to be a "good Anthropocene" as such. The Dark Pedagogy movement (see the essays in series II of this book) respond in this way: try to face the perceived darkness, at the same time, keep or remain sane and "stay with the trouble" in a broken world so as to create feasible solutions from within. From this point of view, it is not advisable to be too optimistic and have un-realistic utopian views of a "good Anthropocene." This ideologically erases—makes one blind to—the real troubles, dilemmas, deep structures, and wicked problems that are pressing, leading only to false hopes and dreams, which are yet but further disillusions and/or blind spots, inaction, and an incapacity to act and recreate or, even more fundamentally, to make necessary radical resistance and change possible.

There are other voices who are optimistic and think that the Anthropocene is an opportunity to create new ways of life, new world understandings, coming into dialogue with more-than-humans, so as to create a much richer life (see especially the essays in series III in this book). While it is true that many young people today grew up with dystopic narratives in mainstream culture, it is possible to find and relate educationally to a literature that possesses active hope. These are narratives that can help young people create hope for the planet, at the same time cultivate their anticipatory imagination, fostering ideas about how to act and create a biophilic society (see especially Chaps. 10 and 12).

Still others propose that it is possible to transition from bad to at least a better Anthropocene, but only if we can succeed by deviating from our current unsustainable monocultural habits and practices. This includes re-creating our educational institutions away from places where students are over-disciplined and controlled, but rather engaged, resilient and open to love and care of the planet. For example, this is the path that is envisioned and enacted by the Wild Pedagogy movement (see the essays in series I of this book), in which a 're-evaluation of all values'is called for, and human 'forgetting'of the more-than-human world is called to account (see for example Chaps. 3 and 5).

What almost all essayists of this book agree on is the bankruptcy of prevailing order of 'control societies.' As the old saying goes: the devil is in the detail. When it comes to details, and basic interpretations of the

Anthropocene situation and adequate pedagogical solutions, most of the essayists disagree with each other, more or less. This should *not*, in our view, be seen as a disadvantage, but reflects the complexity of the issues that call for multiple perspectives we hope readers will find both provocative and stimulating so as to further contribute and foster new and even better thoughts and practices that go beyond what we have dared to think in this book and exist today.

Structure of the Book

We have structured the book into four parts—series—that help navigate the collection thematically. Each series can be read on its own, but also as part of the book's chronology.

The first series, Wild Pedagogies, contains four chapters, which all relate to the Wild Pedagogy Movements, initiated by Canadian scholar Bob Jickling and others. Wild Pedagogies (WP) aims to renegotiate what it means to be human in relationship with the world by engaging in deep and transformational change using educational practices (see https://wildpedagogies.com).

The first chapter in this series introduces the main ideas of WP. Chapters 3 and 4 are independent contributions that relate to and can be viewed as part of the WP movement. Chapter 5 seeks to bring WP into new terrains. Together the four chapters capture the essentials of WP and demonstrate how WP ideas can be used to appreciate more-than-humans in educational contexts. All four chapters respond to the Anthropocene mainly by proposing that new more 'wild' educational practices are needed, which cultivate non-instrumental and non-control-seeking relationships with the world.

Chapter 2 argues that the Anthropocene reminds us of three critical ideas: (1) Earth is in a state that threatens myriad species, including our own; (2) any effective response will require a radical rethinking of ideas and ways of being that run counter to dominant cultural narratives; (3) we live in unprecedented times, meaning that we can no longer prepare students for an assumed world, with confident learning outcome. WP is an educational response to these challenges—and has arisen as concerns about the ways in which control does violence to many species and restricts the possibilities for change in an era of uncertainty. WP is based on at least two premises: first, that modernist relationships with the world must change; and second, that education is a partner in the project. WP argues that the desire for control plays too big a role in education today, and brings too few possibilities for relational engagements within the natural world. As an alternative, WP suggests that teachers should be supported to become wilder, even rebellious, in their practices, and be supported to develop less mainstream, more non-linear, outdoor, and wilder forms of education.

Chapter 3 calls for a shift in Western perception of human and more-than-human relations. The West's propensity to conceive of humans, plants, elements, and land as separate entities merely capable of 'interaction,' limits epistemological and relational possibilities. It is argued that standard (North American) educational practices do little to foster significant relationships between humans and more-than-humans or acknowledge their relationships. As an alternative, Chap. 3 reflects on relational ontology and the need to relearn to love more-than-human-entities, including deep listening to their responses and intrinsic values. This means shifting away from the object-oriented worldview towards more relationally oriented ontologies, whether through forms of new materialism (Bennett, 2010), animism (Stengers, 2012; Bai, 2015), posthumanism (Snaza & Weaver, 2015), making kin (Haraway, 2016) or a scientific recognition of deep interconnectedness of human development (Lieberman, 2013; Narvaez, 2014). Chapter 3 relates this to the nexus of love, and quotes Kimmerer's (2013) important saying: "Knowing that you love the earth changes you, activates you to defend and protect and celebrate. But when you feel the earth loves you in return, that feeling transforms the relationship from a one-way street into a sacred bond" (p. 125). Further, both Chaps. 3 and 5 argue that children have an intuitive entanglement with more-than-humans, but learn today in mainstream education to hide, suppress, ignore, or even unlearn this kind of knowledge.

Chapter 4 demonstrates how nature educational programs can be enriched to help to inspire an appreciation for non-human agency and values in learners. Levels of possible experiences are delineated: Level 1: Learning about non-human beings as isolated objects. Level 2: Learning about non-human beings as interconnected objects. Level 3: Learning about non-human beings as interconnected subjects with agency and interests. It is argued that the most important learning of all will not be the names of creatures or even their broad ecology, but the realization that the organisms being 'studied' are individuals with needs and with value that are independent from any instrumental benefits that humans might derive from them.

Finally, Chap. 5 turns toward children, and explores how it might help educators wild their pedagogies if children—and their capacities for wild thinking—is taken seriously as a starting point for educational activities. Thus the chapter claims that children and adults encounter their worlds in different ways. Children, the chapter suggests, relate to their environments with all their senses, emotions and skills. These relationships position children differently in the world both ontologically and epistemologically. In some senses, their thinking is wild; it isn't corralled or regulated—yet. In other ways, it is argued in the chapter that anthropocentrism is a learned positionality—and that children are taught to be so in the world.

The second series, Dark Pedagogies, contains four chapters, which all emphasize the need to educationally address the dark sides of our current situation and of the world of today. The concept Dark Pedagogies (DP) is used here in a loose sense. Only Chap. 6 explicates and elaborates the DP concept, while the 3 other chapters relate to dark issues and part of the same literature—French philosophy, especially Deleuze and Guattari, new materialism, speculative realism and Timothy Morton and his concept of *Dark Ecology* (2016). Based on this loosely shared frame of reference, we have chosen to bring these four essays under the same umbrella, *Dark Pedagogies.*

Following Morton (2016) Chap. 6 argues that the Anthropocene binds together different temporalities through humans, the planet, and other large-scale entities (called hyper-objects by Morton) into the form of a strange loop. We can no longer see ourselves escaping from being 'cut up' and compartmentalized. The Anthropocene signals to humanity: "Congratulations! You have now become aware of being part of an entity that operates at global scale and there is no way back or out." From this perspective, Chap. 6 argues that the Anthropocene shatters any notion of effectuated intentionality, and brings inherent contingency that leaves us educationally numb, but could also better be countered by approaches that accept the limitations of the human scope of perception and understanding. One could say, following this line of thought, that the Anthropocene makes us ridiculously small and unimportant. We must learn to accept this and form sane responses to the Anthropocene. Thus, education in the Anthropocene should, first and foremost, ensure an openness to the contingent intertwinedness of the world-for-us, the world-in-it-self, and the world-without us as developed by Eugene Thacker (2011).

Chapter 7 develops the concept *Dark Labour* and demonstrates how current mainstream educational thought and practice is pervaded by an industrial, capitalistic, and product-factory worldview that limits the possibilities of overcoming ecological problems of today. In this worldview, the world is seen as a site of potential value to be extracted through labour, including educational labour, linked to an image of human exceptionalism. Where the world is reformatted according to its extractible value for us, it founds a relation to matter that enables the conditions of exploitation and extinction. The world is thus understood within an all-too-human orders of significance and control. This pervasive commitment to production, resides everywhere, including as the orthodoxy of the school of today. Chapter 7 relates education of today to the *Capitalocene*, and argues that modern education is linked to an industrial imaginary. It has throughout its modern development conspired in the exploitation and control of the world by reformatting its value according to its very susceptibility to control; it has also conspired in the production of monocultures and the standard human. The ecological influence of the modern "Educacene" is its implicit 'cheapening' of the world reformatted into institutional value and capital. According to Chap. 7, this also goes for the field of Education for Sustainability Development (ESD), which aims towards the habilitation of an 'optimized productivity', while sustaining the very logic of productivity that today conspires towards the exploitative cheapening of the world. Chapter 7 also argues that the emergence today, in the Anthropocene, of non-human labour in the form of climatological catastrophe advances the understanding that one cannot do what one likes to do with the world; meaning that this alien productivity demonstrates the limits of productions for *organic life*.

Chapter 8 argues that the Anthropocene is not a future disaster that must be prevented. This is the fantasy of extending the Holocene indefinitely. Rather the epoch presents a changed ontology, a new geological and political era, a difference in kind and not degree, marked by the burning of fussil fuels. Chapter 8 also relates the Anthropocene to Moore's (2016) Capitalocene, but with two added features: (1) The Anthropocene understood as a new geological epoch *outdates* capitalism, as demonstrated by Anthropocene science, which studies shifts in Earth Systems far *beyond* the Holocene and the human impact on these systems; (2) The Anthropocene exceeds the geology of our species and should therefore not be equated with the geology of the system of capitalist power as Moore (2015) claims. Chapter 8 suggests framing the educational relevant question of our time as how capitalism hinders or intensifies the state of the Earth System, and what then would be the educational imaginary adequate to this planetary condition. In connection to this question Chap. 8 investigates four paths 'out of the darkness,' so to speak: (1) Ecomodernist humanism, (2) Posthumanists, (3) Posthuman thinkers, (4) Deleuze and Guattari cosmology—and the 'cosmic artisan' as a conceptual persona. All paths are confronted with the dark reality of the Anthropocene, as it is influenced by capitalism. The fourth path is investigated more closely in relation to craft-bioart, biomimesis, and biosensing, where art, technology and biogenetics come together, and seems to be the most promising path if we want to rethink our species relationship to the Earth. Yet, in the end none of the paths seems very promising, all ending up being caught—or risking being caught—within a capitalistic mindset of a world-for-us, yet another form of nonhuman exploitation. Also, the cosmic artisan has been essentially captured by capitalism. Thus, the chapter argues that extreme difficulties for transvaluative change (e.g., movements beyond anthropocentrism and capitalism) persist in the world of today.

The last chapter of series II, Chap. 9, is more optimistic and demonstrates how Dark Ecology, speculative realism and new materialism can be applied to pedagogic and the field of aesthetic learning processes in constructive ways. It is argued that the speculative ideas and concepts of Morton (2016) and others can be used to support and understand in which ways human bodies might be involved in the transition from a well-known human-centred way of relating to the world to an unknown Anthropocene way. The chapter uses the example of an action by Fridays for Future Denmark in front of the Danish parliament to discuss this.

All in all, the four chapters of series II explore in different ways how one can develop new pedagogies that consider the dark sides of Anthropocene reality, and translate ideas of new materialism and speculative realisms into educational thinking and practice.

The third series, Interspecies Inclusion and Environmental Literacy, contains four chapters, which all thematize the possibilities and values of developing relationships to and with more-than-humans within educational practices. Yet, the four chapters move in different directions and base their investigations on different theories and assumptions. What they share is a positive interest in the same issue, and the belief and hope that relationships between humans and more-than-humans can be established, both educationally and as a general value. But how it can be done, differs across the four essays, digging into different spheres—waterscapes, language games, literature, and human-machine entanglements.

Chapter 10 is situated within posthumanist environmental education research, which strives to give voice to humanism's Others. According to Chap. 10, this line of thought is not anti-humanist, since it retains the modernist notions of rights, justice, equality, and freedom. The chapter argues that the human species risks continued failure if nature is not recognized as both a sentient and intelligent creator and conveyor of knowledge, and seeks to decentre the human and explore how more-than-human actors can be invited into interspecies dialogue through rewilding, affiliation, deep listening, being-with, and re-newed connection with more-than-human worlds. The chapter focuses on waterscapes and water literacy as an example of environmental literacy, and examines the pedagogy of entanglements of natural and cultural everyday life within this context, with a special focus on youth engagement. It is argued that the mainstream western management system works against nature, not with it, controlling and managing it according to human economic needs. This has resulted in the planet's diminished capacity to continue to provide the service for *all* life. As a response, the chapter suggests new critical ontologies and pedagogies based on re-visiting and engaging with invitations, and practices from First Nations indigenous pedagogy\ies, which have been less anthropocentric and have a long tradition for working together with and learning from nature. When it comes to waterscapes, it is demonstrated that there is a strong bias within mainstream western culture and education towards an anthropocentric outlook on water (primarily economic) and an overemphasis of cognitive and western scientific approaches to knowledge. As an alternative the chapter presents projects which have developed slow pedagogy and place-based learning approaches that focus on local context, collective learning, interdisciplinarity and cross-cultural learning that convey historical hydrologies, cultural traditions as well as spiritual and ethic-based knowledges. Here students are invited to consider water beyond commodification purposes towards being more "worldly with water" and waterscapes.

Chapter 11 is based on critical-philosophical reflection of human practices and logics in the late Holocene and the beginning of the Anthropocene, and the results thereof—earth-forgetfulness among other things, which calls us to rethink humanity and pedagogy. The chapter argues that what is most lacking in the world today, in the Anthropocene, is the development of dialogical relationships and life communities between humans and more-than-humans, especially, relationships and communities penetrated by mutual love. According to the chapter western educational thought and practice has been pervaded by a cultivation of general concepts, monologue, and monoculture, we-and-they-thinking, and has thereby paved the way for a Holocene mindset, unable to create dialogue and loving relations. Thus, the chapter argues for changing our world understanding from a scenic, monological, objectifying, and human-centred view, focused on resource management, towards a dialogical, loving, and zoë-centred view in which we pay heed to, and try to develop dialogue with more-than-humans. The plea is to join in life communities where both humans and more-than-humans can participate and enjoy, building up concepts and knowledge about each other as irreplaceable unique singular beings. Further, the chapter argues, it is reasonable to demand that educational institutions are changed to become places where we help each other in fostering good interspecies relationships and communities. The chapter, therefore, explores how educational institutions could be set up to enable new generations to gain experiences in trying to form careful communities and relationships, together with more-than-humans. The upshot is a concept of humanity and pedagogy that is not centred on humans alone, but reaches out to our 'life-fellows,' demanding different educational institutions than those of today. Where mainstream educational institutions are made primarily for human needs and only letting them participate as actors in the educational events, the chapter suggests that new educational institutions should be developed in which both humans and more-than-humans can participate, and where both interests are considered.

Chapter 12 is situated within literature studies. According to the chapter *ecocriticism* has shown that literature studies should pay attention to the global environmental crisis. Further, *environmental humanities* have stressed that science alone is not enough to engender a societal transformation of our world needs. Based on this, the chapter argues that one of the greatest challenges facing education in the Anthropocene is to empower young people *to believe* that we can transition to an ecological civilization brought about by the urgency of "the explosion of dystopia" in mainstream popular culture. Earlier, the dystopic imaginary only existed at the margins of mainstream literature. Now, the dominant response to the Anthropocene in literature and film today is dystopian, postapocalyptic, and filled with post-disaster narratives. Students today grow up reading dystopia, playing dystopian games, and watching dystopic films. They grow up hearing that capitalism is unavoidable, even as it destroys the planet. Many young people have, therefore, internalized the belief that imaging a hopeful future is naïve, whereas imaging a postapocalyptic hell is reasonable. Thus, the dystopic narratives seem to erase hope, normalize expectations about dystopic futures and reinforce the belief that ecocide is unavoidable. As a countermovement, the chapter focuses on how non-dystopic literature for young people can be tapped to nourish hope for the planet. The assumption is that hope-oriented anticipatory imagination is a precondition for disrupting ecocide and enabling meaningful change. The chapter suggests *planetarianism* as a term for this anticipatory imagination focused on planet's biocentric future, as a mode of engagement with the issues of climate change in and trough literature for the young reader—and as a *biocentric philosophical commitment to stand up for the planet.* Thus, the chapter suggests that one productive way of engaging with the urgencies of the Anthropocene is through stories that mobilize active hope for the planet and tell stories about how to transit to an ecological civilization. Planetarianism is therefore to be understood as applied hope articulated through stories, that envision the planet as a living entity, imagine a non-ecocidal socioeconomic system, depict disanthropocentrized relationships among humanity and other living beings, and gesture at a biocentric, multispecies future that is worth living for. This keeps alive young people's belief that it is not too late, that we have agency for change, and that even a broken world is worth fighting for.

The last chapter in Part III, Chap. 13, also explores the potentials of literature, but Science Fiction (SF) and stories about Artificial Intelligence (AI). The chapter argues that surviving the Anthropocene is a problem that encompasses the totality of human existence. A pedagogy for the Anthropocene should therefore strive to build people up as holistic problem solvers. Further, the chapter argues that achieving a good Anthropocene requires a shift not merely in political economy and science/technology, but in ontology as well. To turn the Anthropocene around, we therefore need to cultivate richer capacities for being-in-the-world than in hitherto dominating western metaphysics. The chapter explores alternative metaphysics and how these could make room for inclusive political subjects open to human-nonhuman entanglements. To that purpose, the chapter dives into a corpus of fictional texts with AI characters and narrators that can be used as starting points—or illustrative examples—for a poetical ecopedagogy for the Anthropocene. It is argued that reading SF can lead to a deconstruction of one's worldviews, and to the construction of new ones. SF holds this specific potential as it flows around three poles: the world, the text, and the SF world. The de-construction of current mainstream worldviews is, so to speak, immanent to the SF genre. The chapter compares this to the decolonization of thought, when anthropologists unravel indigenous ontologies as different than western ones, and thereby open alternative perspectives on the relation between body and soul. According to the chapter, the SF genre does some of the same. A pedagogy which incorporates SF readings will therefore be able to function as exercises in *perspectivism* and *alternative worlding.*

The fourth series, Critical Rethinking and Future Practices, is more heterogeneous, containing four chapters that are all critical in relation to the current Anthropocene situation and suggest alternative ways to rethink education and shape the future world differently.

Chapter 14 argues that in an educational context the Anthropocene draws attention to challenges pedagogy faces such as human practices, structures and notions that threaten and damage life, landscapes, earth systems and ecosystems. According to the chapter these challenges converge in a transformational task that involve critical thinking, identifying, and addressing what must be transformed. The chapter therefore explores the position of critical thinking in environmental and (post) sustainability education, especially the possibility of a 'critical place-based pedagogy'. It suggests that the ethical grounding of critical thinking may be located within a place-based education that explores the ontological condition of living with other beings. According to the chapter, this is the basis insight brought in from an ethics of proximity (Levinas and Løgstrup). The implication is, among other things, that education in the Anthropocene should include an accommodation of student's experiences and existential concerns of anxiety, sorrow, and loss.

Chapter 15 articulates what the authors—as educators—observe as some of the on-the-ground challenges and opportunities in science education in teaching for sustainability at a rural Midwestern high school. In particular, the chapter outlines the ways that fundamentalist anti-social movements threaten the actual doing of teaching. The chapter explores from this point of view rurality as a site of extraction within global capitalist economies. Further it illuminates fundamental tensions in rural education in the United States. These two contexts lead to a discussion of the daily political barriers that rural teachers face in educating for science. What also emerges from their chapter is that ecological sustainability and the realization of planetary care is overwhelmingly ideological, and that praxis is complicated by geo-political differences in democracy and representation. Yet, the chapter also argues that rural places offer unique opportunities for climate education that have potential to further evolve in favour of the earth.

Chapter 16 argues that according to *the ecopedagogical movement* formed after the Rio-conference in 1992—building on the critical pedagogy of Paulo Freire—the origin of our current Anthropocene crisis can be tracked back to Greek education (Paideia-thinking). The anthropocentrism of the Greek mind stemming from an oppressive slave-based society has allegedly paved the way to the Cartesian dualism between nature/culture and the technical rationality underlying our current ecological crisis. It is argued that this story is partly wrong, insofar it involves an unwarranted claim that classical civilization, as such, must be superseded. Instead, an ecological mindset of today must address our classical heritage, and aim at recovering practices and a mindset which were once made possible by means of slavery, and raise the question whether we can reformulate the conditions of this mindset in a modern world without its oppressiveness. The chapter looks for solutions to our current problems in the conceptual landscape of ancient Greek time, especially in the hostile inclination towards labor and work, and its ideas about *theoria* as a non-interfering spectating attitude towards the world. According to the chapter, these ideas hold a critical potential to thinking and developing a relation to the world not dominated, as today, by the demands of work, productivity, and utility.

Chapter 17 ends the book, by presenting and discussing an un-finished current project that explores *Mycelic pedagogies in the Anthropocene*. The question the project, and the chapter, seek to answer is: how do you lead the way into a culture of re/generative education in the Anthropocene? The educators/leaders of the project (and authors of the chapter) try to change the world locally, in Copenhagen, for the better, through experimenting with and applying ideas proposed by thinkers such as Latour, Stengers, Haraway and others. Yet, as demonstrated in the chapter, their efforts are met by tremendous challenges, among other things by apathy/depression and in-action from the locals they try to involve and interact with (under the pandemic conditions). The chapter shows how difficult, but not necessarily impossible, it can be to do real changes, and how locally and concretely, it is necessary to rethink, resituate and try out new ideas than those one has from the start, and think about how reactive citizens can be transformed into active ones. This involves a shift of focus, vocabulary and concepts. For instance, it is explained in the chapter that the educators were forced to shift focus from a smaller group cartographing their area to the birth of a new pedagogical concept, "the Pollination Academy". This demonstrates nicely how relevant pedagogical concepts must be developed in practice as they cannot be thought-out beforehand.

A Non-conclusive Conclusion: The Diversity of Anthropocene Pedagogies

As it by now clear, this book brings no *final* or unitary answers to the table, but a diversity of Anthropocene Pedagogies, suggestions and new questions and issues. As stated several times, we do not see this as a disadvantage. Problematising the complexity of today's world and its anthropogenic issues, is a landscape in constant flux. Hence, final answers are not possible nor sought in this collection. What we have sought to do is map out the pedagogical problematic, and to develop and re-imagine concepts that are adequate to such a task. It is our unified hope that the sixteen different voices represented in this book can inspire, enrich, and change current educational thought and practice in directions that are worth striving for. To quote the author from Chap. 12, the slogan for this whole book and all it has suggested could be: Anthropocene Pedagogies—*Planetarianism NOW!*

References

Bai, H. (2015). Peace with the Earth: Animism and Contemplative Ways. *Cultural Studies of Science Education, 10*(1), 135–147.

Bennett, J. (2010). *Vibrant matter: A political ecology of things.* Duke University Press.

Braidotti, R. (2013). *The posthuman.* Polity Press.

Chakrabarty, D. (2015). *The human condition in the Anthropocene* (The Tanner Lectures in Human Values). Yale University.

Coole, D., & Frost, S. (2010). Introducing the new materialisms. In *New materialisms* (pp. 1–44). Duke University Press.

Crutzen, P. J., & Schwägerl, C. (2011). *Living in the Anthropocene: Toward a new global ethos.* Yale University.

Crutzen, P. J., & Stoermer, E. F. (2000). Global change newsletter. *The Anthropocene, 41*, 17–18.

Davis, H., & Turpin, E. (2015). *Art in the Anthropocene: Encounters among aesthetics, politics, environments and epistemologies* (p. 416). Open Humanities Press.

Ellis, E. (2018). *Anthropocene*. Oxford University Press.
Emmett, R., & Nye, D. (2017). *The environmental humanities*. The MIT Press.
Fredriksen, B. C. (2020). More-than-human perspectives in understanding embodied learning: Experience, ecological sustainability and education. *FormAkademisk-forskningstidsskrift for design og designdidaktikk, 13*(3).
Haraway, D. (2016). *Staying with the trouble*. Duke University Press.
Kimmerer, R. W. (2013). *Braiding Sweetgrass: Indigenous Wisdom, Scientific Knowledge, and the Teaching of Plants*. Milkweed Editions.
Kopnina, H., Spannring, R., Hawke, S., Robertson, C., Thomasberger, A., Maloney, M., Morini, M., Lynn, W., Zakari Muhammad, N., Santiago-Ávila, F. J., Begovic, H., & Baranowski, M. (2021). Ecodemocracy in practice: Exploration of debates on limits and possibilities of addressing environmental challenges within democratic systems. *Visions for Sustainability, 15*, 44–58.
Latour, B. (2017). *Facing Gaia*. Polity Press.
Latour, B. (2018). *Down to earth: Politics in the new climatic regime*. John Wiley & Sons.
Lieberman, M. D. (2013). *Social: Why our Brains are Wired to Connect*. Crown Publishers.
Lin, H. (2010). Earth's critical zone and hydropedology: Concepts, characteristics, and advances. *Hydrology and Earth System Sciences, 14*, 25–45.
Lysgaard, J., Bengtson, S., & Laugesen, M. (2019). *Dark pedagogy*. Palgrave Macmillan.
Malm, A. (2014). Tahrir submerged? Five theses on revolution in the era of climate change. *Capitalism Nature Socialism, 25*(3), 28–44.
Malm, A. (2016). Who lit this fire? Approaching the history of the fossil economy. *Critical Historical Studies, 3*(2), 215–248.
Malm, A., & Hornborg, A. (2014). The geology of mankind? A critique of the Anthropocene narrative. *The Anthropocene Review, 1*(1), 62–69.
Marris, E. (2011). The end of the wild: Climate change means that national parks of the future won't look like the parks of the past. So what should they look like? *Nature, 469*(7329), 150–153.
McKibben, B. (2006). *The end of nature*. Random House Incorporated.
McNeill, J. R., & Engelke, P. (2016). Climate and biological diversity. In *The great acceleration* (pp. 63–102). Harvard University Press.
Moore, J. (Ed.). (2016). *Anthropocene or capitalocene*. Kairos.
Moore, J. W. (2015). *Capitalism in the web of life: Ecology and the accumulation of capital*. Verso Books.
Morton, T. (2009). *Ecology without nature: Rethinking environmental aesthetics*. Harvard University Press.
Morton, T. (2013). *Hyperobjects: Philosophy and ecology after the end of the world*. University of Minnesota Press.
Morton, T. (2016). *Dark ecology*. Columbia University Press.

Morton, T. (2017). *Humankind—solidarity with nonhuman people*. Verso.

Morton, T. (2018). *Being ecological*. Penguin Books.

Narvaez, D. (2014). *Neurobiology and the Development of Human Morality*. W. W. Norton Publishers.

Oppermann, S., & Iovino, S. (Eds.). (2017). *Environmental humanities. Voices from the Anthropocene*. Rowman & Littlefield.

Oreskes, N., & Conway, E. M. (2011). *Merchants of doubt: How a handful of scientists obscured the truth on issues from tobacco smoke to global warming*. Bloomsbury Publishing USA.

Parenti, C. (2015). The 2013 ANTIPODE AAG lecture the environment making state: Territory, nature, and value. *Antipode, 47*(4), 829–848.

Paulsen, M. (2021a). Cautiousness as a new pedagogical ideal in the Anthropocene. In K. Petersen, K. Brömssen, G. Jacobsen, J. Garsdal, M. Paulsen, & O. Koefoed (Eds.), *Rethinking education in light of global challenges: Scandinavian perspectives on culture, society and the Anthropocene*. Routledge.

Paulsen, M. (2021b). From late Holocene to early Anthropocene educational thinking: Humanism revisited. In K. Petersen, K. Brömssen, G. Jacobsen, J. Garsdal, M. Paulsen, & O. Koefoed (Eds.), *Rethinking education in light of global challenges: Scandinavian perspectives on culture, society and the Anthropocene*. Routledge.

Paulsen, M. (2021c). The plateau of learning in the Anthropocene—How to relate differently to the earth. In D. Kergel (Ed.), *The great book of learning*. Routledge.

Paulsen, M. (2022). *Onto-sympathy, embodied beings and pedagogy in a (post) Antroprocene age*. In The Journal Futures of Education, Culture and Nature - Learning to Become (FECUN), vol 1 (1).

Postman, N. (2011). *Technopoly: The surrender of culture to technology*. Vintage.

Purdy, J. (2015). *After nature*. Harvard University Press.

Scranton, R. (2015). *Learning to die in the Anthropocene: Reflections on the end of a civilization*. City Lights Publishers.

Skrbina, D. (2017). *Panpsychism in the West*. MIT Press.

Snaza, N., & Weaver, J. A. (2015). *Posthumanism and Educational Research*. Routledge.

Sørlin, S. (2017). *Antropocen*. Weyler.

Spannring, R., & Hawke, S. (2021). Anthropocene challenges for youth research: Understanding agency and change through complex adaptive systems. *Journal of Youth Studies*. Published online first Open Access June 2021, pp. 1–17.

Steffen, W., Leinfelder, R., Zalasiewicz, J., Waters, C. N., Williams, M., Summerhayes, C., et al. (2016). Stratigraphic and earth system approaches to defining the Anthropocene. *Earth's Future, 4*(8), 324–345.

Steffen, W., Persson, Å., Deutsch, L., Zalasiewicz, J., Williams, M., Richardson, K., et al. (2011). The Anthropocene: From global change to planetary stewardship. *Ambio, 40*(7), 739–761.

Steffen, W., Sanderson, R. A., Tyson, P. D., Jäger, J., Matson, P. A., Moore, B., III, et al. (2006). *Global change and the earth system: A planet under pressure.* Springer Science & Business Media.

Stengers, I. (2012). Reclaiming Animism. *e-flux journal*, #36. https://www.e--flux.com/journal/36/61245/reclaiming-animism/

Stewart-Harawira, M. (2012). Returning the sacred: Indigenous ontologies in perilous times. In A. McIntosh, R. A. Roberts, & L. Williams (Eds.), *Radical human ecology: Intercultural and indigenous approaches* (pp. 73–88). Ashgate.

Thacker, E. (2011). *In the dust of this planet* (Vol. 1). Zone Books.

Tønder, L. (2020). *Om magt I den antropocæne tidsalder.* Djøf forlag.

Zalasiewicz, J., Waters, C. N., Do Sul, J. A. I., Corcoran, P. L., Barnosky, A. D., Cearreta, A., et al. (2016). The geological cycle of plastics and their use as a stratigraphic indicator of the Anthropocene. *Anthropocene, 13*, 4–17.

Zalasiewicz, J., Waters, C. N., Summerhayes, C. P., Wolfe, A. P., Barnosky, A. D., Cearreta, A., et al. (2017). The working group on the Anthropocene: Summary of evidence and interim recommendations. *Anthropocene, 19*, 55–60.

Zalasiewicz, J., Williams, M., Smith, A., Barry, T. L., Coe, A. L., Bown, P. R., et al. (2008). Are we now living in the Anthropocene? *Gsa Today, 18*(2), 4.

PART I

Wild Pedagogies

CHAPTER 2

Wild Pedagogies: Opportunities and Challenges for Practice

Sean Blenkinsop, Marcus Morse, and Bob Jickling

Introduction

We live in extraordinary times. The stories of our age are being written in mass species extinctions, catastrophic events, and accelerating climate change. It is also a time of social upheaval. Justice movements, such as #BlackLivesMatter, #IdleNoMore, #MeToo, school strikes for climate change (#FridaysForFuture), and the COVID pandemic make it clear that normalized social practices are troubling and inadequate. We cannot continue as we are; the current path is not sustainable. Social distress is increasing, and nature is crying.

S. Blenkinsop (✉)
Simon Fraser University, Burnaby, BC, Canada
e-mail: sblenkin@sfu.ca

M. Morse
Outdoor Environmental Education, La Trobe University, Melbourne, VIC, Australia

B. Jickling
Lakehead University, Thunder Bay, ON, Canada

M. Paulsen et al. (eds.), *Pedagogy in the Anthropocene*, Palgrave Studies in Educational Futures,
https://doi.org/10.1007/978-3-030-90980-2_2

We recognize that the term Anthropocene, which many are using to describe our times, is controversial. It sums up all humans together without separating perpetrators from victims. The Anthropocene does not adequately reflect the complexity of phenomena at play. There is the danger of attempting to predict an epochal change when most major changes—whether we call them tipping points, gestalt shifts, or paradigm changes—only become visible after the fact. Could this be an additional act of *hubris* as we humans attempt to take control of a geostory that is really being co-authored by all earthly forces?

While we are sympathetic to the critiques, we also assert that Anthropocene narratives have significant impacts: they remind us of three critical ideas. First, Earth is in a climatic state that threatens myriad species, including our own. Second, any effective response will require a radical rethinking of ideas and ways of being that run counter to many dominant cultural narratives. Such narratives have been framed in a number of ways—modernist, globalized, westernized, euro-centric, neo-liberal, colonial, Cartesian, anthropocentric—yet collectively these framings are entwined in knitting a resilient status quo that seems difficult, if not impossible, to shake. Third, we live in unprecedented times. Our children and grandchildren will grow up in a very different world. We can no longer prepare students for a "business as usual" world that claims confidence in its learning outcomes. We, as educators and scholars, must be differently as well.

Educational Responses: A Scholarly Ethos

To move cultures from where they are—epistemologically, ontologically, ethically, metaphysically, and practically—we, as educators, researchers, and teachers cannot continue to repeat the same established narratives; we cannot continue to be the *same* people, the *same* educators, and the *same* researchers. We hold that education must be a necessary part of any response that requires such a fundamental rethinking of ideas and practices.

This is not to suggest that there are not already many incredible teachers—across a variety of educational settings—pushing limits, defying the status quo, and persisting in offering radical and hopeful alternatives. We think of them as "rebel teachers" (Blenkinsop & Morse, 2017). Similarly, we join with other researchers—including our colleagues in this book—seeking to break free of normalized scholarly practices that hold us back (Jickling et al., 2018b).

We have called our response to educational demands of our times *wild pedagogies*. In the next section we describe key underpinning ideas of wild pedagogies and include more practical *touchstones* intended as provocations and reminders of what we are attempting to do. But we also ask—through two vignettes of practice—whether such attempts, guided by theory, are enough? When the reimagining and rethinking required runs so counter to cultural ways of being, significant pedagogical challenges in practice are inevitable. Such cultural change cannot be achieved simply through a theoretical shift; ontological alternatives can be difficult to sustain, and "wild" educational experiments may seem fleeting.

Enacting these experiments requires determination and practice. Deep cultural assumptions are often hidden from view in pervasive language choices, hierarchical social structures, and the scope of knowledge and understanding considered neutral. These assumptions constantly and silently work to bend educators back to the status quo. Indeed, these cultural forces can be the "real authorities." How might we meet these challenges and enable productive and hopeful pedagogies? Part of this task must involve naming the challenges and being ready to offer alternative responses.

Consider the challenge of communicating across borders in a transdisciplinary seminar comprised of eco-literary critics, educators, and a curious engineer. Interestingly, such a gathering did take place a number of years ago. However, the engineer, in particular, could not penetrate the polysyllabic words—code, we think, for barriers created through nearly impenetrable dialects (Braidotti, 2019). Sadly, he chose not to return the following day.[1] This incident is a reminder that in academia, we can easily forget how difficult it can be to communicate effectively beyond our comfortable and established bubbles. The point is that this kind of border crossings does not come easily; we must learn how to do it.

In this chapter we will attempt to address this issue by writing in a way that we hope is more inviting and readable—as if we were keeping cross-border allies in mind. The task at hand is formidable and complex. No one will succeed alone. We suggest that a new ethos will require more collaborative research attitudes, generous scholarship, and an assemblage of scholars gathered to build a community of rebel researchers.

Norwegian philosopher Arne Næss puts difficulties in building community this way: "we tend to specialize. With this specialization there is a tendency to feel opposites instead of feeling the complexity of the relations and complementarity" (Næss & Jickling, 2000, p. 50). In a nod to how we might learn to think and do things differently, he recalls his experiences as a young scholar amongst the Vienna Circle of philosophers. In the end, he rejected their logical positivism; however, he greatly admired their searching minds. They were very different personalities, yet they were constantly helping each other, with invitations to try things differently and to look for new meanings. This ethos constituted what Næss called a kind of "research attitude" (Næss & Jickling, 2000, p. 51). Perhaps such a collaborative shift in research ethos, together with a dose of generous scholarship (Russell, 2006), would go a long way toward building the kinds of community that are required for our times.

We aim to open some space for movement in these directions by presenting wild pedagogies as a heuristic—that is, an agent of discovery rather than a rigid framework or plan of action. In this spirit we invite educators and scholars to experiment with our ideas, to try them out in their own places in the world, to look for new meanings, and to suggest ways that this work could be done differently, expanded, and even undone. Similarly, we encourage readers to seek convergences amongst all the chapters in this book, and then, to celebrate divergences as creative opportunities to take excursions together in new directions.

Wild Pedagogies

Wild Pedagogies arises within a complex of concerns about control. These concerns are about the ways in which cultural controls do violence to many and restrict the ways that we can think, act, live, and respond to possibilities for change in an era of uncertainty. We believe that current times require responses that are imaginative, creative, courageous, and radical.

Wild pedagogies is a relatively new term that has simmered since 2014. Initial experiments with wild pedagogies occurred during international gatherings of like-minded educators who sought to explore and expand this idea as an agent for educational change. By 2018, a small group of wild educators and scholars, affectionately called the *Crex Crex Collective*, banded together to publish a provisional gathering of ideas in the book *Wild pedagogies: Touchstones for re-negotiating education and the environment in the Anthropocene* (Jickling et al., 2018b). These heuristic-spirited gatherings have since continued and generated a growing body of literature that has been represented in a number of journal special issues.[2] These works rest upon two key premises. First, modernist relationships with the world must change; and second, education is a necessary, even fundamental partner in the project.

The work of wild pedagogies has been to reclaim language and reconceptualize ideas about the "wild" and wildness. And it has been driven by the frustratingly difficult task of enacting meaningful change, particularly in formal schooling settings (Aikens, 2021). What unites this work is a persistent concern about how issues of control can shape possibilities for change—explicit control, as well as more implicit controls embedded in contemporary language, metaphor, and cultural practices.

Wilderness, Wilding, and Will

Inspiration for wild pedagogies comes from ideas of wilderness, wildness, and will. We are well aware of critiques leveled against "wilderness." We understand that as a colonial tool, it has been used to disenfranchise people and cultures the world over (Bird Rose, 1996; Cronon, 1996). We also recognize that wilderness can be presented in a way that reduces its value to that of a backdrop for human-centered, self-serving, and colonial ends (Stewart, 2004). Yet we have also long known that there is more to wilderness than an absence of people, and a playground for heroic adventures.

Thus, despite its liabilities, wilderness still seems to be a potentially useful concept. There are places where more-than-humans flourish and where humans enter on terms that are more equitably dictated; wilderness is more than just an idea. At the same time, physical wildness is being located and encountered much closer to home—including in colonized urban areas—by those who are looking. Inspired by Deleuze and Guattari's (1994) idea that concepts are constantly being created and re-created, it seems timely to think again about how wilderness can be reconceived. We argue that a robust conception of wilderness does not necessarily rely on disenfranchisement of people from their homelands (Jickling et al., 2018a, 2018b).

In making this renewed case for wilderness, we appeal to Old English etymology. Here the word "wildoerness" can be said to derive from "wil" which in turn can be linked to wild or willed. "Doer" can be linked to beast, and "ness" is linked to place or quality. Putting these together suggests that wilderness can be thought of as a place of wild beasts, or more evocatively, *self-willed land* (Foreman, 2014). When this idea of self-will is juxtaposed against domestication, where "domesticate" is used in the sense of having been brought under control by humans (Livingston,

1994), the inherent agency of wilderness is given weight. Its wildness is celebrated; it informs us, and indeed, it teaches us if we watch, listen, and feel.

In an interesting twist, Norma Kassi, a member if the Vuntut Gwitchin First Nation in northern Canada, reached out to wilderness advocates in a different way. She affirms that there is no word for wilderness in her language; however, she does assert that there is a word for freedom (Kassi, 1996, p. 24). In this offering she suggests that in her culture, freedom overlaps with the best qualities of our own conception of wilderness.

Hints to Norma Kassi's understanding of freedom may lie in the hurt she expresses when humans "manage and study" animals. For example, she has spoken of a caribou that her brother once hunted that had been fitted with a radio collar: "under the collar," she said, "was covered in worms, it was tight. I do not know how the caribou lived, it was skinny and segregated from the others" (Kassi, 1994, p. 215). Kassi's freedom is not a freedom of individualism or economic imperialism. Rather, is seems aimed at an inherent freedom of self-determination and a freedom to flourish—even a kind of intrinsic value.

We acknowledge that wilderness and domestication should not be thought of as absolute qualities; wildness occurs in varying degrees of freedom and will, perhaps along a continuum. Still, for wild pedagogies, it helps to problematize ideas related to control while at the same time acknowledging the wild *agency* of the more-than-human world (Abram, 1996). Given the evolving ecological crisis of our times we suggest that ideas about a self-willed wildness can provide leverage in rethinking human relationships with the more-than-human world in ways of being that are less anthropocentric, less hierarchical, and more equitable for all. In the heuristic spirit of this project, we do acknowledge that this work is far from complete. See for example recent work on wilderness and wilding by Irwin (2021) and Quay (2021).

Wilding of Pedagogy

The desire for control often plays out in our educational institutions in ways that make things measurable, routine, universal, and that work to delineate ways of being. It is made manifest in many ways—often working to push educational practices into particular rationalistic ways of seeing the world. Such worldviews frequently run counter to the lived experiences of educators, learners, and parents, and serve to limit and domesticate educational opportunities. Impulses to push toward more radical reimagining of educational possibilities are tamed. There are too few possibilities for relational engagements within the natural world. The epistemological positioning required for mutual flourishing in a more-than-human world is often absent (See for example, Au, 2011; Jickling, 2009, 2015; Smith, 2016, Spannring & Hawke, in this volume).

Wild pedagogies is inspired by wildness. It represents a desire to let go of an overabundant sense of control, to invite the places we visit to become an integral part of our work, and to respond to provocations in spontaneous, and at times unforeseen, ways. A wilding of pedagogy rests on the premise that an important part of education can include intentional activities that provide a fertile field for personal and purposeful experience without overly controlling the environment and its actors, learners, or educational outcomes.

Problematizing control does not mean aiming for a directionless free-for-all. Rather we wish to challenge existing assumptions, to rethink possibilities, to push open the doors to educational opportunities, to expose the limits imposed upon epistemology, and to embrace the learning opportunities arising from being present to the more-than-human world. Thus, we are interested in how we might start pushing back on domestication and the desire for control in education.

Crucial to any success of wild pedagogies will be making concrete links between ideas and practice—pedagogies on the ground. We need to understand that social systems are often hostile to change, and subject to forces that bend actions back in the direction of the status quo. It is easy to lose sight of progressive, and indeed rebellious aims as we try to work out how change might manifest itself in what we do (Blenkinsop & Morse, 2017). We have been developing what we call touchstones to aid in this process.

Touchstones

In linking theory and practice, the touchstones described below aim to provide reminders, challenges, and a place to return to for educators interested in experimenting with wild pedagogies. They offer questions that educators can ask every day to remind themselves of what they are trying to do in their daily activities. For some, wild pedagogies will provide recognition of what they already do. For others it might inspire a wilding of their practice—providing opportunities to attend to the wildness of places, themselves, and their students in a deeper way.

Importantly, these touchstones are not static. They are provocateurs to be read, responded to, and revised as part of an evolving, vital, situated, and lived practice. What follows is our summary of those touchstones as we currently see them and one sample for each touchstone of the kinds of question posed for practitioners to consider. They rest on a substantial corpus of previous work, yet as this summary shows they continue to grow and change. (Blenkinsop et al., 2018; Jickling et al., 2018b; Morse et al., 2021).

Touchstone #1: Nature as Co-teacher

This touchstone asserts that education is richer, for all involved, if the more-than-human-world is actively engaged with, listened to, and taken seriously (Blenkinsop & Beeman, 2010). At one level this touchstone seems easy to understand and to put into practice. The claim is that the natural world is a vibrant, active, agential place that is worth listening to and learning from. Accepting this claim means that educators will spend more time outdoors and access different pedagogical possibilities. However, this touchstone also has implications for what knowledge is and how learning happens. If nature is embraced as co-teacher then the human is de-centered and learning becomes a shared project that is no longer ever complete or human-based. With this discussion as background, consider the question: How can I invite and provide space for the natural world to be present as a co-teacher in my practice?

Touchstone #2: Complexity, the Unknown, and Spontaneity

Education is richer for all involved if there is room for surprise. If no single teacher or learner can know all about anything, then there is the possibility for unexpected connections to be made, unplanned events to occur, and simple explanations to become more complex. Knowledge, if given space, is wondrously dynamic. This touchstone celebrates the unpredictable as it pushes back against the desire to categorize, limit, and contain. It listens for a diversity of voices, especially those that are marginalized or lost in learning environments where the standardized, the measurable, and the definable are the focus. For educators, this involves risk. Emergent approaches tend to complicate situations and curriculum design can no longer rely solely on desired learning outcomes. The suggestion here is that the world does not work in a clean, predictable, linear fashion and that something important is lost when we assume that it does. With this discussion as background, consider the question: How did my practice today take risks in moving away from full control of assumed ends?

Touchstone #3: Locating the Wild

The wild can be present everywhere but difficult to find. It can be made hard to see by cultural tools, by colonial attitudes, and, in urban spaces, by concrete itself (Derby et al., 2015). This touchstone cautions against the cultural constraints of much of modern public education and the often-present colonial orientations toward the natural world and many peoples. It challenges educators to think about their own privileges, including those related to the more-than-human world. It requires educators to be constantly aware of how language, metaphors, the structures they work within, and the tools they employ, can either challenge or sustain the status quo. It pushes back against the desire to control—both as humans controlling the more-than-human world and as centralized institutions controlling learners and educators. The wild, like freedom, runs contrary domestication and can be located anywhere, in the rural and the urban spaces, but also in individuals and their own acts of resistance. With this discussion as background, consider the question: How can I provide ways to acknowledge the wild and wildness in everyday encounters?

Touchstone #4: Time and Practice

This touchstone acknowledges that building relationships within the more-than-human world takes time and discipline. This touchstone focuses on both the processes and practices involved in building and maintaining these relationships, especially with those denizens who live near us. This process requires significant amounts of time immersed in particular places, listening to the world. For many, this will also require slowing down, changing habits, and listening to our own bodies and those others around us, in different ways. In practice, this requires work and discipline, much the same as developing a meditative practice. It will also take work to develop the one's own pedagogical practices—the how of one's teaching and the assumptions and habits that motivate that our work. With this discussion as background, consider the question: How might I leave space in my teaching to allow for meaningful engagement with nearby places and the beings living there?

Touchstone #5: Socio-cultural Change

We assert that the way many humans currently exist on the planet needs to change. This change is cultural and education is necessarily political player in this process. This touchstone begins with the radical premise that much of current educational practice is anti-environmental. It will not be enough to simply tinker with its edges. These premises, place the teacher in the role of activist, who recognizes that choices made in classrooms have explicit and implicit implications for how learners come to understand themselves and the natural world. This touchstone also recognizes that the future is no longer easily predictable, and that children are not growing into the same kind of world that their parents or grandparents did. Thus, educators need to challenge children to respond to uncertainty with creativity, visions for change, and building of shared community outcomes. With this discussion as background, consider the question: How can I actively make choices that provide students with possibilities for alternative relational ways of being and knowing while not furthering a sense of catastrophe fatigue?

Touchstone #6: Building Alliances and the Human Community

This touchstone seeks to build strong alliances and flourishing communities in a more-than-human world, but also not to forget to build human alliances in environmental and social justice communities. The goal is to push against individualization and alienation while resisting colonial moves to separate marginalized groups and place them at odds with each other (Simpson, 2017). We must listen and learn from each other while creating equitable and flourishing communities. Diverse platforms bring more perspectives to our conversations and can lend support to each other. Through such alliances, educators can learn from others—environmental educators from critical race theorists, community organizers from experiential educators, popular educators from gender theorists and more. And, there is much to be learned from alliances inclusive of the more-than-human world. With this discussion as background, consider the questions, When I think of educational possibilities, which communities do I reach out to? Who is not included? And, who might I add?

Touchstone #7: The Imagination—Limits and Possibilities

This touchstone aims to increase the imaginative range and the creative impetus in wild pedagogical work. Change can only happen when we can imagine alternatives, while also seeing ourselves as capable of acting in new ways. Yet, imagination is not unfettered. The edges of imagination are drawn by complex combinations of culture, experiences, histories, and our own creative practices. Working with wild pedagogies thus requires spotting our collective limits, then finding ways to offer our students and us experiences, encounters, and content that might expand imaginative ranges. As our work involves cultural change, we must extend our own imaginations in ways that penetrate existing cultural frames—including moves beyond current anthropocentric limits. With this discussion as background, consider the questions: Where are the edges of my imagination that limit my ability to create different kinds of education? And how might I expand my own, and my students', imaginative range?

Considering Wild Pedagogies in Practice

These touchstones and the kinds of question arising from them are a first step toward linking theory and practice. In the following section we consider some practical examples and consider further steps. We offer two vignettes of engagements with wilding pedagogy in practice. We acknowledge that these are locally based examples from the authors and that each attempt to work with wild pedagogies will be situated in different local contexts.

In offering these vignettes we include elements of the experiences that appear to tame the experiences and bend learning opportunities back toward the cultural status quo. Significant challenges are inevitable. However, enacting these experiments is exciting work that can provoke change. The question we ask here, though, is can we be better equipped to meet these challenges and enable hopeful pedagogies?

Vignette #1

We are sitting in the sun in a big circle on the well-manicured grass of a large urban park. It is late May on the west coast of Canada. The grounds are immaculate, and we are surrounded by a cascade of flowers, carefully cropped shrubs, and a bevy of unusual, non-Indigenous, tree species. The noises of kids playing team sports floats in the background like a babbling brook. This is pre-COVID picturesque. The group I am working with is made of 25 practicing teachers who are in their third semester of our nature-based experiential learning graduate diploma. This two-year program seeks to support teachers in becoming wilder, even rebellious, in their practices. It also seeks to sustain intriguing concepts such as nature as co-teacher, nature as colonized, and teacher as activist. This is the semester where teachers have been immersed in the ideas of wild pedagogies and are being asked to implement more nature-based lessons in their classrooms. This is the move from theorizing to practicing, to taking the ideas of wild pedagogies and trying them out in real situations with real kids, real families, in real schools. It is also the semester when we start to talk about the challenges, the push back, and our own limitations.

"Any challenges, discoveries, learnings arising from trying to get outside and implement wild pedagogies in your classrooms?" A hush falls. Folks are thinking but also worrying about being the first to speak; maybe everyone else has had a great time and things are going swimmingly. Ben, as he often does, is the first to break the ice and tells us a story of how one child asked why he was lying to the class about nature being in trouble. "My Dad says there is no climate change and that you should stop teaching lies." The group lets that one land, gathering before discussing, but this is tough to hear. Jennifer, always full of extravert energy, dives into the silence and tells a story about when she came upon two students killing ants and disturbing a third student, who was in tears. It was clear that her reflective-self regretted not dealing with the ant death at all. "I haven't really thought about how to deal with death in my classroom and not only that I totally forgot about nature as having rights and that living beings were actually being killed in all of this!" She had, by her own admission, slipped into both anthropocentrism and "old teacher habits" focussing solely on the emotions of the situation, the bullying, and the humans involved. Alyssa is next to speak and in her quiet deliberate way shares with us her sense that she is still "too controlling" to let learning emerge, to trust nature to "teach," to see time pass where it looks like kids aren't doing anything. "I know, in my head, that allowing them to build relationships with the natural world is a good thing. And I know learning is happening, because I have seen it, heard it, and documented it. And yet, there is voice inside my head that still questions this. I hear my colleagues wondering if we are just playing outside and I wonder if the community just thinks I am weird."

As the reader will note through this vignette there is a lot that can happen as teachers experiment with less mainstream, more outdoor, and wilder forms of education. While the teachers' excitement in wilding their pedagogy was profound, such attempts run up against elements of the political, cultural, and individual status quo that can frustrate and even stop the work. Below we highlight some of these in an attempt to be proactively prepared for them.

The politics of a "neutral" education. Ben's example is a pronounced version of the political in the school classroom, and the way wild pedagogies can push against assumed norms. It provides an example of the difficulties of implementing the critical touchstone of *socio-cultural change*. In many ways the assumed "common sense" idea of public education, what it looks like, what and how teachers teach, is embedded in the mainstream. This means that teachers moving in the direction of wild pedagogies are confronted with the politics of that move. Some see this as bringing an agenda to their classrooms—to manipulate or propagandize. This rests in the assumption that teaching is supposed to be from a position of neutrality. Such a flawed assumption arises from the privilege of being ensconced at the center—the status quo. As the old adage goes, the fish doesn't see the water in which it swims. Said another way, all teaching is political, but what are the politics that each teacher brings, and why? Answering this involves careful thought, ongoing community education, deep humility, and a critical eye. The question, then, becomes what are the insights I am seeking to offer.

Cultural frames of anthropocentrism. Cultural frames echo through each response. In Ben's response, we hear how the dominant culture and its assumptions are more apparent to those on the margins or those who are seeking to change that culture. In Jennifer's we hear a specific encounter with anthropocentrism and human exceptionalism—and of how challenging it is to recognize those habits, and then to change them. Through Alyssa we come up against the assumption that the teacher is the expert. It is also assumed that lessons should be organized to achieve measurable learning outcomes, control is held by the teacher, and students are managed. Implicitly, knowledge is the purview of humans, fragmentable, and distributed in bite-sized chunks. For wild pedagogues this process of encountering culture, of wrestling with troublesome habits, and of discovering ingrained ways of teaching is an ongoing accomplishment. Wilding pedagogies offers teachers opportunities to change in ways that reflect expansive expressions of educational freedom and their inner wildness.

The challenge of self-reflection. Many teachers experimenting with wilder pedagogies confront discursive battles within themselves. For Alyssa there is an intellectual commitment to being outside, and to working with the *nature as a co-teacher* touchstone. Yet, there is a culturally generated voice inside her that questions this. Alyssa points to a sense of "weirdness" that is likely a response to a western cultural context that pushes way from the natural world. In her context, it is often seen as soft, weird, or crazy to be an independent, autonomous, human who talks to and cares for the natural world. Another piece of this discursive battle lies in finding ways to trust both the natural world to perform as a co-teacher, and the students to be engaged learners, while seeking to create rich educational encounters. A wilder education requires relational alternatives in practice, and these, in turn, require pro-active trust in learners, places, and teachers themselves.

Vignette #2

We are standing on a shingle bank beside the river, looking out across the river towards water-worn features in the cliff wall on the opposite bank. We arrived in this place via rafts, journeying on the Big River in South Eastern Australia, and the group is made up of final year pre-service Outdoor Education teachers and a group of first year university students. It is a teaching opportunity for the final year students to trial outdoor environmental pedagogical approaches in this place. As part of their studies the pre-service teachers have been considering pedagogies that respond to our times—including wild pedagogies. In particular they have been working with the touchstone nature as co-teacher and the implied question of "how can I invite and provide space for the natural world to be present as a co-teacher in educational encounters?"

The experience begins with the pre-service teachers' invitation to consider the formations and imagine ways in which these geological features have been formed. There are several thoughtful responses from the first years that prompt discussions within the group. The idea here is to be led by the place. As the discussions develop the pre-service teachers add in catchment and geological information, including timelines and ideas about layering and metamorphosis. It is a lively and informative discussion based on learning from the place, and through direct first-hand experience within the place. There is a sense of excitement from the students and teachers alike. Yet, at the same time there are challenges in fully enacting nature as co-teacher. At times, for example, when students head off in a previously unimagined direction, they are in subtle ways returned to the view of "discovering" things about the place. "If you look over here you will see…" *or* "the rock you see here was formed by ...." *In many ways the pre-service teachers are following their lesson plans that were designed to work with the place as co-teacher, but in practice there are continuous cultural temptations to return to teaching about, rather than with, the place.*

In highlighting possibilities and challenges for educators in wilding pedagogical approaches, the vignette above brings into focus some contextual concerns and the critical importance of the touchstone *time and practice.*

Following a plan/avoiding risks. Having a linear lesson plan in place can promote an engaging lesson, but an important question might also be "what opportunities are being missed?" This question is not easily answered, as they may never become apparent without imaginative experiments such as wild pedagogies. What might happen, for example, if we carefully and deliberately set up experiences that welcomed the as yet unknown or unpredictable to occur? What might happen if, instead of working to funnel learners toward designated curriculum objectives, we begin with the quality of the relational encounter and the place itself? How might this look in the example above? Time would be required for students to explore and find things—places and interactions that, on their own terms, draw students' attention. Such points of attention and departure could be discussed using careful language that present materials, forces, and other beings as actively agential. This involves risk; it involves challenging ideas of control, trusting the place and the learners, and shifting cultural ideas about teacher identity (Green & Dyment, 2018).

Sliding into anthropocentric language. The *nature as co-teacher* touchstone asks how might we actively work toward knowledge held by places and other-than-human ecologies? One important response is to avoid sliding into an anthropocentric language of knowledge production. This is difficult work, because it is very easy to fall back into assumed cultural habits as we speak. For example, rather than saying, "the rock you see here was formed by..." we could ask... "if we try to imagine a different timescale, what might these rocks be telling us about how they arrived here and how they continue to influence things and lives in this place?" And critically, then, how might educators respond? Must the conversations lead to a singular compartmentalized knowledge or might it be okay to explore multiple available storylines within a place? To actively promote the *complexity, the unknown, and spontaneity* held within a place? This vignette, and other examples of practice (Blenkinsop & Piersol, 2013) suggest that if we listen carefully, stories may bubble up from outside our peripheral vision. Such stories might be macro—formation/transformation, or micro—spiraling whirlpools that catch insects on currents, or literal. This touchstone encourages students to see things from different angles—to test their own ideas about how things interact. The place might then guide and co-teach through gaining, providing, and becoming with attention.

Politics of knowledge out there. The lesson described in this vignette was considered successful teaching about, and to some degree with, the place. It also revealed an underlying assumption that teachers could ultimately explain the place; that knowledge to some degree exists *out there*, to be explained through a process discovery. Following this assumption, the place becomes a textbook, of sorts. It becomes a place where students can assume to discover a singular and relatively static reality. While this does provide a version of learning with the place, it does raise questions, such as, What epistemological possibilities might be side-lined by this assumption? And, in what ways, might the things, materials, and forces continuously act to produce the place on their own undiscoverable terms? In this way, knowledge might then be considered as situated, partial, dynamic, and necessarily co-produced with the place. We suggest that what are needed are more deliberate acts of pedagogy that place us directly, politically, and relationally within the world.

Concluding Thoughts

We must act differently—we cannot continue as we are—and education must play a role in the cultural change required. David Orr (2017), like many others, calls for serious educational change, because "without exaggeration it will come down to whether students come through their formal schooling as more clever vandals of the Earth and of each other" on the one hand, "or as loving, caring, compassionate, and competent healers, restorers, builders, and midwives to a decent, durable, and beautiful future" (pp. ix–x) on the other. What will it take to nurture caring, compassionate, and competent restorers of the earth? In part we believe this will require shift in scholarly ethos and we turn again to Arne Naess for insight. In the end, he preferred to put aside academic competitiveness, in favor seeking minds, sharing ideas, and a research outlook. At the core of his ethos was "trying to help each other"—to improve each other's work and to find new ways forward.

This book, in itself, steps alongside David Orr's question and toward a research outlook. With wild pedagogies, and with the other chapters presented here, we aim to provoke opportunities for reimagined relationships, enlarged more-than-human communities, and nurture caring and compassionate educators.

Responding to the ecological and social crises of our times, however, will require more. In working with students and communities to enact such cultural change, educators and researchers are called to rethink education, reimagine pedagogies, and, when needed, to fiercely resist the status quo—to be rebel teachers. By framing key underpinning ideas of wild pedagogies, situating them through the touchstones, and then experimenting with our practices, we hope to have offered a way forward that can provide possibilities for each of us to become better educators and allies of, for, with, and in the more-than-human world.

Notes

1. AHRC network, The Cultural Framing of Environmental Discourse, Workshop I, December 2–3, 2010, in Bath, UK. The Network is part of a programme on "Arts and Humanities Approaches to Researching Environmental Change."
2. See for example, *Pathways: The Ontario Journal of Outdoor Education* (2016) *28*(4); (2020) *32*(3); *Journal of Outdoor and Environmental Education* (2018) *21*(3); *Policy Futures in Education* (2021) *18*(3); *Canadian Journal of Environmental Education* (Forthcoming, 2022).

References

Abram, D. (1996). *The spell of the sensuous: Perception and language in a more-than-human world*. Pantheon Books.

Aikens, K. (2021). Imagining a wilder policy future through interstitial tactics. *Policy Futures in Education, 19*(3), 269–290.

Au, W. (2011). Teaching under the new Taylorism: High-stakes testing and the standardization of the 21st century curriculum. *Journal of Curriculum Studies, 43*(1), 25–45.

Bird Rose, D. (1996). *Nourishing terrains: Australian Aboriginal views of landscape and wilderness*. Australian Heritage Commission.

Blenkinsop, S., & Beeman, C. (2010). The world as co-teacher: Learning to work with a peerless colleague. *Trumpeter, 26*(3), 26–39.

Blenkinsop, S., Jickling, B., Morse, M., & Jensen, A. (2018). Wild pedagogies: Six touchstones for childhood nature theory and practice. In A. Cutter-McKenzie, K. Malone, & B. Hacking (Eds.), *International research handbook on childhood nature*. Springer.

Blenkinsop, S., & Morse, M. (2017). Saying yes to life: The search for the rebel teacher. In B. Jickling & S. Sterling (Eds.), *Post-sustainability and environmental education: Remaking education for the future* (pp. 49–61). Palgrave Macmillan.

Blenkinsop, S., & Piersol, L. (2013). Listening to the literal: Orientations towards how nature communicates. *Phenomenology and Practice, 7*(1), 41–60. http://www.journaltocs.ac.uk/index.php?action=tocs&journalID=30103

Braidotti, R. (2019). *Posthuman knowledge*. Polity Press.

Cronon, W. (1996). The trouble with wilderness: Or, getting back to the wrong nature. In W. Cronon (Ed.), *Uncommon ground: Rethinking the human place in nature* (pp. 69–90). W.W. Norton.

Deleuze, G., & Guattari, F. (1994). *What is philosophy?* (G. Burchell & H. Tomlinson, Trans.). Verso.

Derby, M., Piersol, L., & Blenkinsop, S. (2015). Refusing to settle for pigeons and parks: Urban environmental education in the age of neoliberalism. *Environmental Education Research, 21*, 378–389.

Foreman, D. (2014). *The great conservation divide: Conservation vs. resourcism on America's public lands*. Ravens Eye Press.

Green, M., & Dyment, J. (2018). Wilding pedagogy in an unexpected landscape: Reflections and possibilities in initial teacher education. *Journal of Outdoor and Environmental Education, 21*(3), 277–292.

Irwin, R. (2021). Rewilding policy futures: Maori whakapapa and the ecology of the subject. *Policy Futures in Education, 19*(3), 307–323.

Jickling, B. 2009. 'Environmental Education Research: To What Ends?', *Environmental Education Research, 15*(2), 209–216.

Jickling, B., Blenkinsop, S., Morse, M., & Jensen, A. (2018a). Wild pedagogies: Six initial touchstones for early childhood environmental educators. *Australian Journal of Environmental Education, 34*(2), 159–171.

Jickling, B., Blenkinsop, S., Timmerman, N., & Sitka-Sage, M. (Eds.). (2018b). *Wild pedagogies: Touchstones for re-negotiating education and the environment in the Anthropocene*. Palgrave Macmillan.

Jickling, B. (2015). Self-willed learning: Experiments in wild pedagogy. *Cultural Studies of Science Education, 10*(1), 149–161.

Kassi, N. (1994). Science, ethics and wildlife management. In J. Peepre & B. Jickling (Eds.), *Northern protected areas and wilderness* (pp. 212–216). Canadian Parks and Wilderness Society & Yukon College.

Kassi, N. (1996). What is a good way to teach children and young adults to respect the land: A panel discussion. In B. Jickling (Ed.), *A colloquium on environment, ethics, and education* (pp. 33–48). Yukon College.

Livingston, J. A. (1994). *Rogue primate: An exploration of human domestication*. Key Porter Books.

Morse, M., Jickling, B., Blenkinsop, S., & Morse, P. (2021). Wild pedagogies. In G. Thomas, J. Dyment, & H. Prince (Eds.), *Outdoor environmental education in higher education* (pp. 111–121). Springer.

Næss, A., & Jickling, B. (2000). Deep ecology and education: A conversation with Arne Næss. *Canadian Journal of Environmental Education, 5*, 48–62.

Orr, D. (2017). Foreword. In B. Jickling & S. Sterling (Eds.), *Post-sustainability and environmental education: Remaking education for the future* (pp. vii–xi). Palgrave Macmillan.

Quay, J. (2021). Wild and willful pedagogies: Education policy and practice to embrace the spirits of a more-than-human world. *Policy Futures in Education, 19*(3), 291–306.

Russell, C. (2006). Working across and with methodological difference in environmental education research. *Environmental Education Research, 12*(3–4), 403–412.

Simpson, L. (2017). *As we have always done: Indigenous freedom through radical resistance*. University of Minnesota Press.

Smith, W. C. (Ed.). (2016). *The global testing culture: Shaping educational policy, perceptions, and practice*. Symposium Books.

Stewart, A. (2004). Decolonising encounters with the Murray River: Building place responsive outdoor education. *Journal of Outdoor and Environmental Education, 8*(2), 46–55.

CHAPTER 3

The Epistemological Possibilities of Love: Relearning the Love of Land

Estella Carolye Kuchta

Prelude: Learning to Walk in the Dark

I learned how to walk in the dark, first pregnant, then holding a baby. The first lesson of darkness is: Most darkness is not so dark. On the remote, off-grid California mountaintop where I lived, locally and fondly known as "Last Chance,"[1] shoals of stars lit up the night sky. The fingernail moon brightens enough stones to find one's way down the path. The smiling Cheshire cat moon turns bull pine, madrones, spiky agave plant, and trails a milky-blue, so the walk can be less cautious. The generous pancake moon casts shadows beneath the elbowy march of a beetle and sheds pretty dappled light through leaves to the forest floor. A stroll through

For Maxwell and for Last Chance.

E. C. Kuchta (✉)
Department of English, Langara College, Vancouver, BC, Canada

Faculty of Education, Simon Fraser University, Burnaby, BC, Canada
e-mail: ekuchta@langara.ca

M. Paulsen et al. (eds.), *Pedagogy in the Anthropocene*, Palgrave Studies in Educational Futures,
https://doi.org/10.1007/978-3-030-90980-2_3

that darkness is a smiling, ethereal event, playful and surreal. Amidst this shifting beauty, I went out to fetch the firewood, to trek to the outhouse, to take a shower, to fetch a forgotten sweater from the garden, or to travel between our primitive cabins (mostly a series of shacks: a 'main' cabin, a bedroom cabin, a bathhouse cabin, a spare bed cabin, and our primary home—a converted school bus.)

The second lesson of darkness is: It is beautiful. It's an under-appreciated, warm thing wrapping the body in soft, black fur. Morning offers hope. Evening offers respite. But night gives us pause, a sense of being outside of time. Often, I put a hand to my belly and made excuses to step out into it. When I did, joy bloomed inside me. Lesson three: Stepping into the night means stepping into a story told by a dear friend. The groan of a madrone trunk tells of uncomfortable contrasts between daytime heat and night cold. The small beetle now pausing on the trail senses an insect nearby. Pausing to sniff the air, Tango, my blue heeler, helped me notice the scent of cool rocks, tangy pine, and sweet madrone. Occasionally, muskiness hinted that a raccoon or rat was nearby. When I stepped into the night, my senses—the five ordinary ones and the other extraordinary ones—seemed to ask: What is the story here?

Occasionally, however, darkness is complete. If thick clouds hid a new moon, if my partner was away so the lights were off in the other dwellings, if I left my flashlight somewhere else—then night wrapped a black blanket too tightly around me. In those moments, I couldn't see my own body and I had 50 yards of winding, bumpy trail to navigate through the trees. The propane lights inside the cabin were too dim to shed light up the trail. The most immediate threat was that I might trip and hit my head or belly on a rock. But my mind went first to the nocturnal mountain lions in the area and the peculiar, translucent scorpions that ventured out to devour beetles in the night. But we had also seen black widow spiders, rattle snakes, lynx, large bucks, and packs of wild boars on the mountaintop. Fear jammed up inside me—throbbing in my chest—undecided whether to make me freeze, run blindly in panic, or scream, though no human would hear me. That's when I learned the fourth lesson of night: Fear is a terrible guide through the dark. Fear constricts and threatens to scatter needed focus. It divides—me and my unborn baby on one side—and the land and its other inhabitants on the other side.

Beneath the fear, something else, softer and more inviting, called to me. I didn't decide what to do next. I just did it. With a hand on my belly, where my growing baby slept, I *asked* the land to show me the way. I knew this land and this land knew me. Remembering that—feeling it—calmed me. Still on high alert, I reached my sandled toe toward the trail.

At the trailhead, I veered right to make room for the small fir branch leaning over the trail and stepped tall to avoid tripping on the three raised madrone roots. I veered slightly left, my foot crunched dried leaves—too far left. I straightened onto the trail. I paused briefly to listen for Tango. Her silence reassured me. If a mountain lion was near, she would explode with noise. (I believed that on faith, but several years later, she proved it while successfully protecting my young nephew from a night lion.) The map in my mind reminded me where to step around and over the next stumps and roots, and when to bend hard right up to the cabin. I touched the arms of my tree friends, remembering the location and size of each. Soon enough, I reached the school bus, found the matches, and lit a candle. Then I paused and said, *thank you*.

Ontology of the Isolated Individual and the Limits of Standard Education

As numerous scholars have addressed (e.g. Jickling et al., 2018; Louv, 2005), standard North American educational practices do little to foster significant relationships between humans and more-than-humans or acknowledge their existing relationships. My own cultural and institutional education illustrates a few ways children are drawn away from naturally occurring relational ontologies emerging from shared identity, care, and love. Twelve years of standard Canadian education helped me understand moon cycles, cloud formation, and the basics of animal motivations. At church, I learned that animals—including my beloved pet dog—and plants have no souls and are, therefore, inferior and destined to be ruled by humans. Before the age of six, I never doubted that the maple trees in my yard knew me, cared about me, and communicated with me. This sense went beyond object-projection because, unlike a teddy bear, the trees responded to me. They swayed and dropped leaves and, most importantly, talked to me albeit without human words.

One day, however, I arrived home to see my father had felled one and was going at it with the chainsaw. To some extent, my experience in that moment relates to Gilligan and Snider's (2018) assertion that cultural patriarchy demands the loss of some relationships in exchange for broader cultural acceptance. Seeing my horror, my father, a second-generation

Eastern European settler, explained that trees don't feel, can't think, and don't have the consciousness to understand their lives or deaths. Believing him meant denying all I had experienced in my friendships with those trees. It meant abandoning the quiet closeness we shared and intimate sense of abiding mutual care. But not believing him meant that I was witnessing the death—the murder—of one of my close companions—a pain I could not bear. In severing my relationship with all trees, I stepped into an ontological orientation aligned with the dominant North American culture and reinforced through formal education.

Standard North American education orients from a fundamental assumption about the isolation of the individual. Bayo Akomolafe (2020) articulates, "One of the most persistent and sticky habits of perception that has possessed those of us gestating in modern civilization is we tend to see things as separate from each other." This ontology is so deeply rooted and unchecked that even earnest educational attempts toward more wholistic epistemologies and greater interconnectedness tend to orient from it. Scholars and teachers might talk about *reaching out to* the other and *making* relationships, but these actions can only take place if we assume the starting point is that of the isolated individual. For example, standing at the edge of the black abyss that night, I might have, like Richard Kearney (2015), prioritized my sense of touch for "making sense and *receiving* sense from … something *other* than myself," crossing "back and forth between self and strangeness" (p. 104). Kearney's sensual analysis of touch limits understanding of the "other" to what skin can sense and brain can, thus, interpret. This framing orients around the primacy of material bodies—mine and that of the trees, stones, and soil around me. From this perspective, identity is bound to the material body, and only with concerted effort might we perceive the other who is bound within their own material self.

Perhaps more so than any other foundational belief, the ontology of individualism limits the capacity of humans to engage in ecologically ethical and relationally meaningful ways with our more-than-human community. Indeed, this is the worldview of the cultures that have carried us into the Anthropocene. If my perception of my own beingness is primarily that of independence, naturally, I will care for *self* before considering my impact on more-than-humans. Furthermore, I will expect all others, human or not, to prioritize themselves as well, leading to a worldview defined by competition and, thus, hierarchies. From this standpoint, more-than-humans are seen not as relationships but as resources. Ultimately, actions that impact land, water, and other entities within the natural world are seen as justifiable if they provide personal profit. This is the kind of faulty reasoning underscoring the Anthropocene, a time of reckoning with the dangerous consequences of cultural errors.

Can Land *Love*?

Many contemporary scholars are shifting away from the object-oriented worldview and making moves toward more relationally-oriented ontologies, whether through forms of new materialism (Bennett, 2010), animism (Stengers, 2012; Bai, 2015), posthumanism (Snaza & Weaver, 2015), making kin (Haraway, 2016) or a scientific recognition of the deep interconnectedness of human development (Lieberman, 2013; Narvaez, 2014). The field of love research has much to offer this movement. In human experience, love is a central binding agent and catalyzing force with the capacity to spark new life, and thus, new subjects which are inherently already in relationship. Up on the dark mountain, the growing baby inside me, like most unborn babies, was sparked through relationship and through love, and was already in relationship to the mountain prior to birth through the oxygen and nutrients absorbed by me along with accompanying affective experiences. Clarifying Pierre Teilhard de Chardin's theory of love, Ilia Delio (2017) writes, "union precedes being because love is the core energy of evolution and love is intrinsically relational" (p. x). Chardin (1817) explained simply, "love is the primal and universal psychic energy" (p. 4).

Can land *love*? Dominant North American culture says it cannot. What is the evidence for this belief? First, many psychologists would point out that the composition of soil, rock, and flora does not contain the necessary physiological systems required for experiencing love, namely, linked hormonal, neurological, and epidermis systems. It is a typically Western fallacy to assume that intelligence and feeling can only be measured using the human yardstick and human-like faculties. Even Western scientists have begun to understand that the human exceptionalism bias has crippled our capacity to comprehend the vast complexity of intelligence and sensibility possible outside the human sphere (e.g. De Waal, 2016). For example, Peter Wohllenben's research (2015) on trees illuminates sophisticated systems of care, community, and communication. More broadly, the theory of Gaia urges educators and activists to consider the entire earth as a highly elaborate network of living organisms with enormous capacity for response to its human inhabitants (Latour, 2017). Some scholars may feel squeamish about seeing land as "loving" because, to them, this invokes quaint images of tearful trees and other silly personifications. While moving to recognize the agency and capacities of more-than-humans, let's sidestep the tendency to remake them in our own image.

Love has been defined as an action (hooks, 2001) and an energy (Savary & Berne, 2017). Here, theologian Pierre Tielhard de Chardin's theories of love overlap with Potawatomi Botanist Robin Wall Kimmerer's (2013) explanations of love and reciprocity. Teilhard, according to Delio (2017), believed "the *physical* structure of the universe is love" (p. x). Savary and Berne (2017) elaborate, "Teilhard always thinks of love in all its many forms primarily as energy—as the ability to do work. Love is energy because it is able to accomplish things, make a difference, transform people" (xiv). Kimmerer described, "Something essential happens in the vegetable garden. It's a place where if you can't say 'I love you' out loud, you can say it in seeds. And the land will reciprocate, in beans" (p. 127).

Love from land is a highly complex series of interconnected and ongoing processes, that is, energies and actions that are supportive, generous, and sensual. Land loves through interconnected entities of light, soil, and plant life, as well as those of weather patterns, and sensory stimulus. The land on the mountaintop provided me and my unborn baby with black beans, green beans, bell peppers, lettuce, tomatoes, potatoes, herbs, flowers, and herbal medicines along with rain and creek water for washing. We took these forms of love into our bodies, absorbing their goodness and nourishment. This love gave us the physical energy needed to eventually move to other lands, just as the love from other lands had carried me to this one. Food, water, and natural medicines are just a few of the most obvious ways the land expressed love.

Other ways involve the generosity of the endless beauty, familiarity, and welcomeness, a feeling of 'Home.' Sue Gerhardt (2004) explained that repeated emotion experienced by mothers during pregnancy can and does help shape a child's physiological attachment systems, often for life. How might my repeated joyful and intimate encounters with night have imprinted my unborn son? The dazzle of stars? The gift of an armful of kindling on cold nights? The sweet snap of a pea eaten in the moonlight while the San Pedro cactus reveals its yearly 24-hour bloom? The land at Last Chance offered me generosity, beauty, inspiration, friendships, and companionship. Love binds people to land and offers a 'between' space, where ecologically responsible and relationally attuned knowledges can emerge. Kimmerer (2013) explained, "Knowing that you love the earth changes you, activates you to defend and protect and celebrate. But when you feel that the earth loves you in return, that feeling transforms the relationship from a one-way street into a sacred bond" (p. 125). Two decades later and a country away, my son and I still feel a sacred bond with Last Chance.

Although loving bonds to land lend to ecologically ethical epistemologies, standard education actively undermines the relational ontologies that support it. Numerous scholars have identified the ways contemporary North American culture increasingly experiences an impoverishment of love (Sorokin, 1954; hooks, 2001; Savary & Berne, 2017), relationality (Gerhardt, 2010; Narvaez, 2016) and perversions of emotions in general (Illouz, 2007; Cederstrom & Spicer, 2015); however, love impoverishment also applies to the dominant North American culture's relationship to more-than-humans, a relatively unexplored dynamic. Love scholars tend to view plants, animals, and land as objects deserving appreciation and care (Noddings, 1986; hooks, 2001) while eco-education scholars tend to discuss 'relationships' and 'care' for the natural world without mentioning the love inherent within those relationships (Jickling et al., 2018). In essence, ecological education can further deepen notions of relationality from the lens of love research, while love research can map previously overlooked terrains of love relationships with land. As Delio (2017) articulates, "How we love is how we live, and who we love shapes our relationships, communities, our connection to the Earth, and the future of the Earth" (p. xi).

Definition of Relational Ontology

Particular pathways within science, Indigenous education, theology, and environmental ethics converge around the idea of a relational ontology. This orientation, as I define it, begins from the philosophic perspective that an "I" does not exist without relationship; thus, relationships come first, individualism is secondary. Western education, however, is rooted in an ontology of individualism, an understanding of the world where the "I" is central and of primary importance while relationships are secondary. Describing a relational ontology, Ross and Mannion (2012) state, "the world is a domain of relational entanglement" (p. 303) and where "learning is a process of 'attunement' to the meanings that inhere in the relationships that make up the world" (p. 304). Referencing Tim Ingold's theory of dwelling, they describe, "ants, humans, stones and mountains are to be understood as knots in or interlacings of relationship, or a 'domain of entanglement'" (p. 305). In this writing, the relational ontology refers to an ecological worldview, wherein all entities are seen as entangled within relationship and those relationships assume greater importance than the entities themselves.

The relational ontology is supported by contemporary Western science, but education, by and large, hasn't quite caught on. As Darcia Narvaez (2016) asserts, "Perhaps the most misleading aspect within Enlightenment philosophy is the conception of the basic human condition, that of individualism, a conception that is untrue on nearly every level of analysis (e.g. at the quantum level everyone on earth is connected; at the biological level, humans share DNA with virtually every other entity and each person is a community of microorganisms)" (p. 8). From an educational standpoint, having a relational ontology is not an invitation to step into relationship or to "plac[e] relationship at the nucleus of … pedagogical practice" (Glover, 2019, p. 87), for the individual can no more "place" relationships at the center of teaching practice than the earth can "place" the sun at the center of its yearly circuit. Rather, orienting from a relational ontology within education means a recognizing and prioritizing the deeply entangled relationships existing between all things. Understanding the science of interconnection makes it easier to see how the orientation toward individualism has been a choice, rather than a philosophic or scientific fact, as contemporary education tends to imply.

Traditional North American Indigenous education, for example, makes an altogether different choice. In his description of traditional Indigenous education, Gregory Cajete (2016) points out that children first learn about relationships with family, culture, and place (p. 371). Much later, they learn about their individualization, but even then, "a deep understanding of relationship and diversity" are emphasized (p. 372). When European settlers first arrived in North America, several hundred diverse Indigenous cultures existed, each containing unique customs and languages; yet, they all shared deep-seated respect for more-than-humans to whom they felt inextricably bound. The downgrading of individualism and prioritizing of relationship is often evident in these languages which tend to favor verbs and verb forms, rather than nouns. Noun-based languages centralize *things* rather than processes, interactions, and interminglings. Indigenous verb-based languages, such as Potawatomi and Rarámuri, highlight the sentience of plants, animals, elements, and land

(Kimmerer, 2017; Wyndham, 2020). A relationship—in the true sense of the word—with a sentient being will naturally be more involved, entangled, meaningful, and complex than a 'relationship' with a non-living object.

These Indigenous ontological orientations are important for settler educators to understand for a couple reasons. First, due to their own lack of awareness, educators can inadvertently make Indigenous students feel uncomfortably different and unable to 'fit in' with expectations. Imagine the experience of an Indigenous student whose well-meaning teacher assumes children can only begin to understand the natural world through understanding Western-style science. This child's entire relational experience as well as their culture is sidelined, if not discarded entirely. Stan Rushworth (2020), who was raised by his Cherokee grandfather described, "In the classroom was where the divisions took shape, in the mind and heart" (p. 9). (It's not only Indigenous students who might experience this. More on this later.) Secondly, the fact that hundreds of diverse cultures relate to the natural environment of North America similarly, suggests that perhaps the land itself invites certain kinds of interactions and shapes human ontologies. Sheridan and Longboat (2006) explain that the Haudenosaunee (Mohawk) believe all human creativity and solution-finding results from more-than-humans gifting humans with ideas in this shared space of cognition. Referring to the Haudenosaunee creation story, Sheridan and Longboat note:

> As the last being Teharoniawakon created, humans are and remain dependent on all other beings, and whether those beings offer their lives for our nutrition or their sentience for our thinking and imagining, ancient reciprocities continue. We are thankful. Their love of us and our love of them guide our path to that future that replicates and restores antiquity. (p. 366)

Sheridan and Longboat (2006) suggest that "minds and culture mature" with the land and that settler culture is still developing along these lines (p. 366).

Relational Ontologies in Educational Alternatives

Although my own cultural and formal education could not help me navigate the dark night, other kinds of learning did. These learnings were acquired very much at the fringes of mainstream society. After stumbling upon the *Medicine Cards* (Sams & Carson, 1988) in an alternative bookstore, I renewed my relationship to trees and other more-than-humans. One day, I had a profound and memorable conversation with a homeless Indigenous man which encouraged me to trust my own sense of knowing, despite mainstream epistemologies. Later, in an elective class at a California college, I read about the inherent relationality of individuals, families, communities, war, land, and weather in Leslie Marmon Silko's *Ceremony* (1977) and felt the profound, life-altering sense of relief of one whose worldview has for the first time been articulated and validated.

These experiences awakened something in me that had been largely dormant through my elementary and high school years; I began to actively and deliberately enter the shared non-material spaces of humans and more-than-humans. I felt an open, clear, but wordless communication pass back and forth between a neighboring cedar tree and me. I listened to the wind, the rain, and the coyotes with my whole spirit—not just my brain—with an openness to whatever communications might come. Nel Noddings (1986) wrote that a "psychic relatedness lies at the heart" of an ethic of care (p. 1). Although her ethic of care doesn't extend to more-than-humans in any meaningful sense, her astute observations that "all caring involves engrossment" (p. 17) "a 'feeling with' the other" (p. 30) aptly describe my own engagement with the land at Last Chance. With a couple of my human loved ones and this specific land, I shared a "psychic relatedness" where communication was fluid and unencumbered by laborious human language. In other words, it was not always necessary to be in each other's presence to know how the other was doing or what they were experiencing.

Additionally, as a woman pregnant with her first child, I was undergoing a profoundly embodied and relational love experience that inherently stands apart from traditionally masculine realms and delineations of reason and the brain-centered education of schooling. A pregnancy, like the act of stepping outside in the night, is an experience of loosening Western notions of linear time. During pregnancy, one steps outside time into an immediacy that is part inheritance, part the memory and the daydream of love, and part invitation to possible futures. In this sense, it is reminiscent of Sheridan and Longboat's notions of "future that replicates and restores our antiquity" (p. 366); a pregnancy is an embodied way of remembering into the future. Kimmerer (2013) asserted, "For all of us, becoming indigenous to place means living as if your children's future mattered, to take care of the land as if our lives, both material and spiritual, depended on it"

(p. 9). On the mountain, care for the land—meaning, attending to plants in the garden, respecting the homes of scorpions and rattlesnakes, not going beyond the carrying capacity for water use—was a way to care for the land, myself, and my future child. The three could not be separated.

Pregnancy also lent to another kind of knowing, a powerful intuitive sense likely passed down through generations of attuned mothers. I knew my child before he was born—his temperament, his spirit—the same way I knew his younger sister's later. I could impart things to him and he to me. Admitting this in an academic context means taking the significant risk of being written off as dreamer whose feet have long since floated off the ground. This is part of the 'secret' knowing that I'm aware—without having to be told—that I'm not supposed to talk about. Intuitive knowing is an inherent aspect of relational epistemologies, but in Western realms has always been relegated to categories of superstition, wishful thinking, and fanciful imagination. Could this be because those designing the shape of our academic institutions have little or no experience with genuine intuition? At best, they might chalk it up to keen sensory perception operating just below the level of consciousness. It's true that when one views the world from the standpoint of the isolated individual, genuine extrasensory intuition appears literally impossible. However, within the relational ontology, the idea that knowing can be shared between mother and unborn child seems logical, practical, and obvious.

Standard Education Actively Discourages Relational Ontologies

Standard North American education, however, does not allow for this kind of interconnection. "Intuiting" and "talking to spirits" or the "spirits of the land" are activities relegated to fluffy non-scientific sentimentality and nonsense. Children who—whether through family culture or innate sensitivity—have an intuitive entanglement with the lizard, the red tailed hawk, the wind, or even each other quickly learn to hide, suppress, and ignore this kind of knowing. Blenkinsop and Piersol (2013) call these children "incredibly subtle feelers and responders; they seem to soak up everything that is around them, such that their self is porous and wide open" (p. 55). For them, the process of being in school is as much about learning what one is *not supposed to know* as it is about gaining knowledge. They can experience this in their knowledge of other people or more-than-humans. Regarding children's connections to the natural world, Blenkinsop et al. (2018) refer to suppression of ways of knowing as eco-double consciousness, the act of splitting the self into two halves: the half that is acceptable to the wider society and the secret, suppressed self that knows things through deep relatedness and sensitivity to the natural world.

Boys, in particular, learn that deep relationality is problematic, unmasculine, and unacceptable. Girls can get away with intuitive relatedness and seemingly 'sentimental' and 'irrational' connections for longer. Blenkinsop, Piersol, and Sitka-Sage further suggest older boys actively belittle sensitivity to nature "not only to gain a sense of power over 'nature lovers,' but to enforce the cultural norms of hyperseparation" (p. 351). They continue:

> In addition to interwoven layers of sexism, homophobia, anthropocentrism, and a troubling conformity under threat of violence, their gesture insists that the "voice of the world" (Evernden, 1985) be, and remain, severed and silent by the time of "manhood." Those who have not "matured" along these developmental lines and who maintain a deeper relationship are deemed: emotionally irrational, sentimental, effeminate, naive, or queer. (p. 351)

For young boys and for all children sooner or later, intellectual reasoning and pragmatic sensibility are prized and rewarded in the classroom and beyond the schoolyard. By high school, most North American boys understand that it would be more acceptable to admit to each other that they watch porn every night than to admit to communicating with a lizard, the wind, or a tree. Desire is permissible, even admirable, in capitalist culture but love, an ethic of care, and intuitive entanglement are not.

Rather than being seen as a gift, an intelligence, deeply relational, intuitive knowledges of any kind are often painfully suppressed by children trying to gain acceptance. My own son, Maxwell (personal communication, 2 June 2020), described battling between acceptable knowledge and unacceptable knowledge in later elementary school and early high school. He regularly perceived things about his teachers that he felt he shouldn't know. He described the acquisition of this knowledge as "like accidentally reading someone's diary." Now a young adult, he recalls this as a period of time when he was uncomfortably aware of being "different" and was preoccupied by "guilt and embarrassment."

Standard education—at all levels—positions us to become unable to know *what we know*, and in this way, we become fractured individuals, severing parts of ourselves for the sake of greater acceptance (Gilligan & Snider, 2018). In his analysis of comparative epistemologies of nature knowledges, Zwart (2008) warns, "The scientific ego has to learn to be on its guard against alluring images and intuitions" (p. 43). With no explanation or definition, "intuition" becomes a forbidden topic, though one might note the hint of gendered bias as both "alluring" and "intuition" tends to be associated with women, not men or the masculine institutions like the academy. To be clear, the personal stakes of off-hand comments such as these are high. The psychic pain of fractured relationships and the deeper injury to the relational self can ripple out across a lifetime in attachment disorders, loneliness, spiritual hunger, and self-doubt.

The rejection of relational knowledges can be understood by recognizing that *agape* (divine humanitarian love) lies at their core and love in all forms is problematic in patriarchal societies. Gilligan and Snider (2018) explain that patriarchy "forces a betrayal of love and then renders the loss irreparable," arguing, "The sacrifice of love is the thumbprint of patriarchy" (p. 16, 33). The boy whose strong sense of care and deep empathy carried him into seemingly forbidden relational spaces with his teacher, must turn away from the teacher emotionally and pretend the loss of that bond was always inevitable. Although Gilligan and Snider explain this paradigm in human-to-human terms, it can be applied to human to more-than-human relationships as well. After my father's explanation of the inanimacy of trees, I abandoned a very real friendship with the maple trees in my yard and accepted that connection was gone forever. The boy who shared thought-space with the lizard, learning ineffable but visceral lessons about what it means to move, react, and experience the world like lizard, turns away from this relationship and likely ignores the heartbreak caused by doing so. Later, sensing a discomfort from a now unrecognizable source, he may discourage his own children from bonding with the lizards.

Mainstream North American culture allows for love of more-than-humans but with limitations. Pets, national parks, and homesteads, for example, can be loved. However, when that love challenges capitalist goals by turning into protection of rivers, old growth forests, and snowy owls, it can lead to raids and arrests, such as was experienced by the Wet'suwet'en pipeline protestors or even ostracism and violence, such experienced by the Dakota Access pipeline protestors. In contexts like these, claims of 'love for land' are seen as unnatural, unreal, and threatening. When North American companies set up camp in other countries, such as mines in Guatemala or agribusiness in the Amazon, Indigenous protectors of land have been threatened, tortured, and murdered. These protectors are seen as terrorists willfully obstructing the engine of capitalism rather than as warriors acting out of love and orienting from a deeply relational worldview.

The Way Forward: An Ontological Shift in Education

If education is to adapt to a new ecological ethos, defined by care and attunement to our natural environment, a shift from an individual-oriented ontology to a relational ontology is an important first step. Beeman and Blenkinsop (2019) acknowledge, "no reasoned argument will ever fully convince homo mobilis out of one ontological space and into another" (p. 10). However, as Savary and Berne (2017) remind, "Almost everyone agrees that people—and the world—have been changed by people loving and by being loved" (xiv). Radical shifts in one's worldview occur because of relationship and, frequently, because of the love inherent within those relationships.

All love is ultimately a walk in the dark, sometimes on a lovely moonlit night, sometimes through a frightening space of danger. Love can involve risk as well as trust, togetherness, entanglement, and unity. The relationship comes first and differentiation comes later. Even when we 'first meet' the other, we are finding our way back to a togetherness—yet unknown, but always already within existence as a potentiality. If we, in North American standard education, are willing to *regain* this sense, we will all feel less alone, we will better understand innate belonging and the responsibility of interconnectedness that is central to an ecological ethos.

Blenkinsop and Piersol (2013) noted, "as a result of this ontologically different orientation, the more-than-human world speaks to you on a literal level, in its own languages and ways" (p. 52). When I asked the land to show me the way, I remembered the familiar and comforting feel of the path beneath my feet, and with that memory, I felt *pulled* forward as though a force that was beyond me and beyond the land but uniting us both compelled me. I sensed the land sensing me, and that awareness eased my fear. Many months later, with a baby on my hip, I stepped out again into the dense opacity of pure blackness. I didn't have a small orb of a flashlight to bump along a trail and constrict the story to a handful in size, and that fact frightened me. If my baby sensed my unease and started to cry, the chaos of noise would utterly disrupt my ability to sense our way, to sense the shared intuitive space where I could listen for the land to guide me. For his sake, I quieted myself and, as Noddings describes, began "feeling with" the land (1986, p. 30). He remained completely calm and quiet as we trekked through the erased world, the weight of him on my hip suggesting a relaxed mood. He *knew* this land in the cells of his body, in his lungs, with his love, with the love we all shared in that space.

Note

1. I gratefully acknowledge that Last Chance belongs to the traditional territory of the Ohlone people. I further acknowledge that while I write this chapter, I sit on the traditional territories of the Tsleil-Waututh, Squamish, Sto:lo, Stz'uminus, and Musqueam peoples.

References

Akomolafe, B. (2020, January 22). Slowing down in urgent times. *For the Wild Podcast*. Host A. Young. https://forthewild.world/listen/bayo-akomolafe-on-slowing-down-in-urgent-times-155

Bai, H. (2015). Peace with the earth: Animism and contemplative ways. *Cultural Studies of Science Education, 10*(1), 135–147. https://doi.org/10.1007/s11422-013-9501-z

Beeman, C., & Blenkinsop, S. (2019). Environmental end game: Ontos. *Environmental Education Research, 26*(9–10), 1479–1490. https://doi.org/10.1080/13504622.2019.1663792

Bennett, J. (2010). *Vibrant matter: A political ecology of things*. Duke University Press.

Blenkinsop, S., & Piersol, L. (2013). Listening to the literal: Orientations towards how nature communicates. *Phenomenology & Practice, 7*(2), 41–60. https://doi.org/10.29173/pandpr21167

Blenkinsop, S., Piersol, L., & Sitka-Sage, M. D. (2018). Boys being boys: Eco-Double consciousness, splash violence, and environmental education. *The Journal of Environmental Education, 49*(4), 350–356. https://doi.org/10.1080/00958964.2017.1364213

Cajete, G. A. (2016). Indigenous education and the development of indigenous community leaders. *Leadership (London, England), 12*(3), 364–376. https://doi.org/10.1177/1742715015610412

Cederstrom, C., & Spicer, A. (2015). *The wellness syndrome*. Polity.

Chardin, P. T. D., & Universal Library. (1817). *On love & happiness*. Harper & Row.

De Waal, F. (2016). *Are we smart enough to know how smart animals are?* Norton.

Delio, I. (2017). Preface. In L. M. Savary & P. H. Berne, *Teilhard de Chardin on love*. Paulist.

Evernden, N. (1985). *The natural alien: Humankind and the environment*. Toronto, Canada: University of Toronto Press.

Gerhardt, S. (2010). *The selfish society: How we all forgot to love one another*. Simon & Schuster.

Gerhardt, S. (2004). *Why love matters* (2nd ed.). Routledge.

Gilligan, C., & Snider, N. (2018). *Why does patriarchy persist?* Polity Press.

Glover, S. K. (2019). Walking alongside my relations. *SFU Educational Review, 12*(1), 83–95. https://doi.org/10.21810/sfuer.v12i1.614

Haraway, D. J. (2016). *Staying with the trouble*. Duke.

hooks, B. (2001). *All about love*. William Morrow.

Illouz, E. (2007). *Cold intimacies: The making of emotional capitalism*. Polity.

Jickling, B., Blenkinsop, S., Timmerman, N., & Sitka-Sage, M. D. (2018). *Wild pedagogies*. Palgrave Macmillan.

Kearney, R. (2015). What is carnal hermeneutics? *New Literary History, 46*(1), 99–124. https://doi.org/10.1353/nlh.2015.0009

Kimmerer, R. W. (2013). *Braiding sweetgrass*. Milkweed Editions.

Kimmerer, R. W. (2017). Learning the grammar of animacy. *Anthropology of Consciousness, 28*(2), 128.

Latour, B. (2017). *Facing Gaia*. Polity.

Lieberman, M. (2013). *Social: Why our brains are wired to connect*. Crown Publishers.

Louv, R. (2005). *Last child in the woods*. Algonquin.

Narvaez, D. (2014). *Neurobiology and the development of human morality*. Norton.

Narvaez, D. (2016). *Embodied morality*. Palgrave Macmillan.

Noddings, N. (1986). *Caring: A relational approach to ethics and moral education*. U of California Press.

Ross, H., & Mannion, G. (2012). Curriculum making as the enactment of dwelling in places. *Studies in Philosophy and Education, 31*(3), 303–313. https://doi.org/10.1007/s11217-012-9295-6

Rushworth, S. (2020). *Diaspora's Children*. Talking Leaves Books.

Sams, J., & Carson, D. (1988). *Medicine cards*. St. Martin's Press.

Savary, L. M., & Berne, P. H. (2017). *Teilhard de Chardin on love*. Paulist Press.

Sheridan, J., & Longboat, R. D. (2006). The Haudenosaunee imagination and the ecology of the sacred. *Space and Culture, 9*(4), 365–381. https://doi.org/10.1177/1206331206292503

Silko, L. M. (1977). *Ceremony*. Penguin.

Snaza, N., & Weaver, J. (2015). *Posthumanism and educational research* (Vol. 35). Routledge. https://doi.org/10.4324/9781315769165

Sorokin, P. (1954). *The ways and power of love*. Templeton Foundation Press.

Stengers, I. (2012). Reclaiming animism. *E-Flux*. (36)1. https://www.e-flux.com/journal/36/61245/reclaiming-animism/

Wohllenben, P. (2015). *The hidden life of trees*. Greystone.

Wyndham, F. S. (2020, May 20). Personal communication.

Zwart, H. (2008). *Understanding nature: Case studies in comparative epistemologies*. Springer Netherlands.

CHAPTER 4

How Might Self-guided and Instructor-Led Nature Education Serve as a Gateway to Appreciating Non-human Agency and Values?

Joe Gray

Introduction

My goal in this chapter is to explore how nature education programmes can be enriched to help inspire an appreciation for non-human agency and values in learners. Within the broad area of nature education, my focus here is on activities that have a strong weighting towards the practice of natural history, as opposed to, say, those simply providing instruction in outdoor skills such as canoeing down rapids or navigating by the position of the stars.

For a definition of natural history, I draw on the work of Fleischner (2002, p. 11), who described it as "intentional, focused attentiveness and receptivity to the more-than-human world." When practised in this way,

J. Gray (✉)
London, UK
e-mail: joe@deepgreen.earth

M. Paulsen et al. (eds.), *Pedagogy in the Anthropocene*, Palgrave Studies in Educational Futures,
https://doi.org/10.1007/978-3-030-90980-2_4

natural history has the potential to be "a wellspring of informed care for our non-human kin and the habitats that they depend on" (Gray & Noss, 2021). In order to encourage this, it is the role of people involved in the delivery of nature education activities to move beyond serving merely as custodians and communicators of facts and become facilitators of attentiveness and receptivity. This is a point echoed by Barrable (2019, p. 2), who suggested that nature education "be considered more from a point of view of relationality and interconnectedness, of building a meaningful relationship, and less from the point of view of gaining knowledge and understanding."

Relatedly, wrote Weston (2002, p. 41), in considering environmental education more broadly: "We must rediscover ourselves in connection with the rest of Earth: we must acknowledge ourselves as animals, come to feel ourselves as parts of larger living systems after all." The overarching goal of such education, he added, is "to address our disconnection, reverse it, to re-situate us, to welcome us home" (Weston, 2002, p. 41). These comments, I feel, have a particular relevance for nature education, which offers a primary means for students to find their place in a world peopled by innumerable non-human others. In this way, nature education can encourage and underpin an overarching societal purpose of the mutual flourishing of humans and non-humans.

Before turning to describe the structure of the chapter, I will share a final initial consideration, in the form of a quote from Frank Forencich's *Sapience Curriculum* (2020, pp. 1757–1759, 1807–1810).

> *These days, everyone talks about the urgent need to "save the world," but in order to save it, we have to know it. And to know it, we have to spend time in contact with it. We have to feel the rhythms, the textures, the creatures, and the forces that are the very stuff of life…*
>
> *Ultimately, learning habitat is a question of attitude and relationship. To really learn the outdoor world, you'll have to abandon your cultural inclinations toward ownership, domination, and imperialism. You're not there to conquer the mountain, the land, or the river; you're a student, there to experience and learn. Forget what you know and engage your beginner's mind. Open your attention to everything you see and feel.*

Structure of the Chapter

I write this chapter primarily from my experience as both a tutor and a tutee on natural history courses in the UK, which are educational activities aimed at people across the age range that have taken place outside of the architecture of formal education. While speaking from the perspective of a practitioner, and not a scholar, of pedagogy, I do, nonetheless, incorporate ideas from the academic literature, where they have obvious practical ramifications.

The chapter's scope embraces both self-directed and instructor-led modes of education. Let me first clarify that by 'self-directed education' I mean education facilitated through, for instance, the reading of an information board at a nature reserve or the study of an accompanying leaflet. There may be a tendency, as I argue below, for the importance of engagement with such materials to be underestimated within the arena of nature education. Turning to instructor-led education, there is another clarifying point that I wish to stress. Specifically, this mode of education need not be incompatible with the emergence of pedagogical agency among the non-humans encountered or within the students themselves. This is something that I discuss in more detail later in the chapter. Following the sections on self-directed education and instructor-led education, I then go on to provide thoughts on the importance of the language that is used in delivering education, which is a discussion that has implications for both of the broad categories.

In order to structure the various educational possibilities within a framework, I tentatively propose the hierarchy in *Box 4.1*, which I will refer to within the chapter. Running from weaker to stronger possible experiences as a student of nature, the hierarchy is not a validated scale but simply crystallizes my own thinking as a tutor on nature courses and a creator of materials for self-directed learning. In the chapter, I speak of 'achieving' the different levels in *Box 4.1*. I use this word very loosely and do not wish to imply that these levels be considered as explicit goals for nature educators to set their students, and that a box be ticked at the end of a session and a certificate handed out. In the spirit of John Holt, an important critic of the approach to education typified by formal schooling, I suspect that the best learning experiences are likely to be those free from any incentives or disincentives that are external to the activity (Dickerson, 2019). In other words, they are activities that learners undertake because they strike them as being worthwhile in their own right.[1]

Box 4.1 A Hierarchy of Possible Experiences as a Student of Nature

- Level 0: No meaningful engagement with the material or activity.
- Level 1: Learning about non-human beings as isolated objects.
- Level 2: Learning about non-human beings as interconnected objects.
- Level 3: Learning about non-human beings as interconnected subjects with agency and interests (including through receptivity to non-human beings as teachers).

Self-directed Nature Education

In the context of nature education, self-directed learning can be inspired, supported, and guided by a whole suite of possible materials. These include:

- **on-site materials**, such as information boards and leaflets that people might choose to engage with after having already made the decision to visit a nature reserve or similar place;
- **off-site materials**, such as large books and streamable documentaries, which are best suited to home-based learning;
- **portable aids**, such as field guides and nature-identification apps, which are designed to assist people spending time outdoors to learn more about the world around them;
- **engagement stimulators**, such as articles in weekend newspapers with suggestions for nature-based activities, which might prompt people to undertake self-directed activities, as individuals or in groups (*e.g.*, family units);
- and, of course, the **natural elements present in the site**.

Through this broad range of materials that can inspire, support, and guide self-directed learning, there exists the potential to engage with a far larger audience for nature education than is possible through instructor-led programmes. For one thing, there may be people who find instructor-led sessions less appealing, for a variety of reasons. These include, but are not limited to, time constraints relating to personal circumstances, a reluctance to engage in group-based activities, and concerns with ease of access (*e.g.*, the terrain may be difficult for someone who uses a wheelchair, for instance, if 'universal design' has not been practised at a site).

In addition, since different people will have different preferred means of engaging with educational activities, it might be that certain materials for self-directed learning offer something that some people would struggle to gain from an instructor-led session. For instance, an information board illustrating the ecological linkages between different species within a habitat may, for some people, offer a more compelling depiction of interconnectedness than a verbal description from an instructor could, helping such learners to achieve a Level 2 experience (see *Box 4.1*).

Furthermore, some people may find it easier to achieve a deep level of engagement with non-human beings, including in the role of teacher, while away from other humans and not inhibited by the structure of instructor-led education. And this is something that might facilitate the achievement of a Level 3 experience.

Despite these potential advantages for self-directed learning in nature education, at least for some individuals, it is my experience that there is a tendency, in considering the delivery of outdoor environmental education, for a focus to be placed on instructor-led sessions. Building on a point that I raised in the Introduction, I suggest that both scholars of pedagogy and people involved in the delivery of nature education should be careful to avoid underestimating the importance of this mode of learning. Interestingly, it might just be that the temporary changes to the way in which people access educational opportunities during the Covid-19 pandemic, with a shift away from instructor-led sessions, leads to an increased focus on this important area in the near future.

With this discussion complete, I now turn to consider ways in which self-directed learning activities can help foster an appreciation of non-human agency and values. I shall start with an example activity that I consider to be unlikely to get most users past a Level 1 experience. As part of a new woodland project in my local area, an arboretum of native species was planted with trees arranged according to the ways in which humans use the materials obtained from them, such as for furniture making or as medicines. One section includes species whose wood is used to create objects for making music, which is a case of nature's instrumentalization, in two senses of the word.[2]

The objectification of non-human beings is similarly reinforced by educational materials with a strong focus on 'ecosystem services' such as flood prevention and carbon sequestration.[3] While learners might gain a certain appreciation of the ways in which the organism being discussed is connected to the wider system, and thus achieve a Level 2 experience, such narratives are not helpful in moving to Level 3. This is not to say that anthropocentric considerations should necessarily be entirely excluded, but, at the very least, they should not be dominant if materials are to support learners in enriching their understanding of non-humans as beings with agency and interests. As Molina-Motos (2019, p. 10) observed:

> *[A]n ecocentric perspective of [environmental education] includes humanistic intentions without a species supremacism and integrates them into a broader aspirational and pedagogical project.*

The idea of rallying against human supremacy is reinforced in the following comment by Sitka-Sage et al. (2017, p. 31):

> *[W]e advocate that it is incumbent upon environmental educators to supplant master species metaphors and practices that perpetuate an image of the world as "ours" to remake according only to our desires—even in urban centers.*

As a direct challenge to 'master species metaphors', there are various ways in which materials can help learners achieve Level 3 experiences. One is to encourage *sympathy* by highlighting threats to particular organisms' thriving—either in the habitat in question or in the wider landscape—as long as this is not then diluted by presenting the threat of loss primarily in terms of the potential implications for humans. So, for instance, rather than positioning the plight of pollinating insects as being principally a threat to agricultural systems, it could be described as being bad for the insects themselves and the ecosystems of which they are an integral part, which then foregrounds their intrinsic value (*e.g.*, Mathews, 2016).

Another way is to encourage *empathy* for organisms by describing facets of their lives, such as the challenges of parental care, that have clear parallels with our lives as humans. Here, in place of traditional third-person, scientific descriptions of organisms, consideration might be given to adopting first-person prose (*e.g.*, "As a great crested newt, I depend on this pond for rearing my young. If humans release fish into the pond, my young will not be able to grow and survive.").

A third way is to present suggestions for activities that might help individuals deepen their recognition of agency and interests in non-human others. Such activities could be as simple as spending time observing nature. In this way, the centrality of the author of the materials in the role of teacher is dissolved, and the non-humans themselves become integral to the learning process. As a demonstration of the potential of simply observing nature, here, Challenger (2021) has noted the following in regard to her experience of spending time watching wild rabbits:

> *What emerged through observing these animals running about after each other, avoiding dangers, seeking food, or just resting and looking into the distance, was the transparency of the uncountable motivations of their lives. What my mind responded to was not a rational calculation of feeling or interests but the observable movements of need.*

In spite of the thoughts presented above, it is important to note that even materials suited only to Level 1 and Level 2 learning experiences might still help individuals on their educational journeys to an ecocentric awareness of the life around them. No educational experience occurs in isolation, and traditional descriptive information might work synergistically with other resources from which an individual is learning in parallel. Indeed, in a short piece in *The Ecological Citizen*, co-written with fellow ecocentrist, Ian Whyte, we reflected how, against a background of philosophical study into deep-green thought, we both found that nature walks with a field guide for a companion were instrumental in shaping our own worldviews (Whyte & Gray, 2020, p. 119):

> *Against a background of ongoing deep-green study, we have found that accentuating the experience of walking in wild places with the information in these books has done something remarkable. Out of objects it has forged subjects, subjects that are imbued with meaning and value and that have independent concerns.*
>
> *Through this shift, our own worlds have changed. New relationships and value centres have become evident everywhere. And the realization has followed that we, too, are part of the immense and integrated new whole […]*
>
> *[F]ield guides offer a window into local-scale diversity, connections, complexity and beauty, and there follows an inevitable conclusion: Everything intertwines. And thus, gradually, one realizes that all life is one's equal.*

Instructor-Led Nature Education

The first thing that I want to say about my experience of instructor-led nature education is that too much attention may be given to helping learners name a creature to species level or acquire other *mere* facts. These may help individuals to achieve Level 1 and Level 2 experiences (the latter might be the case if the facts presented aided learners in seeing the organisms as being part of an interconnected whole), but they are not necessarily conducive to Level 3 experiences. While the same point that I made above about no educational experience occurring in isolation applies here also, I do think that an important opportunity is being missed if little attention is paid to helping individuals transcend the notion of non-human organisms as objects.[4]

In the case of educational activities aimed at young children, there may be a particular temptation for instructors to feel that their primary role is to help their students find insects or amphibians, say, and then learn the names of everything they encounter, seizing the opportunity to cram facts into the developing minds in front of them while the holes in their brains are those of a sponge and not a sieve. But to do so may lessen the young children's chance for non-instrumentalized learning (*i.e.*, learning not primarily motivated by external incentives or disincentives (Dickerson, 2019)), as their experience might be too similar to that which they are used to in a classroom. In this latter setting, motivations for education may be to prepare for a test or to meet the criteria for receiving a certificate, say, rather than learning for the sake of it.

Instead, instructors can choose to let the young students be the discoverers, and offer them only broad reassurances that their attempts to put a name to the beings in front of them are on the right lines. Let us say that a young girl asks the instructor: "Is this a ladybird?" It is okay for the instructor to simply respond with an encouraging "yes," rather than feeling the need to tell her: "Well, actually, this is a seven-spot ladybird."

In keeping with the overall theme of this chapter, the most important learnings of all will not be the names of creatures or even their broad ecology, but the realization that the organisms being 'studied' are individuals with needs and with a value that is independent from any instrumental benefit that humans might derive from them. Here, instructors must move beyond the idea that non-human beings are principally 'things' to learn about—which, in my experience, is the dominant paradigm of nature education—and give learners the opportunity to unlock realizations of agency and interests in non-human others.

In this regard, the ways in which other beings are encountered is of much significance. Weston penned the following passage about finding wild beings *within* the classroom setting, but the essence of what he was saying applies very much to outdoor education too (Weston, 2002, p. 50):

> *It should be very clear that I am not speaking of bringing spiders or other insects into the classroom as exhibits, in bottles or tanks, appropriated and confined for our scientific or merely curious inspection. This is a philosophical experiment, not Show and Tell. The aim is to attend to how it changes our sense of this space when we discover such Others already present, co-inhabiting this space we were so sure was only our own, elusive but independent, on much more equal terms.*

In the outdoor setting, equipping students (from across the age spectrum) with collecting pots for temporarily confining an encountered organism is one practical means for them to bring a community of other lives into their personal spheres without risking damage to these beings. Here, handling live organisms with care and respect is an important way through which an appreciation of their interests plays out in real time. Similarly, it can be edifying for students to be instructed to keep a respectful distance from organisms who might find their proximity stressful. Enjoying the presence of others from a distance is another way to enact an appreciation of their interests.

Nevertheless, the importance of the tactile element in encountering certain organisms should not be underestimated. For, just as one cannot truly sense the fragility of human babies until one takes them into one's clutch, there is nothing like placing a bush-cricket, say, on one's palm for appreciating their delicate beauty. Furthermore, it is inevitable that students will interact to some extent with the non-humans whom they encounter, and the relationships inherent in these interactions can be helpful in turning objects of study into subjects leading their own meaningful lives. Thus, while it may be regrettable if students cause, say, a bird who is feeding in a tree to take flight by getting too close, the interaction at least reaffirms that the students and the bird are of the same world and evidences the existence of a real and meaningful inter-relationship.

For younger children participating in activities that involve collecting live organisms in pots, I find that the most important experience of all almost invariably comes at the point in the session when it is time to return the captured beings to their homes. Some youngsters will have an inclination to want to keep the organisms in the pots and take them away with them. Here, the lesson is that just as we humans have a home where we feel safe and can access all the things we need to live, such as food, so do all these non-human beings. The exercise of releasing the organisms back into their own habitat can be a very powerful one. A similar phenomenon can occur in pond-dipping, where there should be a particular urgency in returning the organisms encountered back to their aquatic environment.

Other, more direct, ways of giving students an opportunity to discover agency and interests in the non-human beings whom they encounter would be to set exercises that might encourage this. Weston (2002, p. 49) suggested: "Try, for instance, to think of some familiar and specific aspect of 'our' world from the perspective of specific other animals." Another exercise would be one of simple focused listening, where students are asked to try to tune in to the sounds that the birds around them are making and to imagine what messages they might be attempting to convey.

In one session that I facilitated, I talked about bird song and calls, starting with *intra*-species communication and moving on to *inter*-species communication. Then, I asked participants to consider if they had ever tuned into something that a bird was saying. My idea was to present this as a natural continuum from inter-species communication, with humans just being another species, and I had my own example ready. Remarkably, the first person to answer gave the very example that I was going to share, which was the alarm call of a blackbird in his back garden who was threatened by the approach of a domestic cat. He was able to get into the garden in time to save the bird (as was I in my case). I know that the blackbird was not calling to me specifically, but the communication was effective nevertheless, and the connection between us as two beings sharing a particular place was powerful.

Such activities offer means of 'de-centring' the tutor and empowering nature as a 'co-teacher', which is something that Jickling et al. (2018), in an article on 'wild pedagogies', described as an educational *touchstone*. Receptivity to non-humans as teachers, they argued, "involves carefully listening to available voices and will at times involve actively needing to de-centre the taken-for-granted human voice and re-centring more-than-human voices" (Jickling et al., 2018, p. 162). Now, a willingness to accept non-humans as teachers does not automatically move education away from an anthropocentric framing.[5] Relevant here, I feel, is the legend relating to Robert the Bruce in which he is said to have learned of how perseverance can lead to positive outcomes by watching a spider struggling to make a web but eventually succeeding. The positive outcomes that he is said to have had in mind were military in nature and thus distinctly human. This observation notwithstanding, de-centring the tutor is certainly a way to enrich the educational tapestry and enable students to embed ideas of agency in non-human others. Another educational touchstone that Jickling et al. (2018) described in their treatment of wild pedagogies is the enabling of uncertainty and the relinquishing of control. Here they suggested "pedagogical shifting toward emphasis on creating diverse and stimulating learning environments with generative questions" (Jickling et al., 2018, p. 164).

For older students, including adults, there is likely to also be value in explicitly including a discussion of the difference between anthropocentric and ecocentric standpoints within nature education programmes. Cocks and Simpson (2015) argued that this will help to "shed light on alternative experiential possibilities." They also noted that even if some outdoor educators "do not believe that ecosystems possess intrinsic value, it still may be a notion they would want their students to be aware of" (223). One way to do this would be to incorporate ethically challenging issues into the educational experience, rather than shying away from them. So, where, for instance, there are 'feral' populations or 'invasive' species present in an area, the ethical dilemmas that arise from their presence could be directly discussed. Where these populations or species are encountered as real individuals (as opposed to more abstract collectives), a realization of their agency and interests may be fostered. In turn, this might just play some role in stewarding humanity towards a culture that strives for solutions to thorny issues in conservation that are mindful of the moral standing of every individual. (For an insightful discussion of this issue, within the context of horses in the Australian Alps, see Jukes (2020).)

Nature educators should also be mindful of the way in which they describe the settings in which educational activities take place. Here, Sitka-Sage et al. (2017) cautioned against describing places for outdoor education as a playground, or even as a setting for exploratory play and learning,[6] as this may hinder participants' ability to realize that these places are agential co-teachers. Relatedly, Jickling et al. (2018, p. 162) observed: "No longer is the environment an important backdrop upon which learning happens, nor is it simply something to be interpreted solely by adult humans, but it might become an active member in teaching and learning."

Finally, and perhaps most importantly of all, nature educators should be open, where possible and meaningful, to incorporating Indigenous teachings into the activities that they facilitate. Zimanyi et al. (2020), for instance, described how they incorporated Indigenous teachings into an early-childhood programme focusing on water that is taught in forests and meadows on the banks of the *GabeKanang Ziibi* (the Humber River), outside the Canadian city of Toronto. Their programme promotes, for instance, the Anishinaabe teaching that *Nibi* (water) is the blood of *Aki* (the Earth). Such approaches may offer a powerful means of opening students' eyes to agency and interests outside of humanity's sphere.

Choice of Language

Just as the broad-brush details of educational activities discussed above are important in guiding possible educational experiences for students, so may be the finer points of the specific language that is used. In Box 4.2, I present some suggestions for language that creators of materials to support self-directed education and those involved in instructor-led activities may wish to consider.

Box 4.2 Considerations for Language Choice Used in Materials to Support Self-directed Nature Education and Instructor-Led Nature Education Activities

- Describe seeing or hearing non-human others as 'encountering' and 'meeting' in preference to 'spotting' or 'recording'. (The former are warmer and imply a relationship.)
- Use 'who' in preference to 'that' when describing non-human life (*e.g.*, "the newts who lay eggs in this pond" rather than "the newts that lay eggs in this pond").
- Avoid the objectifying term 'it' in referring to non-humans. Instead, intersperse the use of 'he' or 'his' and 'she' or 'her' (*e.g.*, "Can you hear the robin calling to defend his territory?"), or construct sentences inclusively with plurals and use 'they' or 'their' (*e.g.*, "With slugs, you may notice how they like to come out on wet nights.").
- Use the term 'community' to mean not an exclusively human group but the interconnected web of human and non-human beings in a particular place.
- Avoid language that positions nature as being a resource for humans above all else.
- Avoid language that separates humans from nature (*e.g.*, using 'animals' to mean non-human animals).
- Avoid language with potential for demonization, such as 'pest' (the value judgements inherent in such language may skew learning experiences).

Concluding Remarks

Above, I have offered suggestions—drawing both on my experiences as a tutor and tutee on nature education courses and on the academic literature—for how self-guided and instructor-led nature education activities can be shaped so that they have the potential to serve as a gateway for learners to appreciate non-human agency and values. In doing this, these activities can encourage and underpin an overarching societal purpose of the mutual flourishing of humans and non-humans. Key themes to emerge in the chapter are the importance of self-directed nature education, the dangers of instrumentalized learning experiences (as opposed to learning for learning's sake), the potential benefits of de-centring the tutor, and the possibility for reinforcing inter-species inclusion through an openness to non-humans as partners in pedagogy.

I will finish with a call to action. Specifically, as the challenges facing human and non-human Earth citizens escalate—in a period of time so different from what has come before it that some scholars now consider it a new geological epoch, the Anthropocene—there has never been a greater urgency for re-evaluating the praxis of nature education.

Notes

1. Instructors on nature courses and creators of materials to support self-directed learning may wish to bear this point in mind in shaping their approach to supporting learners.
2. Here, if we turn to consider instructor-led nature education ahead of the main section that covers it in this chapter, it is worth noting that, in the case of this arboretum, a skilled tutor would still be able to help students discover agency by looking at the trees as individual living beings and exploring the community of denizens that make each one a great temple of life.
3. Even literature primarily expounding the virtues of nature-rich places in regard to human mental health—controversial though this is to challenge in today's world—may serve to perpetuate the objectification and instrumentalization of non-human organisms.
4. In certain cases, such as in specialist education aimed at upskilling people for the task of conducting biological surveys, there may be a need for a strong focus on simply transmitting facts and providing instruction in techniques. But in educational activities of a more general nature, this same requirement does not exist.

5. One can imagine a cartoon, if and when the concept of non-human teachers becomes mainstream, in which a rabbit is running away from a human who is holding a notepad and pencil. The caption is: *It isn't enough to try to eat me? Now you want me to be your teacher too!*
6. I wonder if activities that involve young students converting, say, a wild scattering of dead wood on a forest floor into rudimentary shelter-like structures might not be followed by a reverse exercise to restore the habitat to its pre-activity state? This would enact a respect for the habitat as habitat rather than a mere outdoor classroom.

References

Barrable, A. (2019). Refocusing environmental education in the early years: A brief introduction to a pedagogy for connection. *Education Sciences, 9*, 61.

Challenger, M. (2021). Sounding out other species. *The Ecological Citizen, 4*, 159–162.

Cocks, S., & Simpson, S. (2015). Anthropocentric and ecocentric: An application of environmental philosophy to outdoor recreation and environmental education. *Journal of Experiential Education, 38*, 216–227.

Dickerson, A. (2019). *John Holt: The philosophy of unschooling*. Springer International Publishing.

Fleischner, T. L. (2002). Natural history and the spiral of offering. *Wild Earth, 11*, 10–13.

Forencich, F. (2020). *The Sapience Curriculum: Teacher's guide for an age of turmoil* (Kindle ed.). Exuberant Animal.

Gray, J., & Noss, R. (2021). Fostering a love of the living world—Or, the need for a grand revival of natural history. In J. Gray (Ed.), *Visions for a post-Covid world: Defining a radically new normal* (pp. 45–59). Dixi Books.

Jickling, B., Blenkinsop, S., Morse, M., & Jensen, A. (2018). Wild pedagogies: Six initial touchstones for early childhood environmental educators. *Australian Journal of Environmental Education, 34*, 159–171.

Jukes, S. (2020). Thinking with a landscape: The Australian Alps, horses and pedagogical considerations. *Australian Journal of Environmental Education*. https://doi.org/10.1017/aee.2020.26

Mathews, F. (2016). From biodiversity-based conservation to an ethic of bioproportionality. *Biological Conservation, 200*, 140–148.

Molina-Motos, D. (2019). Ecophilosophical principles for an ecocentric environmental education. *Education Sciences, 9*, 37.

Sitka-Sage, M. D., Kopnina, H., Blenkinsop, S., & Piersol, L. (2017). Rewilding education in troubled times; or, getting back to the wrong post-nature. *Visions for Sustainability, 8*, 20–37.

Weston, A. (2002). What if teaching went wild? *Philosophy of Education, 2002*, 40–52.

Whyte, I., & Gray, J. (2020). Field guides as a gateway to appreciating more-than-human concerns. *The Ecological Citizen, 3*, 119.

Zimanyi, L., Keeshig, H., & Short, L. (2020, April 19). Children make connections to Aki (Earth) through Anishinaabe teachings. *The Conversation*. Retrieved February 2021, from https://theconversation.com/children-make-connections-to-aki-earth-through-anishinaabe-teachings-133669

CHAPTER 5

Where the Children Are

Marianne Presthus Heggen, Bob Jickling, Marcus Morse, and Sean Blenkinsop

Polar bears are drowning. Children rage. And education is, so far, failing to provide a clear pathway out of our ongoing ecological and social crises. We need change—different thinking, different relationships, and different solutions. But where to turn? In this chapter we turn towards children, for two reasons. First, the "wonderment" with which they often seem to encounter the world. This wonderment can baffle us, but in this chapter, we ask, in what ways could the children be encountering and thinking

M. P. Heggen (✉)
Western Norway University of Applied Sciences, Bergen, Norway
e-mail: mph@hvl.no

B. Jickling
Lakehead University, Thunder Bay, ON, Canada

M. Morse
Outdoor Environmental Education, La Trobe University, Melbourne, VIC, Australia

S. Blenkinsop
Simon Fraser University, Burnaby, BC, Canada

M. Paulsen et al. (eds.), *Pedagogy in the Anthropocene*, Palgrave Studies in Educational Futures,
https://doi.org/10.1007/978-3-030-90980-2_5

differently—knowing the world in different ways? And in what ways might they be offering possibilities for pathways we are missing?

Second, we set out to consider children as qualified *beings* (James, 2011), and to see children as *citizens*, performing their citizenship in political, yet different ways (Grindheim, 2017). Looking at citizenship as an enacted process opens space for other contributions to citizenship—not only human. We do not suggest that children, as citizens, are responsible for the environmental acts of society. Rather, we think that their actions may contribute towards more equitable, relational and caring futures, and thus be worthy of note. In other words, they might be enacting examples of ecocitizenship (Heggen et al., 2019). We also see the ethical obligation in listening to the children's contributions. After all, it is their future.

This chapter begins with the premise that children are often positioned differently, perhaps less anthropocentrically, in the world, both ontologically and epistemologically. In some senses, their thinking is wild; it isn't corralled, regulated, or enculturated yet. By following their lead, we ask what might be revealed through their embodied experiences and emotional encounters with the world? We wonder, too, about the political and pedagogical implications of taking children seriously. How might this help educators wild their pedagogies, wild themselves, and provoke cultural change? (See also Chap. 2 in this book).

Writing as a Method

The methodological approach to this chapter might, at first glance, be described as a thought experiment offered through writing. We write to figure things out (Richardson, 2001). Our method, of presenting evidence in response to appearances and ambiguities in the world has a long history stretching back to the Renaissance philosopher Montaigne. Through his art of essaying, he wrote to complicate ideas, to undo easy explanations, and to seek new, if incomplete conclusions.

For Montaigne, the essay is a gathering of experiences, the sensuous and 'mute life' of the physical body, the concrete realities of material life. He refused positions that were not grounded in experience, in the flesh and blood of real life, and the comfortable answers of the status quo. For us, this original form of the essay is appealing. It offers space to explore a diversity of responses and theories and to explore a different explanatory space. It also requires us to not just intellectualize this discussion, a move which has often historically been one of the steps towards anthropocentrism.

For Montaigne, appearances matter, and experiences are to be trusted (Bontea, 2008). But this raises questions of how, and how far? To understand these questions Montaigne often incorporates associations from his own work and that of others. For him, the essay attempts to connect seemingly random events, ideas, and explorations from different realms of experience. We take this approach here as we attempt to meet the children where they are.

In what follows we present two vignettes, or keyholes, of children engaging with their world. We each then offer some speculative responses and noted dissonances to these keyholes. Following Montaigne in our discussions, we gather experiences, ideas, and dissonances into new explanatory pathways that might, we hope, point towards more equitable and just ways of knowing and being in the world.

Keyholes

The two vignettes of children's play—seen as if looking through a keyhole—reveal glimpses of a day in the life of pre-schoolers. They describe everyday situations children seem to enjoy, and the pre-school teachers seem to value. Something intriguing seems to happen here, something that is difficult to decipher. Looking at children's contributions is not new in early childhood research, yet we question, how do these children viscerally encounter the world? How might these encounters be different from those who have already been fully 'educated' in their cultural context? And what might we learn from these encounters?

Naming the vignettes keyholes reminds us that we are only catching a small glimpse into the children's experiences, and even then, making assumptions in our interpretation. We acknowledge, too, that both vignette keyholes are situated in particular cultural settings. Although we do not always hear directly from the students in theses vignettes our attention was attracted, in part, by the mystery and wonder of things unsaid. Rather than a reflected reality, we are seeking available possibilities—a thought experiment through writing. By trying to meet these children where they are, we hope to gain some insight into their ways of encountering the world. We attempt to draw together our combined writing, and to take this discussion beyond the realms of early childhood.

The keyholes are set in Norway, where it is common for pre-schools to hike in local nature areas. These areas are visited and revisited, and may be located just outside the fence, or within a couple of kilometres. Hikes to such areas also include opportunities for explorations. The children know these areas, they feel comfortable there, and often know what they wish to do. Our vignettes return you to May 2020 to an urban pre-school at the outskirts of the city centre. These children hike weekly in their nearby community.

First Keyhole: Excited by the (Seemingly) Unexciting

Early one morning, a group of toddlers start the day's hike. The goal is a small grove of trees with a little stream running through, a place the children seem to love. Just a few meters from the gates, in the middle of the asphalt, a boy stops next to a manhole cover. He looks down and walks decidedly towards some nearby rocks. He finds some rocks of the appropriate size and throws them down the manhole. The cover has both small and large holes, and not all rocks go through. There is a distinct "plop" when the rock hits water. He tries again, but this rock is too big for the holes in the cover, he finds another rock and throws that instead. He walks back and forth, finding new rocks, throwing them towards the cover, finding new rocks, and… After many repetitions, his teacher asks: "Shall we follow the others?" The hike continues for two or three meters. By the side of the street are some interesting slugs. The boy stops intrigued and observes them closely. Three days later the same group leaves the pre-school for another hike, and another boy stops by the manhole, and begins throwing strands of grass, small leaves and sticks.

Reflections from Marianne

In addition to the everydayness, what strikes me in this story is the way that play was started and lead by the boy, without interference, in a kind of free play. The children decide without adult interference what, and how, to explore. There is a multitude of dimensions evident in this small story. As adults, we tend to let our interests lead our insight; a physicist will see different learning outcomes from a musician when 'evaluating' this boy's learning. But the boy might gain many other things which we don't recognize. I am intrigued by how we don't realize or recognize what the children gains, and that we do not know why they do these activities. Maybe we try too hard to analyse an activity that is simply meant as playing, as having fun?

The child's activity resembles other educational activities with small children, as when using a sorting box. Yet, a sorting box does not provide the variety of sounds, smells, textures, qualities, weights, etc. as the things the children throw into the manhole. By repeating the activity in different conditions, the children experience differing things, changing some factors, but not others. By making a sorting box, we believe we can be confident the children have experimented with shapes; and perhaps, with colours. But the factors that change are limited. What are the consequences of this simplification and flattening of experience?

Reflections from Marcus

I am struck by my own initial response—to override the situation in some way. I might ask the child if they are ready to go or what they are doing. In doing so, I might manage the situation and move things along. In my own way, I am setting out pedagogically to think of how I might help or guide the child. I recognize also, that I could let the child continue to play. To experience the world on their own terms in a form of free play. I am challenged also to wonder, though, what is actually going on here for the child. How self-directed is it really? The child is not alone; they are engaged in and with the world around them. If I hold back my urge to consider human directed influences on play and coming to know the world, how might I reconsider the child's experience in the moment? Might it be that these rocks, spaces, textures, and sounds actively guide the child in the interactions? I have a sense, also, that by intruding into the experience I might also break the magic of this moment.

Reflections from Sean

One thing this story recalls is the way adults can create quick thought pathways, determining that which is noteworthy. My sense is that part of this is a quite helpful response to a world filled with stimuli. Otherwise, one might spend the time stopped short in front of our gates dropping rocks through a manhole while our ice cream melts for lack of attention. And yet, I wonder about what is lost in this split-second programming of what one attends to? And how we learn what to notice and what not to? How are these systems developed over time and what role do teachers and culture play in their development? In some ways the cultural norms of focusing on the hike, getting to the chosen destination, picking the key fragments of knowledge that will lead to the preferred outcome are not yet part of this two-year-olds vernacular. Given that everything is unfamiliar, they are moving in a world where their attention is freely requested.

I am intrigued, also, by the idea of place-making tools. Vygotsky suggests, in his conversations about cognitive tools, that our birth cultures offer us tools that help us make sense of our world. Tools such as language and story allow us to frame our world and belong with our kith and kin. Yet the tools are not simply there for us to use and put down at will, they shape us as we use them. I wonder, given our long relationship with the natural world and particular places therein, if there might be another layer of tools being offered by the natural world, with similar effects? What if a child's focus on the spatial, the sensations of holding and dropping rocks, and his dedication to task are all part of the tools the world offers us, and which could shape us into ways of becoming place-bound?

Reflections from Bob

The curiosity and the intensity of the activity are interesting. The engagements seem silent. What could be going on? Is it purely sensuous, the plop and the physical manipulation of the stones and later, sticks and leaves? Is it even possible that there is something going on here that is outside of linguistic understanding? And if so, what would that mean? I am drawn to this line of questioning by my own interests in understanding and expanding the ways that learning, knowing and thinking can be perceived. There is something within our group's experiences, often working with learners outside of classroom settings, that drives this curiosity.

I am provoked by Jan Zwicky's work in lyric philosophy. It seems to arise from an interest in propositions that manifest themselves in the world, yet fall outside of logical structures of interpretation, or cannot be captured in words. Part of Zwicky's lyric philosophy is to embrace those understandings that arise suddenly and affect us as sensuous beings with bodies and emotions. If you are a musician, think about the wondrous performance that can occur when jamming is elevated to a perfect understanding amongst a group. Such experiences seem to suddenly announce themselves, not as a collection of logically linked parts; rather, they arrive in a moment as whole understandings. Zwicky calls these whole understandings gestalts. They are neither rational nor irrational; they are *arational* in that they elude adequate capture in words. Could it be that the manhole cover, the rocks, the plops, and the water were singing to the children in their own way?

Second Keyhole: A Squirrel Isn't Enough

Another day, a group of three- to five-year-olds hike from the pre-school. One of the teachers shares her story: While I watch children playing in the stream, some ask if we can go into the woods at the hiking site. They know they must bring an adult. The forest grows on a steep hill with its roots protruding from the ground in places. Several of the 5-year-old children, both girls and boys, want to go and I let them, as long as they stay within my field of view. They talk, and I can hear them looking for mushrooms and studying everything they find on and in the ground. They repeatedly ask if they can go further up. After a while, another teacher arrives to look after the children in the stream, and I can follow the children up the hill, allowing them to ascend further. I spot a squirrel looking for food in the trees and try to get the children's attention. They look up, but do not see the squirrel as they are more interested in something they have found. It takes me a few tries before I realize that only I find the squirrel exciting. The children are more interested in the ground. They dig with sticks and inspect various stones and plants. Suddenly someone shouts "I found a fungus in the ground!" He is clearly excited, and the other children gather around him. Everyone agrees that "yes, that's a fungus". I'm pretty sure it's not a fungus they've found, but do not object and wait to hear what they think. They ask if we can take it to the pre-school to find out what the fungus is, so I put it in my pocket. Back in the pre-school, the mysterious fungus was not prioritized, and it was forgotten. A week later, the children ask for the fungus, but it had unfortunately disappeared. We agreed that the next time we go to the same destination, we will go looking for a new one and take the time to investigate it better together.

Reflections from Marianne

In this keyhole, the children initiate and sustain the activities, fulfilling the adult criteria of play. As in the first story, play here seems to be filled with sensory aspects—the thrill, the engagement, perhaps the feeling of flow, of forgetting time and space, of simply "being". One thing that intrigues me in this vignette is the children's resistance. The children stray from the teacher. Observing children play in nature, I often see children straying away from the others, seemingly seeking room to investigate by themselves. In this example, the children collectively resist the teacher as she tries to direct their attention.

It was the children who initiated going to this part of the forest. Through the place, the mushrooms and the forest itself, the surroundings contributed to the play. I am drawn, then, to the active role nature takes as a playmate. If children get these kinds of playful experiences again and again, will they lead to an intimate knowledge of nature? The statement that children playing in nature learn to know nature and will take care of it is both worn and disputed. Still, when children play with each other, they might become friends. Can looking at nature as a playmate refine aspects of the children's play? Can looking at the children's play in nature be seen as socialization into, and within, a more-than-human world? What would such a socialization be like?

Reflections from Marcus

This keyhole raises my desire to redirect attention. If the intent of the educator is to direct attention towards the world in a way that engages and guides children's learning—why might I, and presumably the educator in the vignette, feel a compelling urge to control attention? The child's attention is already held in the moment, with a seeming sense of fullness of engagement with/in the world. What might be missed in redirecting the children's attention? Although I might have a positive intent, it might also limit the experience of knowing the world through direct encounter, in ways that defy calculated, logical, and conceptual understanding. Ways we might consider *child-like*. I am often left to wonder at the quality of attentiveness paid by children to natural things. In these descriptions, it is seemingly the stones, sticks, or fungus that, at least in part capture the attention and guide the encounter.

Reflections from Sean

This keyhole has me thinking about the assumed scarcity of time and about how that plays a role. We see it in the desire to move the child along from something that appears to provide limited educative return on to "the good stuff." In the disappearance of the dug-up object, 'of limited curricular value', although this is framed as forgotten, it is clearly remembered by the children. Again, my sense is that the adult is shaped by a set of cultural norms related to how learning works, to what is being learned, and to the "important signposts" for learning.

This leads me into wondering about the "stuff" of knowing. That thinking, idea generating, and learning is built out of stuff (rather the way good soil is built out of compost) and that part of what is happening in both these stories is, in fact, "stuffing". There might not be obvious learnings attributable to this digging in the forest seeking mushrooms, but there is a gathering of experience, encountering, testing, and exploring that can be part of the "stuff" of other encounters. That can be built upon, flexed, added to and that at this age, and maybe all ages, the more the merrier. Perhaps it is an error to impose a linear concept of learning and valuation to encounters. Who knows, maybe this encounter with dirt and life or holes might be helping to position the child in the world in important ways.

Reflections from Bob

It is interesting to notice how seemingly normal it is for these children to go outside of the school grounds. The children know what is happening, they know how to conduct themselves, and they have a sense of what they wish to do. But they also appear to have an enormous capacity for spontaneity. They are interestingly attentive, in ways that appear less contaminated by other aspects of culture. I can't help imagining that conversations with older children and teens could include jokes, singing popular songs, and conversations about cultural phenomena and artefacts. But then, would these older students be so distracted if excursions like this had continued to be part of their everyday experiences throughout their schooling? Or, might it be that the desperate move to belong to the human peer group arises because there is no underlying sense of belonging to a place in the larger world?

The children do not seem interested in the squirrel. Is it hard for them to see the squirrel? Or is it because they have already found something interesting—and see the squirrel as a distraction? Or is it the intimacy and physical engagement, the digging with sticks, that captures their attention? Whatever the answers, it does appear that the children were determined to engage with their environment differently, in a sensuous, embodied way. Perhaps interesting parts of play are the experiences gained, the gathering of stuff, and maybe even responses to the agency of the place itself that happen when the adults urge to control the activities is loosened.

Resonances and Dissonances

Some resonances and dissonances run through our reflections, as we are all struck by the children's sensuous and spontaneous engagement with their perhaps agentic, surroundings. Likewise, we problematize the adult urge to look for clear and obvious aims the children gain in these situations. This makes us wonder; what is lost if the teachers with their pedagogical and cultural tools had redirected the attention, created thought pathways, flattened and simplified these multidimensional situations? To be clear, however, we are not claiming that teachers should not interact pedagogically with their students. Accepting children as different learners also means accepting that it is the responsibility of adults to meet children's contributions, to be present and to follow their lead, but also to protect the children, and to guide the children. Rather, we explore what children might give us—or what is won—if we loosen our urge to control their learning and if we listen to them, if we seek to meet the children where they are.

A Wonder/Wander

In our own Montaigne-esque style, we have gathered responses to these vignettes—an aggregation of associations. We have attempted to capture some of the vividness of possibilities that children might have with and in the more-than-human world, while acknowledging the enigmatic nature of how these experiences appear to each of us. It is now time, in the tradition of the essayist, to explore these interpretations.

Montaigne was somewhat unusual in Western history in that he was directly involved in the public political life of his society and very privileged, yet at the same time, perhaps because of his upbringing, he was deeply concerned with everyday life and embroiled in examining and critiquing 'the system'. His essays are, thus, more than just a description of a lived life; they also search for ways in which lives are worth living. They are a sweet spot in humanism that is more generous with conceptions of proof, such that they can reach more fully into the experiences of life, beyond the limits of rigorous science and accepted cultural rules of thought (see e.g., Bontea, 2008).

Montaigne's essays were not the polished perfectly formed arguments that we aim for today. Rather, they were explorations, wanderings with ideas, with the "what ifs" of open-ended discussion. For in French, "essay" is exactly that, a try, an attempt, an adventure for writer as well as reader. It is in that spirit that we respond to these vignettes and reflections, these witnessing's of young children's interactions, prioritizations, engagements, play-ins with the world "under open skies". In this chapter, we seek *to follow the children's lead and ask; What if the children have other, less anthropocentric, knowledges? And what if we were to support these other knowledges in education?*

What If the Children Have Other Knowledges?

The first *what if* we are proposing is an epistemological adventure, a wandering with knowledge. We have each been struck by seeming dissonances between the adults' suggestions and the children's priorities. On the one hand, adults can tend to be immersed in an epistemology that is progress-based, linear, fragmentable, ordered, and transferable, whereas children may appear to be *resisting* adult agendas for potentially indeterminable but intriguing reasons. We wonder why, as Jan Zwicky considers: "are we so deeply susceptible to the charms of epistemological security? Partly because as a species we enjoy the exercise of power for its own sake; we have a penchant for controlling things." (Zwicky, 2019, p. 95). It is a loosening of control, then, that intrigues us.

A Gestalt of Knowing

Meeting the children where they are opens challenging possibilities. We wonder if there may be learning experiences that fall outside of linguistic and a particularly rationalized and logical knowing. We wonder about gestalts of knowing that appear as intact wholes, rather than built from component parts. The children's activities in these examples, together with their teachers' responses, may be showing us that some of their most engaging learning is difficult to predict, plan, or control. This leads us to wonder: What if there are other forms of knowledge outside of the ones prioritized by mainstream educational systems.

Perhaps the most disturbing possibility is that this form of knowledge might crumble and decay under analysis. That is, when we disrupt children's play, or ask seemingly harmless questions like 'what are you doing?' Our logic-oriented inquiries might undermine the learning itself. Zwicky remind us that our inclination towards logical "rules of thought" are aimed to "secure against prejudice, superstition, and whim" (2019, p. 95). These aims are in a tension with any underlying proclivity to accept unexamined gestalts, in this case the capacity of children to wonder at phenomena presented by manhole covers and mushrooms.

Zwicky asserts an over reliance on "rules of thought", and that these alone, may deprive us of thinking's most significant dimension:

> The first and fundamental aim of thinking is to understand, to discern the lineaments of reality. The correction of mistakes accompanies this

> discernment; it is not achieved by shackling thought to ensure that mistakes are never made.
>
> There is no series of steps we can implement to precipitate gestalts in all audiences. Real thinking does not always occur in words; it can decay under analysis; its processes are not always reportable. This means that real thinking is in some sense wild: it cannot be corralled or regulated. But it is also the only access humans have to the experience of insight, to moral and mathematical beauty, to ontological vision. (Zwicky, 2019, p. 95)

This is risky business for educators. We are challenged to ask what is lost when thinking is reduced to just thinking in words, to relying on a particular form of rational logic alone, to see the world only when reduced to its constituent parts through analysis. How might we throw light on alternatives that increase capacity for thinking with and in the world? Here we return to the proposition laid out near the beginning of this chapter: is it reasonable or insightful, to deprive learners, young and old, of access to forms of understanding. How might we throw light on alternatives open to them, to increase their capacity for thinking with and being in the world? And we wonder if wilding capacities for thinking can enrich educational possibilities for understanding, and herein lies an enormous challenge. Yet, in Zwicky's words, "Where the danger lies, there too lies meaningful life" (2019, p. 95). And we wonder if bringing such child-like, holistic, and connected knowing into practice, to balance more dominant, linear, individualized, fragmented, anthropocentric, and separating ways of knowing, can be a pathway towards doing things differently, towards being differently in the world. Could this be an important piece of building the kinds of relationships that might allow all beings to better flourish, to respond to crises of alienation, extinction, anxiety, and loss of biodiversity?

Knowing with/in the World

> Imagination is understood to be a quality of mind in settler culture. In Haudenosaunee/Mohawk tradition, the same quality is understood to be animal and spiritual helpers manifesting their presence in one's life … [The settler conception of] imagination dominates where fear of the unknown, uncertainty of memory, and placelessness thrive. (Sheridan & Longboat, 2006, p. 365)

Sheridan and Longboat add a compelling piece to our gathering of ideas. First, there is the suggestion that the settlers' western epistemological concepts of knowing, thinking, and imagining are not the sole purview of the isolated human mind; that this way of thinking of the imagination is in fact deeply extractive, colonial, and anthropocentric; and that there are other conceptions out there. Epistemology is culturally framed and there are different ways understanding and, by extension, encountering knowing. Second, there are threads that reach from this assertion back into our own reflections. They connect with our concerns about culturally specific and adult learning agendas which may be "settling" into the lives and minds of small children. Sheridan and Longboat point to beings who understand themselves as part of a larger community of knowing and being—beings who are receptive to gifts being proffered by their "animal and spiritual helpers" and by rocks, fungi, slugs, and places.

There are also links here to Zwicky's positioning of a different kind of knowing emphasizing holistic, non-linear, differently reasoned, and languaged forms. Such relational forms might be understood if they were gifts from other knowers. And in time, through play, encounters in "flesh and blood", diversities of time spent in place, these gifts might lead to an 'old-growthing' of minds. A knowing needed for this uncertain world:

> Haudenosaunee minds are composed not just of visible ecological domains but also by numinous qualities of those domains that, allowed to mature, emerge, and encompass the old growth of their traditional territory. Old growth minds and cultures mature, emerge, and encompass the old growth of their traditional territory. Haudenosaunee minds are congruent with their traditional territories but more important, Haudenosaunee *are required* to accomplish that symmetry in accomplishing their authenticity. (Sheridan & Longboat, 2006, p. 366)

Bildung with/in the World

A third conversation that resonates with this more relational, non-languaged, less anthropocentric, non-analytical way of knowing is the concept of *Bildung*. The act of Bildung occurs when our skills are coupled with a reflexivity around what we are doing, and why, in encounters with *someone*. In the reflective sense, Bildung is thought of as both a critical exploration of one's being in the world, how one relates to oneself and the surrounding world, and as a transgression of existing order (Paulsen,

2021). There are many interpretations of the word, but in general, it represents the fulfilment of human potential. In a simplified version, it has been framed as "what is left after we have forgotten everything we learned" (Ellen Key, in Steinsholt & Dobson, 2011, p. 7).

Bildung has no beginning or end. It is part of all life, not only education, and not only with children and young people. It is neither linear nor compartmentalizable in the ways of more conventional Western education; as such, it stands as an alternative to an understanding of an uncoupled analytical world (Jensen, 2000). It allows us to think of learning in ways that acknowledge what we previously called the "stuffing" in learning. Bildung implies that we exist in the world, that we are not just something in spirit. This leads us to children's sensual and phenomenological meetings with nature. In the same way that we are not free when meeting other humans, we are not free when meeting nature. Perhaps the keyhole vignettes suggest that in the children's explorational, relational, and complex play, nature takes the form of an active 'someone' during formative encounters. The concept of Bildung is often framed as cultural formation. But what if we think of Bildung not as a cultural formation, but as a natural-cultural formation? What are the possibilities for this *someone*, or *something*, who supports the development of our reflexivity to be inclusive of a more-than-human world?

What If We Were to Support This Different/Other Knowledges in Education?

Our second wandering/wondering *what if* asks what education becomes if the different knowledges we point to exist. How might they be supported? How might this other kind of knowing be sustained, nourished, and given space to develop and flourish? Might meeting students where they are provoke further consideration of alternative non-languaged, embodied, and relational ways of knowing?

Play as Encounter

When we observe children playing, we can attribute a sense of freedom or wildness to their actions. In a school setting for example, we might describe the moment as involving 'student-directed' activity (or *free play*), along a continuum with more adult-directed activities (*guided play* or *teacher play* e.g.). But what if we consider student-directed play as a relational encounter with/in the world, that it is not solely the child that makes decisions? What if the child's play with materials, animals, sounds and shadows, for example, are viewed primarily as encounters, as formative exchanges amongst humans and myriad relational beings? In other words, what if this play is not simply child-centred, or even human-centred, but is a learning encounter with the world and its ways of knowing? In considering the possibility that material objects might guide a child's experience, Rautio (2013) suggests:

> To appreciate also the momentary and the seemingly unguided in children's everyday lives ... we would have to embrace the thought that teachers—those who invite, guide, support and steer us—can also be other than human beings. Tiny black inanimate pebbles can invite us into interaction by virtue of existing, guide the nature of this interaction by virtue of their physical form, support our activity through lending themselves to be investigated and engaged with. (p. 402)

What if, in other words, there is a way of knowing the world that is not just driven by human subjects? What if the children know the world as a place that asks questions, actively engages, and demands attention? What if it is not just the children directing their own attention but a world that seeks them out and requests it of them? Thought of this way knowledge, and ways of knowing, might be supported in different ways.

Paying Attention/Orienting to the World

It is, perhaps, not by chance that the two keyhole examples we have chosen are set out of doors. There is an everydayness about the descriptions, but also an attentiveness in the descriptions which we might easily take for granted. Such attentiveness is commonly observed in play outside. What is it with such engagements that draw in young children in this way? It is not to suggest that such attentiveness does not occur in a multitude of contexts, including indoors with man-made objects, but children can often appear intensely engaged with natural objects/settings. Might there be something in it when James (2009) suggests that because 'natural things have not been designed to fulfil any human purpose and so there typically seems to be more to them than can be comprehended in instrumentalist or functionalist terms... so natural things can invite attention' (p. 108)? In this way, time spent in settings not designed for humans might offer opportunities to attend to the world in ways that places us directly within the world through attentive engagement. Such engagements can include both paying careful attention, and also being reciprocally open to the intrusions of the world.

Merleau-Ponty (1968) describes such experiences as potentially involving a paradox of being both the 'seer' and the 'seen'. This shared encounter of coming to know the world in this way troubles individualism, isolation, and anthropocentrism. It places us directly in an expanded more-than-human world. Merleau-Ponty describes the reversibility of such perceptions (seeing and being seen, touching and being touched) with an example of touching the world while realizing that we are touched by the world at the same time. There is a profound and inescapable *openness* to the world; it intrudes into us and we intrude into it: 'the seer and the visible reciprocate one another and we no longer know which sees and which is seen' (p. 139). Merleau-Ponty argues that the simultaneous intertwining of the two perceptions is at the heart of being in the world.

Wildness and Education

At the heart of this chapter has been the idea of control—an idea arguably deeply entwined with narratives of the Anthropocene. Responses to the keyholes involve a recognition of the desire to exert an analytical and deliberate control. In loosening our desire to control situations, what might be gained? Throughout this chapter, and inspired by the keyholes, we suggest that an enlarged range of epistemological possibilities for learning with the wildness of the world might enable alternative knowledges. We have turned to the children as the starting point for this chapter; yet we are also suggesting that such alternative possibilities for knowledge must be broadly available within education beyond early childhood. This assertion rests on the premise that current overriding concerns with analytical, calculative, and fragmentable knowledge is failing us. Such knowledge is valuable, but it is not complete and alone it appears to lead to serious injustices. If we are to work with students in becoming caring and compassionate humans within a more-than-human world, then we must deliberately offer a broader range of experiences that allow us to be differently in the world. It is perhaps, in part at least, through meeting the children where they are that we gain inspiration and insight into these possibilities.

In conclusion, we return to Montaigne's conviction that we should trust our perceptions of experiences; but how far should this trust reach? The emergent convergence in our thinking provides a triangulation of sorts. Yet, it is certainly not strongly enough to satisfy traditional notions of scholarship. Perhaps a better measure lies in the readers' responses to, and recognition of, our wanderings. Do they strike you as familiar? Do they reflect shared insights that you feel? In short, do they resonate? For now, we assert, that there is verisimilitude in resonance.

If we hold that at least one important aspect of play is the releasing of control over our actions, constrained by adult supervision, curricula, cultural expectations, and in this case, scholarly expectations, then this research is playful. And this the writing is too. There is risk in this playfulness, but in it, there are also rich possibilities for meaningful futures.

Acknowledgements We would like to thank the children and the nature surrounding them for letting us look through the keyholes and get a glimpse into their world. We would also like to thank the teacher who authored keyhole two for letting us learn from her story.

References

Bontea, A. (2008). *Montaigne's* on physiognomy. *Renaissance Studies, 22*(1), 41–62.

Grindheim, L. T. (2017). Children as playing citizens. *European Early Childhood Education Research Journal, 25*(4), 624–636.

Heggen, M. P., Sageidet, B. M., Goga, N., Grindheim, L. T., Bergan, V., Krempig, I. W., Utsi, T. A., & Lynngård, A. M. (2019). Children as eco-citizens? *NorDiNa: Nordic Studies in Science Education, 15*(4), 387–402.

James, A. (2011). To be (come) or not to be (come): Understanding children's citizenship. *The ANNALS of the American Academy of Political and Social Science, 633*(1), 167–179.

James, S. (2009). *The presence of nature: A study in phenomenology and environmental philosophy*. Palgrave Macmillan.

Jensen, A. (2000). *Naturverdfilosofien og vegledning i friluftsliv*. Norwegian University of Science and Technology.

Merleau-Ponty, M. (1968). *The visible and the invisible: followed by working notes*. Northwestern University Press.

Paulsen, M. (2021). Bildung & technology: Historical and systematic relationships. In D. Kergel, M. Paulsen, & J. Garsdal (Eds.), *Bildung in the Digital Age* (pp. 1–25). Routledge.

Rautio, P. (2013). Children who carry stones in their pockets: On autotelic material practices in everyday life. *Children's Geographies, 11*(4), 394–408.
Richardson, L. (2001). Getting personal: Writing-stories. *International Journal of Qualitative Studies in Education, 14*(1), 33–38.
Sheridan, J., & Longboat, R. H. C. (2006). The Haudenosaunee imagination and the ecology of the Sacred. *Space and Culture, 9*(4), 365–381.
Steinsholt, K., & Dobson, S. (2011). *Dannelse: introduksjon til et ullent pedagogisk landskap*. Tapir.
Zwicky, J. (2019). *The experience of meaning*. McGill-Queen's University Press.

PART II

Dark Pedagogies

CHAPTER 6

Action Incontinence: Action and Competence in Dark Pedagogy

Jonas Andreasen Lysgaard and Stefan Bengtsson

Introduction

This chapter revisits the action competence approach in the context of the Anthropocene (Pétursdóttir, 2017). The action competence emerged in the 1990s as an effort to critique instrumentalist pedagogical efforts that stress behavior modification and has, over the last three decades, been a strong voice in ongoing debates within research and practice linked to environmental and sustainability education, arguing for *informed action* and the role of *free will* of the learner (Mogensen & Schnack, 2010). Our return to the action competence approach is aimed to make a contribution to its revitalization, where we "revitalize" action competence by drawing out some implications of the Anthropocene for thinking competence and educations contribution to the development of action competence.

J. A. Lysgaard (✉)
Aarhus University, Aarhus, Denmark
e-mail: jonas@edu.au.dk

S. Bengtsson
Department of Education, Uppsala University, Uppsala, Sweden

M. Paulsen et al. (eds.), *Pedagogy in the Anthropocene*, Palgrave Studies in Educational Futures,
https://doi.org/10.1007/978-3-030-90980-2_6

This chapter is structured in the following manner. First, we will return to the classics outlining the action competence approach in order to substantiate certain key assumptions about the human subject as learner, the conception of competence, as well as the role of action. Second, we will reflect on what notions of temporality, spatiality and causality are at play in these conceptions. Third, we aim to draw insights from the discussion of the Anthropocene in order to rethink these notions and action competence.

Accordingly, the effort of this chapter is not only to revitalize the action competence approach, but also to engage in a broader reflection on the conditions and possibilities of developing the subject's ability to shape and enact the future in the Anthropocene. The foci of reflection that we will apply in the second and third part of this chapter will be on how action competence relates to a notion of individual, human causality that is the main or exclusive focus of the approach, as well as a relating to a past (human historicity) as reference point for future competence. The reflection is here to interrogate: what kind of dialogues with the future are we entering into through education?

The Action Competence Tradition

Action competence as a concept of *Bildung* (formation in and through education) is associated with being able—and willing to—become a qualified participant in democracy, where commitment and commitment in relation to the position one has chosen to take is emphasized (Schnack, 1994). The development of the concept was based on a critique of what was considered a tendency toward narrow and problematic approaches focusing on behavior modification in environmental and health education in the late 1980s and 1990s (Schnack, 1994). Narrow and deterministic perspectives on the means and purposes of education, often derived from behavioral psychology, were by Schnack and colleagues portrayed as the opposite of critical pedagogy inspired understanding of education as *Bildung*, where students, as critical subjects, are to be fostered by developing their capacities and ability to take part in critical, formative, and open Bildung-processes (Schnack, 1994).

The concept of action competence itself originated in the late 1980s with the Danish researcher Hans Jørgen Kristensen arguing that: "The question of what students should learn in school with a view to their further path into the next century can be answered with the fact that they must acquire and develop an action competence." (Kristensen, 1987).

Kristensen further emphasizes that competence to act is not about shaping children and young people into a given society (deterministic perspective), but rather about what "children and young people must learn in order to be able to help shape their own and others' future" (Kristensen, 1987).

This argument is, as argued above, rooted in a Bildung-tradition (derived from the German verb *'bilden,'* i.e., to shape, to form), which emphasizes general education and democratic formation, and which can be distinguished from "*Ausbildung*" understandings, where the school's task is primarily education in order to qualify for participation in working life (Klafki, 2011). The *Bildung*-tradition, especially in the form of its reinterpretation in post-World War II Germany, stresses the danger of deterministic approaches to education, where especially Klafki´s later work aimed to highlight their role of developing critical approaches to education that strengthen learners' ability to openly engage with the content and direction of education (Klafki, 2010, 2011).

With regard to the specific content of such an action competence oriented education, the action competence approach shares strong similarities to Klafki's critical-constructive pedagogy, and his formulation of the societal 'key problem' (epochal key problems) that education must relate to (Klafki, 2010). Such key problems are for example, questions of peace, the environment and inequality, which Klafki describes as universal in a given epoch, in that they are relevant for all human beings. . It is at the same time the confluence of different individual formation processes around these key problems that aligns, at least partially, the process of individual Bildung and that of humanity. The basis of this *Bildung*-approach and the resulting curricular outlook is not, as in other European traditions to education, based on or derived from specific academic disciplines or specific academic knowledge and skills content. It is rather a general educational and curricular perspective (*Allgemeine Didaktik*), which must be considered in relation to several of the school's subjects and topics. Thus, the disciplinary educative traditions and subject-specific contents of education are subsumed to the primacy of a comprehensive educational outlooks (*Bildung*), its purposes and the notion of coherent self as that which is shaped in educative processes. This inheritance from the Klafkian approach to Bildung-oriented critical and constructive education sciences, together with Klafki's emphasis on the democratic educational values—self-determination, co-determination and solidarity—is central to the formulation of the concept of action competence (Klafki, 2011).

With its linkages to German Bildungs-oriented didactics as well as an emphasis on the importance of democratic educational values, the action competence approach shares similarities to other existing approaches to education for sustainable development, such as the *Gestaltungskompetenz*-approach (Haan, 2008) in Germany or the pluralistic tradition in Sweden (Englund et al., 2008).

Participatory Democratic Education

The action competence approach was part of a wider reaction toward notions of seeing education as an intrinsic and direct part of economic growth, and a growing focus on the notions of state reliance on education in order to remain competitive in a globalized world (Vare & Scott, 2007). In the late 1980s and 1990s as researchers at the late Royal Danish School of Education Sciences (1856–2000, now the Danish School of Education, Aarhus University) witnessed a growing focus on environmental issues and ties to the new concept of sustainable development (Breiting et al., 1999). This concept was popularized through the Brundtland report and started to seep into educational perspectives on environmental challenges (UN, 1991). Sustainable development established a new, global framework for understanding environmental challenges, but also marginalized former more localized approaches to environmental education (Poeck & Lysgaard, 2016). The specific focus on environmental, ecological an nature-related issues that framed environmental education (EE), were, during the 1990s, expanded to also tap into social and economic issues as well, through the emerging concept of education for sustainable development (ESD) (Læssøe, 2020). This was a time before for example, the comparative programme for international student assessment (PISA, 2002 and on) had impacted Scandinavian educational research and practice, but tendencies to align state and market priorities and use education as a tool to reach such priorities were evident and growing also in Denmark. The notion of ecological modernization put specific links between market driven incentives and solving environmental challenges together (Dryzek, 2005; Hajer, 1995). Organic food made its mainstream debut and education would again be at the center of a discussion of the core value that a society and the education it supports should hold at its center. This greatly influenced the development of the concept of action competence in the north of Copenhagen as a team, forming around Schnack and other key researchers, set out to critique the educative potential of then current approaches to linking education with sustainable development (Dahl et al., 2011).

Karsten Schnack, as part of the efforts to emphasize the foundation for the development of the action competence approach argued:

> There are two different perspectives on education. In the first case, the educational target is of a formative nature, since concerned with a specific critical way of relating to life (…). In the second case, the educational target is limited to acquiring a set of specific behavioural patterns that can help to solve environmental issues here and now. (Breiting et al., 1999)

This understanding of education and the competing approaches and positions it entails, is not only well known throughout continental discussions of the purpose of education, it also opens up for a distinct and normative take on the values embedded within education. The entailing critique of a less nuanced understanding of education as primarily a producer of behavior draws on critical pedagogical perspectives from Klafki. It echoes the heritage from critical theory and the will to engage critically with democratic values through education (Mogensen & Schnack, 2010; Vare & Scott, 2007).

Such an overall understanding of the two outlined approaches to education, and the proposed shortcomings of a focus on behavior modification instead of critical formation became the foundation of the development of the action competence approach. It emphasized the importance of empowered teachers and pupils that were able to navigate questions of power and avoid being manipulated to meet the needs of others. As was argued as part of the project examining environmental education in the Nordic countries:

> Behaviour regulation displays the most overt use of power. In itself, there is no harm in that. However, the democracy perspective makes it all-important who exerts such power over whom, in what capacity, and to what extent such behaviour regulations are fair and responsible to weaker citizens. This is precisely what action competence is for—enabling individuals to take part in the use of power, and hence behaviour regulation, in a critical manner. This is the life blood of democracy. (Breiting et al., 1999)

While somewhat grand and sweeping such statements underline the critique of using education as a tool for implementation of specific behavior, especially without the knowing consent and critical engagement from the learner, mimicking ongoing current critique of the use of "nudging" approaches to solving pollution challenges (Bessant et al., 2015). The critical and formative perspectives must be part of education, according to Schnack, but not only linked to the individual needs and efforts to navigate learning and educational approaches. Another inheritance from the critical theoretical backlog, as well as traces to critical pedagogy is the specific social and political outlook of education according to the action competence approach. While this in itself is adding complexity, the action competence approach adds to this by not only focusing on current social and societal issues that need to be addressed, but also the state of potential futures societies.

> Possible solutions and actions need to be considered in this societal perspective. It stands to reason that having environmental problems solved remains first and foremost an adult responsibility. All the same, it is the schools' responsibility to encourage and prepare their students, enabling them to reflect critically and take part in debating future environmental problems from a societal perspective. (Jensen & Schnack, 1997)

This emphasis on not only current, but also future challenges and the role of young pupils and students as emerging and future adults links the action competence approach liberal education. Schnack and Jensen here draw on a well-established Danish educational take on liberal education:

> In liberal education, over and above insight in a sphere of knowledge, there lies the fact that a criterion has been established for utilization of that knowledge, that one has accepted a responsibility for how, when and for what one will use this knowledge. (Jensen & Schnack, 1997)

Understanding the foundation of action competence and the specific challenges that shaped the development of the concept, frames it as an effort to introduce critical theory infused emphasis on Bildung and specific societal notions of liberal education. This underlines the importance of addressing environmental challenges as not only localized and something that can be dealt with through behavior modification, but as something that should be understood as part of the social construction of central societal issues that goes beyond limited and opaque use of behavior modification. Action competence thus invests heavily in the role of the individual learner as the normative critical key character in the continued Bildungs-process of individual and society:

> Action competence implies that you will include normative arguments and views in a discussion of what constitutes the relevant issue, and what alternatives and visions can be suggested. This also serves to indicate that when dealing with environmental issues, it is necessary to reflect on normative aspects. In other words, that debating ethical issues makes sense. (Jensen & Schnack, 1997)

An understanding of the core take on the role of education in the action competence approach is seemingly driven by an emphasis on an idealistic kernel, understood as the combination of grounding normativity in how an issue is constituted in an argument. That argument is that a form of normativity is grounded in human will. Thus, the discursive constitution of the issue and the vision for its solution by the human subject "constitutes" the relevant issue.

The Notions of Action and Competence

Based on such an explicit normative ideal of the role of education in society, the group of researchers settled on the concepts of "competence" and "action" in order to describe their Critical-*Bildung*-infused approach to environmental education. While "Competence" in the 2020s reek of the aforementioned efforts by comparative efforts such as PISA and TIMMS to create quantifiable data on the skills and competencies of young people, the concept of competencies did not carry these connotations in the late 1980s and 1990s.

> Developing action competence becomes a formative ideal in a democratic perspective. At best, 'competence' should evoke associations to something about being able to (and wanting to?) to be a competent participant. (Breiting et al., 1999)

The critical potential of such a concept and the democratic undercurrent of engaging with a participatory understanding of competence were only strengthened by linking it with the notions of action:

> 'acting' needs to be read into the entire complex of distinctions concerning behaviour, activities, habits—and hence actions. Strictly speaking, actions

> may well consist of the same movements as kinds of behaviour, yet are invariably characterised by being conscious, reflected, and targeted. (Schnack, 1993)

Action, in the understanding of the action competence approach, differs from simple execution of tasks, as the use of the notion of action in this constellation implies that it is *conscious*. No random faffing about at the will of an old school teacher barking out instructions. The individual learner substantiates her actions through competence and thus moves beyond the limited horizon and potential of deterministic behavior:

> Related to an action, there will always be a conscious making up of one's mind, while this is not necessarily the case with a behavioural change which could be caused by pressure from other people (e.g., a teacher or peers) or by other influences such as advertisements. (Jensen & Schnack, 1997)

Accordingly, education is to foster conscious action in contrast to getting the learner to carry out unconsciously a predetermined action. Transforming the battle cry of the Enlightenment "Dare to know!" (Kant 1789) into "Dare to be conscious of your action!" The action competence approach conceives of education as a means to assure that the learner becomes aware of its action.

Hence, the core idea manifests itself as being both idealistic, beautiful and bordering on the naïve:

> (We) must understand and explain actions by referring to motives and arguments, rather than to mechanisms and causes (Schnack, 1977). Perhaps, this is expressed most succinctly by the term of intentionality. Actions are intentional. (Schnack, 1993)

What can be seen to be at stake is a logical contradiction as we might ask us: Is there something such as unconscious action, maybe an action that is caused by something that I am not aware of or that was not my attention? The last sentence of the quote above can be seen to suggest that we should not consider such an act an action, as they have to be intentional. It is also here that the core paradox of the "enlightened" action competence can also be seen to be rearticulated in new shape: it puts forward that knowing or, in our case not being intentional, is already *self-imposed* (Deligiorg, 2005; Shell, 2009). Hence, if the learner in engaging with "making up of one´s own mind" (Jensen & Schnack, 1997) is then that making up of one´s mind must be a posteriori to an a priori intention to not making up one´s own mind; that is, the intention to be not intentional. The initiators of action competence perspective can here be imagined to chant along the lyrics of *The Hives*' classic *Sturm und Drang* Anthem *Hate to tell I told you so*: "Do what I want 'cause I can and if I don't, because I wanna." The issue we aim to address by pointing to the paradox at the core of this reasoning is to show that there is no escape from intention and non-intention. This would already have to be based on the intention to not have an intention. This weird logic can be seen to be articulated in the following classification of environmental action into two main categories:

> Environmental actions can be grouped into two main categories: (i) actions which directly contribute to solving the environmental problem that is being worked on: (ii) actions whose purpose is to influence others to do something to contribute to solving the environmental problem in question (indirect environmental actions). (Jensen & Schnack, 1997)

Accordingly, there needs to be something as an intention to let oneself be influenced for category (ii) of environmental action to be a possibility. What is the reason for this intention to let oneself be influenced? Following Kant, we might assume that the action competence approach might put forward "laziness" and "cowardice" (Deligiorg, 2005).

Hence, there is a quest for action competence, that of the quest of meaning (Jensen & Schnack, 1997) and potentially laziness and cowardice. Yet, meaning and the quest of meaning can be found to lie in the exposition of injustice and inequality. Consequently, we can see the brave heroes of action competence to engage in the quest of meaning by banging their breastplate, and raising their competence dripping swords in a salute: "To Justice and Equality!" gathering the troops summoned by critical theory and critical pedagogy.

Recruitment to the ranks and fight for the cause requires, not only acceptance of the intention of intention, but also commitment to the greater cause. As always, this critique and the development of the action competence concepts needs to be historically contextualized as part of a reanimation of critical theory and pedagogy in the then changing times which saw a growing emphasis on behavior modification, which would develop in the following decades policy driven obsession with all things quantifiable within global education agendas. In the 1990s, there still was a strong intent to insist on the importance and power of dwelling on the democratic ideals of education:

> The action concept implies a deliberate commitment in the acting person—that you have considered the matter and decided to act. Often, the behaviour concept will not encompass this aspect. As a result, 'behaviour' and its derivative concept 'behavioural modification' private in the shorter and/or longer term? Thus, whenever we talk about modifying student behaviour as an element of environmental education, this tends to signal an education paradigm based on prescriptions and behavioural modification, rather than on democratic elements such as participation, dialogue and co-influence. (Jensen & Schnack, 1997)

These above outlined ideas of action and competence, as imbued within the action competence approach, are not only heavily invested in the idealistic kernel of critical theory and critical pedagogy, but also, today in the third decade of the second millennia, sound as faint echoes of a time where these ideals could still be fought for, without meddling in the ever-changing complexity of notions such as sustainability in times of SDGs and global climate crisis. Today the critical agenda of action competence might seem a bit quaint, mirroring the defeated and paralyzed troops of critical pedagogy, the Left as well as progressivists haunted by fatalism as faint echoes of a lost time.

Yet, back then, action competence was not only part of an ongoing battle over the merits of education in relationship to sustainability and environmental challenges. It also pointed toward the future, toward our challenges of 2021 and beyond. Intentionality was not only caught up in the moment, but also related to future challenges and how the student would engage with these, yet unknown, obstacles:

> This democratic perspective for action competence implies that the concept as such is not context defined, in the sense that it points towards specific action possibilities or views of our future society. All the same, it is prescriptive, since concerning our obligation to relate to issues in an impartial and critically responsible manner, and to base our actions on whatever answers we find—thus participating in developing a democratic, equitable, and sustainable society. (Jensen & Schnack, 1997)

The quote can be seen as to appeal to *Bildung* at its purest, seeing education as a process that would not only emphasize the challenges of navigating current societies while drawing on principles and insights from the past, but also reach into yet unknown future societies:

> …in order for environmental education to qualify students to tackling future environmental issues, a comprehensive, reflective, and critical approach is needed. (Jensen & Schnack, 1997)

Again, the foundational concepts of critical theory and critical pedagogy are brought to the fore as the omnipotent competence of critical thinking are underlined as the way to deal with the challenges at hand:

> As critical thinkers we are engaged in a continual process of creating and re-creating our personal work, and political lives. We do not take our identities as settled; rather, we are aware of the scope of development in all areas of life. We see the future as open to our influence. We regard the world as changeable through our own individual actions and through collective action in concert with others who share our commitment to broader political and social changes. We do not accept the idea that because the things are the way they are now, they must always be this way. And we do not think that we (or anyone else) have the ultimate answer to life's ambiguities and problems. But we do have confidence in knowing, that those things in which we believe, and the actions we take arising out of these beliefs, spring from a process of careful analysis and testing against reality—in other words, from critical thinking. Brookfield 1987 in (Breiting et al., 1999)

And this is where we arrive in our analysis of the original take on action competence: As a formative *Bildung*-ideal, deeply embedded within the critical continental traditions embodied by twentieth century critical theory and aspects of the thinkers associated with the Frankfurt school. Action and competence are entwined with the potential of critical intentionality that invokes the possibility of true democratic participation. Not only was this action competence approach created as a bulwark against tendencies toward behavior modification that seeped into education throughout the 1990s, but it also speaks to the future and aims to establish action competent learners that can continue to critique and intentionally act in order to mitigate the future challenges that they, and the societies they will inhabit will face. As we stated in the beginning of this chapter, the action competence approach has had widespread impact and implications around the world, for many reasons not in its origin country of Denmark, but across Icelandic, New Zealand, Norwegian and South African educational research and curriculum development the concept of action competence pops up, albeit in new forms and contextualized versions, but still harking back to the roots of the ideas of the original action competence approach outlined above.

Educational Spatiality, Temporality and Causality in the Anthropocene

The action competence approach, in its outlined classic form, or one of the newer forms (Carlsson, 2020; Olsson et al., 2019), can still be viewed as both foundational work and something that speaks to and makes contributions to current critical positions within environmental and sustainability education research and practice. Engaging with the approach of action competence, we, in light of what has become to be labeled the Anthropocene and the great Acceleration of global crises, aim to return to the potential routes and paths of action competence, and to revitalize and develop it further through constructive critique. In particular, we aim to address in our critical engagement the notions of spatiality, temporality and causality in action competence in order to illustrate how these notions could be fruitfully re-conceptualized into notions of action and competence in the Anthropocene.

As Morton puts forward in his work on the notions of hyperobjects and dark ecology, we find ourselves living in times where space and spatially are no longer what they used to be (Morton, 2013, 2016). The Anthropocene, following Morton's observations, binds together or twist different temporalities though humans, the planet and other large scale entities into the form of a strange loop, which we can no longer see ourselves to escape from, or realistically cut up and compartmentalize (Morton, 2016). In 2021–2021, we found ourselves inhabiting spaces twisted by a hyperobject called Coronavirus (Bengtsson & van Poeck, 2021). Whereas action competence grapples with notions of *neutralizing* environmental issues (Jensen & Schnack, 1997), Morton insists that this framing of "the environment" as something that can be delimited, addressed and potentially neutralized (as in no longer representing a threat or challenge), is both overly naive, but also a lack of recognition for the weird scale that our planet and its inhabitants operate on.

> There you are, turning the ignition of your car. (...) Every time I start my car (...) I don't mean to harm earth (...). My key turning is statistically meaningless. But go up a level and something very strange happens. When I scale up those actions to include billions of key turnings (...) harm to earth is precisely what is happening. I am responsible as a member of this species for the Anthropocene. (Morton, 2016, p. 8)

What Morton can be seen to highlight is that individual human experience of action cannot delineate the positionality or spatiality of that localized action. The example of turning the key of the car and the global warming it produces highlights that there is something happening that is not accessible in the reduction to acting "here and now." The action of turning the key of the car is both my action and not my action, it is both me turning the key and humanity turning the key. The Anthropocene can be seen to signal to humanity "Congratulations! You have now become aware of being part of an entity that operates at global scale, and there is no way back or out."

Action in the Anthropocene, for example the turning of the key, does not confer to the environments that we can perceive, access or understand. Instead of space, or the spatiality of entities over there is, through the Anthropocene, drawn in, turned and twisted into constant and direct links to everything, from the smallest organisms to the state of the thin circumventing layer of oxygen cushioning us from the interstellar void. Action competence can be seen to keep this closeness of entities at bay by "recognizing" global injustice and the "interconnectedness" of regional or even inter-continental environmental issues (Breiting et al., 1999). Yet, it is keeping these entities at bay by separation and distance. Identifying environmental issues through engagement with our local environment, producing solutions and intentionally applying them in order to neutralize the problem is now inherently impossible in the twisted weird interconnectedness of the Anthropocene, globally linked sustainability crisis and the breakdown of planetary systems such as the climate. The extremity of the Anthropocene and entailing global and planetary weird closeness and stickiness of wicked issues such as climate change or space debris pollutants is overpowering educational aspects of identifying and cutting out specific environmental challenges through critical insight, competence and finally action (Morton, 2013).

How do we address the link between Morton's example of turning the key and the 418 carbon dioxide molecules per million that "currently"

(April 2021) float in our atmosphere? How do we understand the weird connectedness that our physical existence imparts on us? How can we conceive critical competence to engage with such vast links that do leave precious little room for the age-old favorite question of journalists and engaged teenagers alike: "but, what can I do?" You can *do* a lot, but to address comprehensively, exhaustively and reasonably the environmental challenges in anything resembling their extended and weird form is well out of reach, as spatiality no longer confines to ideas of neatly packaged or delineated things perceived, the community, the nation state or even humanity (if there ever was something identifiable as such).

Probing into the action competence perspectives on temporality, links with past, present and futures, only continues to muddy the waters of clear-cut understandings of both action and competence. Operating with a historical perspective can teach us about the immediate Great Acceleration that has brought us the calamities framed as sustainability challenges of global warming. By going just a tiny step further, and including the whole history of the human species and its 300.000 (or so) year span, can easily dismiss any clear-cut efforts to learn from the past, and apply it to the present in order to engage the future with open eyes. According to strands within deep history research (Sørensen & Eskjær, 2014), we, as individuals and societies, might face dramas and tragedies, but as a species, or as the species we might evolve into, our current wicked problems represents business as usual as untold hardships are served to us, as they were to our ancestors, so often (apparently) on the brink of total annihilation. In the context of current development of A.I., we might take John Scalzi's (2010) short story, "When Yoghurt Took Over," as an entry point to contemplate the temporal and historical aspect that deep history research can be seen to address. In the short story, scientists develop, through the most advanced technology, a new type of bacillus used for fermenting yoghurt. The experiment, while seeming to have failed, turns out to have produced a sentient and highly intelligent species of yoghurt that communicates to the scientists: "We have solved fusion. Take us to your leaders." Using humanity as a springboard, the yoghurt develops spaceflight ending the story with the observation: "Life from Earth is going to the stars. It just may not be human life." Hence, we might ask ourselves if the future seemingly intended will be a 'human future'?

The question relates to the causality at play in intention. Is the intention to be located temporally in the past, and where is it to be located? Foucauldian and Deleuzian analysts have been working since the late

1980s with these perspectives, but something new might be at play here (Gołebiewska, 2004; Semetsky, 2003). What we are suggesting is that the causality of intention might be located in the future anterior. Stretching and expanding notions of linear temporality as found by critical pedagogy, we suggest to rethink the temporality of action competence to be located in an engagement with the future, where that future is bringing itself into being. Competence is in this sense a backward realization of the future in the past. Let us take the history of space flight as an example to illustrate. While the US space mission to the moon has been significantly shaped by the work of Wernher von Braun, who has during his childhood been shaped by the work of Jules Verne and the story "From Earth to the Moon," we might be tempted to read the Space mission as influenced by the fiction of Jules Verne. Yet, we might consider the fact that Jules Verne also wrote "Journey to the Centre of the Earth," a fictitious future that did not came "true" (Weingardt, 2011). We might argue that similar to Lorenz work on attractors that Wernher von Braun and others have been influenced by a strange attractor, an attractor that from the future influenced the development of the space program.

The Anthropocene opens up for the weird loops of future attractors imbuing efforts to reconceptualize past and current actions, but also challenges the spatial delimitations as the extreme interconnectedness leaves nothing in a vacuum. No learner is an island, but neither is any given action, thought or educational activity relocated within endlessly intertwined temporality and spatiality. Causal understandings of the links between competence, intention, action and consequence should thus not be approached from limited anthropocentric efforts to pin down and control such processes, but approached from the opposite direction: from the void, where everything and nothing resides in the endless strange loops of the Anthropocene. Such an approach, from the void, renders the action competence, as infused "competence" and "action," impossible remnants from an anthropocentric twentieth century.

The action competence approach, as envisioned by Schnack, Breiting, Mogensen and others in the 1990s, and since then popularized to a large global audience, represented a critical take on how to ensure that links between environmental issues, sustainability and education did not lead to simple notions of empty vessel pedagogy—merely filling the empty heads of children and young people with qualified information on the challenges at hand. More was, and is, needed, in the form of critical thinking and entailing intentional actions. However, as we have tried to demonstrate above, notions of "intentional action" and perceived insight into the consequences of spatial and temporal perspectives, clash with the immense complexity, intertwinement and weird relations imbued in the idea of the Anthropocene. The action competence approach might teach critical thinking and informed action, but still relies on an increasingly impossible conceptualization of causal links between knowing, acting and entailing consequences. Action and competence are powerful concepts of the enlightenment and twentieth century critical thinking and pedagogy, but in the Anthropocene, neither concept of action nor competence can no longer be considered ours alone. The Anthropocene signals here to the "competent" learner that the "natural world" of objects is not the passive background waiting for the human subject to set *things* into motion. Accordingly, competence and action are bend and meddled with by what we considered as non-subjects, like the authors sneezing and potentially oozing lethal amounts of virus all over the keyboard. We wonder if it was "our" competence to round up this paragraph and address the editors' request for clarification or something else´s competence to spread its genome that was happening at "this moment"?

As learners, students, teachers, researchers, citizens, we start to become aware of that we do not "own" or are increasingly unable to impose limits on concepts, objects or phenomena and their entanglement with the world, past and future. Like *Dune's* guild navigators in training we do in the Anthropocene become exposed to spice-induced visions were we become aware of relations of our actions, as they travel and unfold through time, space (and potentially other dimensions) without fully grasping their consequences or being able to control their outcomes. In the Anthropocene, our expanded awareness troubles our previously held notions of the subjects ownership of both action and competence. As Thacker argues, we no longer have access to thinking the world, or our place in it as prospective of reaching equilibrium or healing a broken planet, environment, economy and sociality:

> The world is increasingly unthinkable—a world of planetary disasters, emerging pandemics, tectonic shifts, strange weather, oil-drenched seascapes, and the furtive, always-looming threat of extinction. In spite of our daily concerns, wants, and desires, it is increasingly difficult to comprehend the world in which we live and of which we are a part. To confront this idea is to confront an absolute limit to our ability to adequately understand the world at all (…). (Thacker, 2011)

The Anthropocene's expanded awareness of the "competent" learner bring to the fore the problem of thinking action in terms of effectuated intentionality, calling, as we argue, for a rethinking of traditional understandings of the role and potential of education. The metaphor of exposure to spice as a way to think the Anthropocene as process of becoming aware or accustomed to is by us seen as suitable as it points out the mutative aspect and break in thinking that we see is happening as well as required. Exposure to the Anthropocene unveils to us the anthropocentrism of the teleological way of thinking action and competence as expressed by the early action competence approaches. Exposure to the Anthropocene renders us aware of actions being caught up with and shaped by things radically non-human. To become exposed means to attunes thinking of competence to a beyond of the confines of human will and intention; our increased awareness highlights how outcomes and intentions of actions are shaped by incontinence. Exposure to the Anthropocene is by us seen as to render aware of the incontinence, *as lack of voluntary control over action and intention as well as the lack of self-restraint when intending or acting*. Any apparent competence and intentionality, not only in and of our actions, but also in educational activities (International Coastal Cleanup day, Fridays for Future, Eco-school, Whole school etc.) are dispersed beyond an anthropocentric web of delineated understandings of spatiality and temporality. To paraphrase action competence: "in itself, there is no harm to that," but it drastically underestimates and limits the impact and potential of engaging with education in light of the Anthropocene (Breiting et al., 1999).

Thus, action, intentions, competences happen, but they are shaped by lack of voluntary control and self-restraint. They are incontinent in that any links to intentionality, cause and effect are more based on selective (intentional and empirically limited) interpretation than anything resembling true control over the outcome of a situation. This should, however, not lock us as educators into an impasse. While initial exposure to Anthropocene could leave us initially at a stage of mutation that focuses on the limitations of the human scope of perception and understanding, we argue for even greater exposure to the Anthropocene and a resulting mutation that can help us to transcend modernist anthropocentric lullabies of critical theory and critical pedagogy.

Action Incontinence: Dark Pedagogy in the Anthropocene

We suggest exposing ideas of intentionality and human control over actions and their consequences to the notion of always withdrawn nature of any given objects, actions or phenomena (Harman, 2011; Lysgaard et al., 2019). Inspired by emerging thoughts on speculative realism, as proposed by Thacker, it can be argued that action competence and the educational approaches and strategies that have formed much of our educational system continues an intimate relationship with the world that we have put into words and meaning (Lysgaard et al., 2019; Lysgaard & Bengtsson, 2020; Thacker, 2011). This is the anthropocentric part of Thacker's division of the cosmos, or the world into three parts. The *world-for-us*, which is the world as we intend and interact with it, which we interpret and to which we grant meaning, and which is defined chiefly in terms of our intended relationship with it; the *world-in-itself*, which coexists with the world-for-us, resists or ignores our attempts to mold it, and is primarily accessed through scientific inquiry and technological intervention, and finally the *world-without-us*, which does not and cannot coexist with the world-for-us because it is the subtraction of the human element from the world, and is therefore spectral and speculative (Thacker, 2011).

The action competence approach can be understood as a human undertaking to insist on a gradual expansion of Thacker's world-for-us, while insisting on a critical relationship with the world-in-itself, that throws challenges and surprises at us, as it is slowly digested and tamed by human insight and knowledge and turned into the world-for-us.

Arguably, however, we do not even have full access to what we can term the world-for-us. As the links between Jules Verne and Wernher Braun above illustrate, the depths, twists and *conflation* of what we try to conceive as the world-for-us, through education or otherwise are already deeply intermingled with the world-without-us. The becoming aware of this we have labeled above the Anthropocene. Yet, once exposed we can increase this exposure and our mutation by rendering ourselves sensible to how the world-without-us bursts through the world-for-us seemingly everywhere. Our metaphorical excurses, traveling between different planets or introducing nuclear holocaust as a real potential end to all of humanity relates intimately, almost too intimately, with the world-without-us as it bursts through into the world-for-us. To engage with such questions through education there is a need to expose these breaches of the world-for-us. Breaking with notions of temporality and spatiality that maintain the status of the world-for-us as delimited scene of the narrative of human history unfolding as expressed by action competence is the starting point for exposing ourselves through education to the Anthropocene.

Here we point toward notions of Dark Pedagogy as ways of accelerating our mutation and widening the sensitivity of action incontinence. We are using the term dark in "dark pedagogy" to denote both an ontological position and an emotional, affective resonance. With regard to the ontological position, 'dark' aims to infuse educational thought with emerging realist philosophical perspectives on nature's 'great outdoors;' that is to say, sensitivity to how the world-without-us bursts through, in order to mutate the action competent subject in the face of the nonhuman dark. The darkness of the Anthropocene is uncanny in the German sense of ***unheimlich*** denoting a loss of homeliness and familiarity. In this way, darkness highlights a feeling of loss of place-based identity and culturally safeguarding context. In the encounter of the dark and uncanny, we argue we are encountering a confluence between the world-for-us and the world-without us.

It is hard to deny that the exposure to the Anthropocene and for example the climate crisis tell the competent subject something of profound significance concerning its self-ascribed status as subject and the world-for-us. Exposed to the Anthropocene, the action competent human of good conscience mutates to, first, doubt their status as subject and as well as the flatness of the world-for-us. Mutation sensitizes emotionally and experientially the action competent human to the grip of the *dunkel*, to borrow another German expression. Twilight obscurity ("*Dunkelheit*") renders ourselves strange, it contaminates from a non-specific moment or renders always-already unfamiliar not only the world but ourselves and how we conceived ourselves by appealing to our intentionality, voluntary control and agency.

Consequently, when exposing our-selves to the grip of the *dunkel*, we immerse ourselves into the world-*X*-us rendering, at least momentarily, exposed and vulnerable to that exposure and engagement with it in action (Lysgaard et al., 2019). "X" marks here the "here" and "now" that as we would like to argue already is saturated or conflated with the spectrality of darkness as *Dunkelheit*.

That leaves to question of how such a Dark pedagogy educational approach and an engagement with action incontinence relates to the twentieth century concept of action competence and could transform it into something else. Here we return to action competence and the question pondered Jensen & Schnack by the end of the seminal article from 1999:

> *The question can be asked whether it is possible (or desirable) to aim at a situation where all action are done on the basis of acquisition of a thorough insight and consequent decision making within the spheres in question?* (Jensen & Schnack, 1997)

Taking that question as an invitation to immerse ourselves into the world-x-us, we can tease out something dark and uncanny in the question, that is we expose ourselves to the implications of "acquisition of insight." Where or to whom/what does the insight belong to if it has to be acquired by the addressee of the question, that is the human subject? Action incontinence engages with the premise that question that is it engages with acquiring thorough insight and reflects on what to do with that insight. Yet, it holds that neither insight nor decision is voluntary but exposed and vulnerable to something that is not will or conscious control. The thoroughness of action incontinence draws the subject into the acquisition insight and the temporal and spatial scales that surpass its ability to arrive at "a bottom," hence there is no solid foundation or basis for decision. Yet, there is in this lapse into this abyss of acquisition of insight the possibility of distancing from basis and decision, that is the possibility of estrangement from basis and decision. Accordingly, action incontinence dives into the dark of world-X-us to expose and alter the competent subject, a fall that however not safeguarded by the safety net of desire nor decision.

References

Bessant, S. E. F., Robinson, Z. P., & Ormerod, R. M. (2015). Neoliberalism, new public management and the sustainable development agenda of higher education: history, contradictions and synergies. *Environmental Education Research, 21*(3), 417–432. https://doi.org/10.1080/13504622.2014.993933

Bengtsson, Stefan; Van Poeck, Katrien (2010) What can we learn from COVID-19 as a form of public pedagogy? - In: European journal for Research on the Education and Learning of Adults 12 (2021) 3, S. 281-293 - URN: urn:nbn:de:0111-pedocs-235802 -https://doi.org/10.25656/01:23580.

Breiting, S., Hedegaard, K., Mogensen, F., Nielsen, K., & Schnack, K. (1999). *Action Competence, Conflicting Interests and Environmental Education*. DPU.

Carlsson, M. (2020). Handlekomptencebegrebet på arbejde i Uddannelse for Bæredygtig Udvikling. In J. Lysgaard & N. J. Jørgensen (Eds.), *Bæredygtighdens Pædagogik*. Frydenlund.

Dahl, K., Læssøe, J., & Simovska, V. (2011). *Essays om dannelse, didaktik og handlekompetence*. DPU.

Deligiorg, K. (2005). *Kant and the culture of enlightenment*. State University of New York Press.

Dryzek, J. S. (2005). *The politics of the earth: Environmental discourses*. Oxford University Press.

Englund, T., Öhman, J., & Östman, L. (2008). Deliberative communication for sustainability? A Habermas-inspired pluralistic approach. In S. Gough & A. Stables (Eds.), *Sustainability and security within liberal societies* (pp. 29–48). Routledge.

Gołebiewska, M. (2004). Intentionality and the "Being-in-the-Language" of Michel Foucault. In A. Tymieniecka (Ed.), *Does the world exist?* (Vol. 79)). Springer.

Haan, G. d. (2008). Gestaltungskompetenz als Kompetenzkonzept der Bildung für nachhaltige Entwicklung. In I. Bormann & G. D. Haan (Eds.), *Kompetenzen der Bildung für nachhaltige Entwicklung: Operationalisierung, Messung, Rahmenbedingungen* (Vol. 23–43). Verlag für Sozialwissenschaften.

Hajer, M. (1995). *The politics of environmental discourse*. Clarendon Press.

Harman, G. (2011). *The quadruple object*. Zero Books.

Jensen, B. B., & Schnack, K. (1997). The action competence approach in environmental education. *Environmental Education Research, 3*(2), 163–178. https://doi.org/10.1080/1350462970030205

Klafki, W. (2010). *Studien zur Bildungstheorie und Didaktik*.

Klafki, W. (2011). Die bildungstheoretische Didaktik im Rahmen kritisch-konstruktiver Erziehungswissenschaft. In H. Gudjons & R. Winkel (Eds.), *Didaktische Theorien* (pp. 13–34). Bergmann & Helbig.

Kristensen, H. J. (1987). *Skolen i fremtiden: tværfaglighed og grundlæggende kundskaber*. Gyldendal.

Læssøe, J. (2020). Bæredygtighedsbegrebet og uddannelse. In J. A. Lysgaard & N. J. Jørgensen (Eds.), *Bæredygtigheden Pædagogik*. Frydenlund.

Lysgaard, J. A., & Bengtsson, S. (2020). Dark pedagogy – Speculative realism and environmental and sustainability education. *Environmental Education Research, 26*(9–10), 1453–1465. https://doi.org/10.1080/13504622.2020.1739230

Lysgaard, J. A., Bengtsson, S., & Laugesen, M. H.-L. (2019). *Dark Pedagogy*. Palgrave.

Mogensen, F., & Schnack, K. (2010). The action competence approach and the 'new' discourses of education for sustainable development, competence and quality criteria. *Environmental Education Research, 16*(1), 59–74. https://doi.org/10.1080/13504620903504032

Morton, T. (2013). *Hyperobjects*. University of Minnesota Press.
Morton, T. (2016). *Dark Ecology*. Colombia University Press.
Olsson, D., Sass, W., Cincera, J., Sund, L., Sund, P., Pauw, J. B.-d., … Gericke, N. (2019). *Bringing perspectives of action competence together in education for sustainable development*. Paper Presented at the European Conference for Educational Research.
Pétursdóttir, Þ. (2017). Climate change? Archaeology and Anthropocene. *Archaeological Dialogues, 24*(2), 175–205. https://doi.org/10.1017/S1380203817000216
Poeck, K. V., & Lysgaard, J. A. (2016). The roots and routes of Environmental and Sustainability Education policy research. *Environmental Education Research, 22*(3).
PISA (2002) Programme for International Student Assessment (PISA) PISA 2000 Technical Report. OECD Publishing.
Scalzi, J. (2010) When the Yogurt Took Over. Whatever Blog.
Scheie, E., & Korsager, M. (2017). *Nordisk arbeidsmodell for Undervisning for bærekraftig utvikling*. Retrieved from Oslo.
Schnack, K. (1977). Humanisme – Livsanskuelse og menneskesyn. In F. Nielsen (Ed.), *Pædagogisk teori og praksis*. Borgens Forlag.
Schnack, K. (1993). Action competence and political Bildung. *Didaktiske Studier, 2*.
Schnack, K. (1994). Some further comments on the action competence debate. *Studies in Educational Theory and Curriculum, 12*, 185–190.
Semetsky, I. (2003). The problematics of human subjectivity: Gilles Deleuze and the Deweyan Legacy. *Studies in Philosophy and Education, 22*(3), 211–225. https://doi.org/10.1023/A:1022877506539
Shell, S. M. (2009). *Kant and the limits of autonomy*. Harvard University Press.
Sørensen, M., & Eskjær, M. F. (2014). *Klima og Mennesker*. Museum Tuscalums forlag.
Thacker, E. (2011). *In the dust of this planet*. Zero books.
UN. (1991). *Agenda 21*. Retrieved from Paris. https://sustainabledevelopment.un.org/content/documents/Agenda21.pdf
Vare, P., & Scott, B. (2007). Learning for a change: Exploring the relationship between education and sustainable development. *Journal of Education for Sustainable Development, 1*(2).
Weingardt, R. G. (2011). Wernher von Braun and the race to the moon. *Leadership and Management in Engineering, 12*(1), 26–36.

CHAPTER 7

Dark Labour

Jason J. Wallin

For Francois Laruelle (2013a), the project of Western philosophy perpetrates what he refers to as a "monopoly on the Real". For what has become an orthodoxy within Western philosophical thought, Laruelle contends, is its claim to the truth of reality and the subsequent 'standardization' of thought and experience relative to such 'facticity'. Instances of Laruelle's assertion can be plainly detected within the annals of Western philosophy. In Platonic philosophy, for instance, the Real is constituted from the ideal world and its copy, for Kant, the Real consists of the phenomenal world-for-us and noumenal thing-in-itself, in Badiou, the Real is correlated to the mathematical 'fact' of set theory, and so on (Culp, 2016; Laruelle, 2013b). What insists as a general stratagem across these instances, Laruelle contends, is the presupposition of reality in the first instance and demonstration of its 'givenness' to philosophical reflection in the second. In comic fashion, a preponderance of philosophical exegesis finds the Real 'given' to its particular mode of reflection *always-already* prefigured in a fundamental 'decision' concerning the Real. This scenario articulates both the 'monopoly on the Real' that Laruelle detects at the heart of the

J. J. Wallin (✉)
University of Alberta, Edmonton, AB, Canada
e-mail: jjwallin@ualberta.ca

M. Paulsen et al. (eds.), *Pedagogy in the Anthropocene*, Palgrave Studies in Educational Futures,
https://doi.org/10.1007/978-3-030-90980-2_7

Western philosophical project, and so too, the problem by which thought and experience become standardized according to a founding axiomatic sufficiency of the Decision that subtends philosophy as a limit-thought.

Educational Monopolies

The order of philosophical thought detected in the work of Laruelle poses a compelling challenge for the field of educational thought and research. For where education harbors an albeit 'economized' fidelity to philosophy, it nevertheless labours in relation to certain decisional presuppositions that 'standardize' its thought and action. Such a claim seems patently obvious in the field of educational research, which has today reached 'peak standardization' if but for its fidelity to the orthodoxies of philosophical inquiry detected by Laruelle (2010). For it seems today that the 'standard' image of educational research coincides with the Western philosophical legacy of first naming the Real and subsequently substantiating access to reality via its analytic or reflective tools. In a bald claim to the Real, the intelligibility of educational research today hinges on such dominant performative traits as explanatory function (*this* is how the world works), the hermeneutics of 'deep meaning' (*this* is how things are), and therapeutic functions for making us feel better, smarter, and wiser (*access* to the Real) (Thacker, 2018, p. 28). In part, the problematic of such automatic commitments are elucidated in the work of Weaver and Snaza (2016), who identify in the ubiquitous perpetuation of educational 'methodocentrism' a lingering commitment to the narcissistic presupposition that the world corresponds to our thinking about it.

Despite an emerging program of *more-than-critical* approaches to educational thought and research, the 'standard' commitments of the field nevertheless persist. In part, such persistence stems from the axiomatization of that 'standard' philosophical mode articulated by Laruelle, wherein the intelligibility of educational thought becomes hinged to its performative claim on the Real. A majority of scholarship in the field presupposes this orientation to the world as a confirmation of its labour, where that 'standard' mode of philosophical thinking articulated by Laruelle (2010, 2013a) constitutes a 'genetic' model for thinking-action itself. Here, the persistence of 'standard' philosophy within education is buttressed by the fact of its disappearance into the automaticity of 'common sense'. So it goes that the program of 'standard' philosophy is today 'economized' as a stealth fidelity in educational thought and of the myriad stealth commitments that populate the landscape of educational thought. There is perhaps none as pervasive as the presupposition that education and educational thought inherently labour in fidelity to *production*.

The Facticity of Production in Education

The birth of the modern school in the model of Industrialization stands as perhaps the most overt exemplification of education's commitment to *production* in its standard definition. As the field of curriculum theory (Pinar et al., 2000) has explored, the conceptualization of education in the early twentieth century assumed production as a master signifier according to which the processes of schooling were subsequently shaped. The highly influential work of Fredrick Winslow Taylor (1911), for instance, makes clear that the intent of education is bound to its ability to produce an image of life commensurate with the values of industrial scientific management. The school is not only produced according to the presupposed Real of the Industrial model and its factory worldview, but becomes a fulcrum for the production of an ostensibly efficient and rational socio-political order adapted to factory routine and its hierarchical ordering of life. The primacy of production too subtends the instrumental thinking of curricularist Ralph Tyler (1949), whose seminal works in curriculum study and design assert that educational labour ought to produce a world subject to human mastery and of subjects patterned in the values of industrial management (i.e., progress, achievement, efficiency). Glossing the history of Taylorism and the descendants of Taylor (i.e., Tyler, 1949) in this manner, the problem with production might simply pertain to what production *produces*. As many curricularists have argued, the problem of Taylorism extends from the particular social-mental relations it both produces and simultaneously renders unthinkable. While the critique of Taylorism and its image of what education *ought* to produce remains a crucial aim of educational research, the stealth fidelity to production as a general presupposition of educational thought remains an entrenched orthodoxy. For while the aim of production ought to remain a fundamental issue in educational thought, it is one occluded by a more fundamental decision regarding production as a genetic precursor of education in the first place. Emblematic here is the recurrent curricular question of "what knowledge is of most worth", where critical emphasis is levied upon the value of knowledge production, but rarely the problem of knowledge production in the first instance. It is here that we might return to the thinking of

Laruelle (2013a, 2013b) and recast the presupposition of production as an instance of the "monopoly of the Real" of which the Western philosophical project stands accused.

If we can accept the claim that production founds a central presupposition in educational thought, then its patterning of the field ought to become a focused concentration of educational analysis. This posed, the 'givenness' of production has today been 'economized' within educational thought and research, appearing ubiquitously as a foreclosed characteristic of what it means to educate and be educated. Cursory familiarity with the landscape of educational research is sufficient to observe its commitment to production, if but via the fashionable order-word of 'aims', 'outcomes', and 'outputs' by which institutional thinking and its privileged forms of research are today governed. The avant-garde developments of the field remain as thoroughly committed to production, albeit of an ostensibly different order often emblemized by the more fashionable order words of 'newness' and 'creativity'. And while the image of production appears more starkly in Taylorism's overt standardization of the field in the model of Ford's assembly line, the presupposition of education's fidelity to production seems remarkable as a genetic condition of the field's overarching limit-thought. Even in the ostensibly avant-garde cabals of arts-based research and its aims to short-circuit the 'standard' mode of scholarly exegesis persists a fundamental fidelity to 'alternative' production and signification. Elsewhere, scholarship in the field of education disappears into production and its supplementary connotations of 'use' and 'helpfulness'. Here, the pulsional motors of the field have been absorbed within the expectation that educational thought ought to be oriented to 'practicality'. Such expectations of production are buttressed in their contrast to the problem of 'unhelpful' and 'useless' scholarship that undoubtedly characterizes the mode of thinking plied in this essay. For where production constitutes a presupposition of the educational Real, 'unhelpful' thoughts already appear to court the unfashionable scenarios of resignation and abolition anathematic to its core presuppositions.

The Transmutation of Production and Negation of the Negative

The status of production in education is of crucial importance in the contemporary era, particularly in that production today coincides with a series of enmeshed problematics foremost amongst which are its integration within the capitalist political economy and investment within the motors of anthropogenic climate change. If the decision of education writ large commits to production as a fundamental orientation to the world, it does so on at least two fronts. In the first, the implicate commitment to production in educational thought might be seen to parallel the transformation of production as it is today immanent to capitalist political economy. As Baudrillard (2017) argues, the transmutation of production precipitates a fundamental reformatting of material relations. Where for Baudrillard (2017) the idea of production in 'pre-modern' societies is yet entwined to the problems of excess and scarcity (i.e., potlatch, kula), the transmutation of production under capitalism occurs by way of delinking from the immanent reversibility of production. This scenario today intersects with the problems of what Jason Moore (2017) has dubbed the Capitalocene, a designation that divests the attribution of climatological change to 'humans' in general (the Anthropocene) by foregrounding the profound role of capitalism in accelerating present ecocatastrophes. Deterritorialized from its 'symbolic enchainment' to the world, production becomes 'liberated' from its material imbrication and accelerated as an engine of ostensibly interminable creation (Deleuze & Guattari, 1987). Yet, as the work of Jason Moore (2017, 2015) patently demonstrates, the transmutation of production within capitalist political economy establishes in tandem with its jouissance of consumption the conditions of material exploitation and extinction. Delinked from its pre-modern 'enchainment' to the affordances of material relation, production is relinked to the body of Capital itself. The significance of 'things' herein become absorbed under the sign of Capital, wherein the consignment of matter to its monetary exchange-value ensures the fate of the world to extinction. As Dawson (2016) develops, the accumulative drivers of capitalism today intersect with the horror of exterminism that he calls the *Necrocene*. Where production once referred to a metaphysical order rich in "symbolic reciprocity", it is today reterritorialized upon the seemingly inextinguishable engines of Capital as its basic universe of reference.

It would be folly to disregard the ways in which contemporary education and in particular, the field of curriculum study is born in defiance to the machinations of capitalist political economy. Indeed, the reconceptualist moment of curriculum theory assumed as a central tenet of its praxis the articulation of thought and action resistant to the perceived negativity of capitalism. Yet, for its indictment of capitalism as a solely negative regime, certain vectors of critical curriculum theory have been largely unable to rehabilitate negativism as an orientation for thought (Noys, 2010). Instead, fidelity to 'hope' and 'affirmation' seem to constitute a default presupposition for thinking in the field today. Indeed, a cursory examination of scholarship in the field of curriculum study corroborates this general allegiance, where the referential universe of 'hope' significantly outpaces on all fronts the consideration of negation as a potential orientation to educational thought. While alliance to 'hope' today characterizes an orthodoxy in the field of curriculum study, such allegiance is accelerated in the field of education writ large, where 'hope' and the happy affects it aims to mobilize constitute a dominant orientation for praxis. In unexpected filiation with capitalist political economy, the sense of obligatory happiness proffered in much educational scholarship conspires to enact its own form of deterritorialization by which production becomes delinked from the "accursed share" or reversibility of catastrophe (Bataille, 1991). It is in this formatting of educational thought that the very field of production has not only become delimited, as we find when only 'good' or 'helpful' scholarship constitute worthwhile scholarly pursuits, but cut-off from its immanent reversibility and therein committed to the affirmation of productivity exclusively. At this level of surface simplicity, the field of educational thought is marked by the delimitation of production as it is today reformatted according to its beneficent albeit oblique connection to the teleology of freedom, in which we find education automatically correlated to liberatory functions.

Education in the Mirror of Production

The transmutation of production in early capitalism directly intersects with the reformatting of education in the Industrial era, during which education becomes founded in mirror resemblance to an image of production given by political economy. As Baudrillard (1975) develops, production is reformatted under early capitalism and reoriented in its labour upon a standard idea of value. Yet, as previously explored, such value is marked by its abstraction from material reference and reterritorialization upon the abstract body of Capital itself (Deleuze & Guattari, 1987). In *The Mirror of Production* (1975), Baudrillard remarks on the consequence of production as it is remade in equivalency to the transcendental signified of Capital. As climate change scholarship concerned with the Capitalocene today demonstrates, the reworking of production commensurate with political economy establishes conditions for the material exploitation of matter and its processual 'cheapening' via forced equivalence to Capital itself (Moore, 2015). As Baudrillard (1975) contends, the problem of production's reformatting is not simply an effect of its reterritorialization under the transcendental signified of Capital, but the ways in which such reformatting reorders the significance of our labour in relation to the world. The mutation of production within early modern capitalism establishes for Baudrillard a model in which the world becomes implicitly given to value. Problematically, this presupposition augments the very aims of labour, where today it is a 'given' fact of education that the world is but a site of potential value to be extracted through labour. Further to this problem, the extraction of value through educational labour is today made to reflect in an image of human exceptionalism. It has become the accepted aim of education to find 'our' value by labouring on the world and to realize the meaningfulness of such labour as a connotation of our significance (Baudrillard, 1975). Here, the value of 'things' becomes commensurate with their reterritorialization upon a standard model of production that accords things value in the first place, and in the second, imagines the extraction of such value as an index of human significance.

The model of production subtending educational thought and research intersects directly with the complex problem of climate change. For where the world is reformatted according to its extractible value for us, it founds a relation to matter that enables the conditions of exploitation and extinction. As we see throughout the field of educational thought today, the value of the world is everywhere hinged on its rediscovery within all-too-human orders of significance and control. For despite the field's ongoing interrogation of epistemology and the ideological organization of the school, a more pervasive commitment to production maintains in the very presupposition that the locus of our concern ought to pertain in its first instance to education's 'worth' correspondent to the human *episteme*. This orientation is of course endemic to the aim of modern education, which fundamentally commits itself to the coordination of the world upon the metrics of institutional knowledge, where as so many students find, the world is *always-already* 'given' to its intelligibility within the official meanings of the institution. The Real is here disappeared in its auto-production as meaning and annexed by a conceptualization of labour that dramatizes again the extraction of the world's worth on behalf of its exchange-value as institutional Capital. For overwhelmingly, the labour performed in schools no longer refers to the material world, but to the extraction of its value on behalf of evoking abstract 'signs of labour'. That is, the labour often enacted in schools today is, on one hand characterized by interminable overproduction, and on the other, the ostensible inconsequence of such labour in relation to broader social and mental ecologies (Guattari, 2000). As Baudrillard (1975) develops, the referent of labour has become largely delinked from its material consequence, functioning as but an affirmation of the very model of production with which labour has become forcibly enjoined.

The genetic order of production that inheres curriculum thinking undoubtedly informs upon macro levels of structural and subjective labour. Less overt however, is production's investment within the neuronal or affective register of institutional life. As Larsen (2010) develops this, political economy is today operative as a genetic probe that draws life toward its resemblance in the pulsional motors of production. Institutional life is everywhere today drawn into conformity with the orders of interminable restlessness, agitation, and alienation from its labour. For Larsen, the index of life's affective annexation into political economy figures in the modern myth of the zombie. The modern image of the zombie, Larsen argues, has not only come to emblemize the alienation of labour as it has disappeared into its 'standard' models of production, but illustrates the contemporary estrangement with death that underscores an obsession with productive restlessness. For ultimately, the zombie figures as less a signification of death than its continual deferral. Larsen argues that the strangeness of the zombie redoubles the strangeness of the present, where the perpetual agitation of life into activity has circumvented the paradigm changing force of 'endgame' thinking. Here, Larsen argues that zombie figures in the articulation of our incapability to rethink the future from under its genetic contraction with capitalist production and the ceaseless agitation of life rallied under its genetic ordering of reality.

PRODUCTION'S ROLE IN THE MAKING OF THE 'EDUCACENE'

The model of production assumed in education conspires with what might be called the '*Educacene*', or rather, the legacy of education and educational thought implicate to encroaching planetary ecocatastrophe. Whereas the Anthropocene indicts 'Man' as the primary agent of contemporary climatological change, it is via such alternative signifiers as Jason Moore's (2015) Capitalocene that a more adequate understanding for the antecedent conditions of climate catastrophe might be habilitated. As the generic idea of the Anthropocene is supplemented by a host of corollary precursors such as the Capitalocene, so too might it be linked to the 'standard' image of education as it has come to function as a precursor of our contemporary climatological crisis. Extending from its birth in the Industrial imaginary, education has throughout its modern development conspired in the exploitation and control of the world by reformatting its value according to its very susceptibility to control. In this orientation, education has become the operational counterpart of what Kerian Suckling (2014) has termed the 'Homogenocene', in that education has throughout its modern development conspired in the production of monocultures and the coordination of human biodiversity upon the image of the 'standard human', a term that Laruelle (2013b) uses to connote the highly patterned and non-expansive definition of the human that suffuses Western philosophy and its mobilization within educational thought. Yet, what is perhaps the more profound ecological influence of the 'Educacene' is its implicit 'cheapening' of the world reformatted into institutional value and Capital. The 'Educacene' might in this way be thought as a fulcrum for encroaching ecocatastrophe, in that its 'monopoly on the Real' both accelerates the deterritorialization of labour from its material conditions and perpetuates the conceit that the world is what we think it.

The problem of education's model of production thus posed, it would seem inadequate to insist upon something that might be dubbed the 'end of productivity'. For despite hopes that the mounting force of critical theory would conspire to conscientize the embattled subject of neoliberalism and overturn capitalist political economy, the pulsional motors of capitalist extinction and ecocide persist in their accelerative contagion. As Baudrillard's (1975) indictment of Marx in *The Mirror of Production* aims to demonstrate, the Marxian conceptualization of nature as the dialectical counterpart to human labour might be understood as complicit in the disappearance of the 'natural' world under the sign of human productivity.

For despite the critique of political economy advanced by Marx, the enchainment of human significance as it made counterpart to the extraction of value from the 'natural world' reproduces the very teleological conquest of the 'natural' world Marx's revolutionary thought attempts to critique (Poster, 1975). The problem of production will find remedy in neither the unlikely 'end of productivity' or in the modulation of production toward some more beneficent end. As Baudrillard (1975) argues, it is the very genetic order of productivity and its presupposition of our relation to the world as value that requires transmutation. Here, a secondary 'critical' aim of habilitating what might be called a more 'productive productivity' seems dubious in that the augmentation of production's teleology sustains the primacy of production in the first instance. Such a scenario is endemic to the field of Education for Sustainable Development (ESD), which aims toward the habilitation of an 'optimized' productivity while sustaining the very logic of productivity that today conspires toward the exploitative 'cheapening' of the world.

On the Nightside of Production

Advocacy for an implausible 'end of productivity' or the optimization of productivity through its marriage to some nobler moral teloi today seems inadequate. Yet, the persistence of productivity remains a central problematic for educational thought and curriculum study in that it constitutes a genetic image for how education *ought* to go. 'Standard' productivity today constitutes a reactive habit informing upon education's relation to the world, yet one that has long passed into indistinction with the very project of what education is and how it functions. It is an understatement to suggest that the transmutation of production and value remains an important task, yet one fraught with difficulty in that the transmutation of production and value in educational thought and practice necessitates first altering the very presupposition of the Real maintained in the motors of production 'itself'. Yet, it occurs that the transmutation of production and value is already underway and neither in the form of a more beneficent mode of production or optimal model of productivity enjoined to futures in which 'standard' productivity is maintained as a primary vehicle of human preservation. In contradistinction to the anthropic vectors of production, the transmutation of productivity already underway is of an order that no longer refers to the human or its presumed control over a planet 'given' to its labour. As it always has been, but made more evident by the contemporary climatological catastrophe, production has another side that might be dubbed the labour of the inhuman.

The labour of the non-human/inhuman/ahuman remains a sublimated part of Anthropocene thinking, particularly as the Anthropocene implicitly suggests the primacy of 'man' as a central environmental agent. While the impact of particular humans upon the environment is of course well substantiated in 'Anthropocene' research, it is equally undeniable that the inhuman forces of planetary change have reshaped both the very idea of the human and so too, the image of human futurity on this planet. Such a conceptual shift is in part identified within Haraway's (2015) concept of the *Chthulhucene* as it seeks to describe the 'weird' biological-cultural-political-technological imbrications of human and material life to question what it might mean to live alongside other 'things', temporalities, and the myriad non-Western names and concepts through which the world and our relation to it might be relaunched. Where Haraway's *Chthulhucene* aims toward the articulation of new generative possibilities for human life in its relation to the inhuman, we want to propose another order of inhuman labour as it intersects with the forces of ***absorption*** and ***disappearance***. For while it is clear that the contemporary conditions of climatological change have transpired new images of ecological enmeshment and 'messiness', less overt is the function of inhuman labour as a remote or noumenal order of production that today ratchets open the Real by way of processes that absorb and disappear the 'monopoly on the Real' extending from the genetic model of production presupposed in contemporary thought (Laruelle, 2013a, 2013b).

Inhuman Reversal

Contravening the 'standard' model of production and its presupposition that the value of the world is to be 'liberated' through its extraction by human labour, inhuman labour heralds the event of imminent 'reversal' that exemplifies the horror of climate change. The force of 'inhuman labour' is prefigured throughout contemporary climate change research and its new image of the planetary Real (see jagodzinski, 2018). In distinction to the presupposition of anthropic control given by 'standard' education, climate science today suggests that the effects of climatological transformation have today assumed their own momentum. The release of glacial ice as a result of rising global temperatures produces a positive feedback cycle that accelerates rising global temperatures. Disintegration of plastic waste by wave action and wind abrasion fuels the oceanic and atmospheric presence of toxic microplastics. Bioaccumulation of antibiotic agents within waterways catalyzes the mutation of bacteria into so-called 'superbugs'. While the list goes on, such instances of planetary changes intersect with the image of a planet as it withdraws from the presupposition of human predestination and control. The image of planetary transformation articulated in climate science research emblematizes an image of the Real *without-us*, or rather, a Real indifferent to the primacy of human production that persists at the core of educational thought and action. Herein, the labour of the inhuman suggests another order or direction of production remote to the model of educational productivity dominant in the field today.

In its first instance, the labour of the inhuman demonstrates that production is finite. In distinction from the model of educational productivity and its presupposition on the interminability of production, the labour of the inhuman rejoins production to its material consequence. Such consequence is implicated within such micro-productions as the ubiquitous use of tumble dryers as they present an object (i.e., micro-plastic residue in lint) that exceeds economic models and recuperation (jagodzinski, personal communication, June 9, 2021). Where education founds the conditions whereby the material world is rallied to the side of man, contemporary climatological challenges actualize the return of the object and its implicit threat to the reign of anthropocentrism. Here, the finitude of production finds expression in the event of climatological change writ large, where the presupposed exceptionalism of universal 'Man' is upended by the ramifications of its ecological exploitation. The caveat advanced in this formulation pertains to the way that orthodox education presupposes the inexhaustible extraction of value from the world it aims to represent. Implicate to the logic of this 'standard' educational process is the decision of the world's 'givenness' to productivity and anthropic significance. The imperiled status of orthodox education might here be situated upon the very idea of production, in that the emergence of non-human labour in the form of climatological catastrophe advances the understanding that one cannot do what one likes with the world. The conceptualization of the world as it is *for-us* now lapses into obsolescence, and is superseded by the return of an alien productivity that today demonstrates the limits of production, *but for organic life*. As Colebrook (2013) avers, after man, the orders of non-human productivity will continue if but toward their own horizons of existence and extinction.

The Inhuman Transmutation of the Educational Real

Baudrillard (1975) argues that the value of things is today overdetermined by their forced reflection in the model of political economy. In distinction to such overdetermination, inhuman labour might be seen to short-circuit the automatic relation of value to Capital. For where value has become conceptualized as an extractable commodity and corollary reflection of human vitality, 'inhuman labour' articulates an 'endarkened' process of labour that refers to neither the human or the assumed value of the world as coextensive of human labour. Rather, the inhuman labour emblematized by such events as climatological change founds a new image of value but of an ontological order in which labour is rejoined to its material imbrication. Where the model of productivity rife in education presupposes an ontological hierarchy in which the human deterritorializes the world 'from above', inhuman labour reconstitutes a flattened ontology through which the restless drive for value becomes relinked to such contemporary problematics as ecocide and extinction. This is to derail the myth of interminable production reified in contemporary education, for if anything, inhuman labour marks our mutual enchainment with the affordances of the material world. Where education's standard model of production overdetermines value in relation to human productivity, climatological change short circuits such correlationism by demonstrating the discontinuity of the world's educational production with the Real of inhuman labour. That is, where orthodox production presupposes that the world is *for-us*, the labour of the inhuman articulates the prospect of a world *without-us*, or rather, a world that withdraws from its submission under the *all-too-human* metrics of value (Thacker, 2011). For what education writ large has yet to realize is that *there is another world, and it is this one* (Eluard cited in Wark, 2014).

The expansion of the Real palpated by inhuman labour introduces a more-than-critical indictment against the monopoly perpetrated within contemporary educational fidelities to 'standard' production. For where the Real of education is founded upon an alliance with production and its resemblance in the ambit of human desire, it is via inhuman labour that the unabashedly anthropocentric image of productivity is dilated. The inhuman transmutation of the Real from under its all-too-human conceptualizations is significant along two fronts. First, the Real of inhuman labour suggests that our relation to the world as it is organized and rendered intelligible within education is not *equal* to the world. Second, the Real of inhuman labour suggests that there persists within the educational 'monopoly of the Real' a noumenal world withdrawn from representation. Where education remains wedded to the presupposition that its labour is a corollary of the Real, the noumenal force of inhuman labour incites an unthought-of dimension that outstrips the conceptual apparatus of the field. In the face of this conceptual obsolescence, education might be jolted from its 'monopoly on the Real', and so too, the presupposition that its modes of production correlate to the 'truth' of reality. If anything, the labour of the inhuman advances an insoluble problem for educational thought that will remain unexhausted by the answers levied upon it. This is to rejoin the educational obsession with production to its inadequacy relative to the 'night side of thought'.

Where education habitually espouses its apex of becoming in the model of productivity, inhuman labour demonstrates the limitation of such transcendental adherence. For where experience is often reduced to the accumulative effects of production and their mobilization under the bald rubric of 'use' or 'goodness', the inability of education to encounter the noumenal world of inhuman labour demonstrates that the project of education has become but a bulwark against the 'horror of the Real' (Baudrillard, 1975). Emblematic of civilizational end, education writ large doubles-downs on all-too-human modes of production in obfuscation of education's encroaching obsolescence. As we learn today, virtually all experience can be expunged of its negativity and rallied to the side of production. As is the contemporary fashion of the field, doubt, uncertainty, and failure are expurgated of their negativity and reformatted in fealty to production. Yet, the enigmatic forces of inhuman labour persist as a virulent excrescence to educational experience, functioning as an index of the field's hubris and lack of imagination (Baudrillard, 1975). For where obsessive productivity and its 'optimization' seizes the field, the labour of the inhuman continues to deepen the conditions of negativity recalcitrant to the ostensibly positive contents that productivity is forcibly made to elicit. As climatological research makes abundantly clear, the transformation of the planet is against both our interest and the idea of interest in general (Sexton, 2017). Relative to the inhuman transformation of the planet, the idea that we will produce our way out the encroaching environmental catastrophe seems but a pathetic twinge of self-esteem (Brassier, 2007).

Conclusion: 'But What Will We Do?'

Ultimately, the labour of the inhuman emblemized by climatological change might be understood as less a concept of production than an event of *absorption*. For, the event of climatological transformation implicates the educational project in its entirety, absorbs by way of its implosive energy, education's very presupposition of the Real (Baudrillard, 1975). Inhuman labour not only extinguishes the conceptual moorings of the field but also founds a conspiracy against education's perpetual orientation to the representation and mastery of reality as it is founded in productivity. Perhaps more profoundly, it is by dint of inhuman labour and its impersonal mode production that the representation of the world *for-us* is retroactively annihilated, particularly so as the realities of extinction become increasingly salient. Where educational productivity continues to conspire with the machinations of political economy, it is the labour of the inhuman that absorbs value along a more alien horizon for thought and action. Here, the knee-jerk reaction, 'but what will we do?' stands as an indication that there is no recourse for thought but its conscription to productivity. It is moot to suggest that resignation and abolition are maligned in the field, for their general absence is testament to the singular commitment to production that education assumes as its aim and teloi. Yet, the force of absorption intimate to the labour of the inhuman suggests that the articulation of *other* orientations to the educational project might today be necessitated, if only to remit educational thinking from its automatic inscription under productivity.

References

Bataille, G. (1991). *The accursed share: An essay on general economy* (Vol. 1). Zone Books.

Baudrillard, J. (1975). The mirror of production. Poster, M. (Ed., and Trans.). Telos Press.

Baudrillard, J. (2017). Symbolic exchange and death. Grant, I. H., (Trans.). Sage Publications Ltd.

Brassier, R. (2007). *Nihil unbound: Enlightenment and extinction*. Palgrave Macmillan.

Colebrook, C. (2013). *The death of the posthuman: Essays on extinction, Vol. 1*. Open Humanities Press.

Culp, A. (2016). *Dark Deleuze*. University of Minnesota Press.
Dawson, A. (2016. Extinction). New York: OR Books.
Deleuze, G., & Guattari, F. (1987). In Massumi, B. (Trans), *A thousand plateaus: Capitalism and schizophrenia, Vol. II*. University of Minnesota Press.
Guattari, F. (2000). In I. Pindar, P. Sutton, & (Trans.). (Eds.), *The three ecologies*. The Athlone Press.
Haraway, D. (2015). Anthropocene, Capitalocene, Plantationocene, Chthulhucene: Making Kin. *Environmental Humanities, 6*(1), 159–165.
Jagodzinski, J. (Ed.). (2018). *Interrogating the anthropocene: Ecology, aesthetics, pedagogy, and the future in question*. Palgrave Macmillan.
Larsen, B. (2010). The zombies of immaterial labour: The modern monster and the death of death. *e-flux*, No 15. https://www.e-flux.com/journal/15/61295/zombies-of-immaterial-labor-the-modern-monster-and-the-death-of-death/
Laruelle, F. (2010). Philosophies of difference: A critical introduction to non-philosophy. Gangle, R. (Trans.). Continuum.
Laruelle, F. (2013a). Anti-Badiou: On the introduction of Maoism into philosophy. Mackay, R. (Trans.). Bloomsbury.
Laruelle, F. (2013b). Principles of non-philosophy. Smith, A. P., & Rubczak, N. (Trans.). Bloomsbury.
Moore, J. (2015). *Capitalism in the web of life: Ecology and the accumulation of capital*. Verso.
Moore, J. (2017). The capitalocene, Part I: On the nature and origin of our ecological crisis. *The Journal of Peasant Studies, 44*(3), 594–630.
Noys, B. (2010). *Persistence of the negative: A critique of contemporary Continental theory*. Edinburgh University Press.
Pinar, W.F., Reynolds, W.M, Slattery, P., & Taubman, P. (2000). Understanding curriculum. : Peter Lang.
Poster, M. (1975). Introduction. In Baudrillard, J., The Mirror of Production, Poster, M. (Ed & Trans.). Telos Press.
Sexton, J. (2017). On black negativity, or the affirmation of nothing: Jared Sexton, interviewed by Daniel Barber. Retrieved October 13, 2017, from https://www.societyandspace.org/articles/on-black-negativity-or-the-affirmation-of-nothing
Suckling, K. (2014). Against the anthropocene. Retrieved January 3, 2015, from https://blog.uvm.edu/aivakhiv/2014/07/07/against-the-anthropocene/
Taylor, F. (1911). *The principles of scientific management*. Harper.
Thacker, E. (2011). *In the dust of this planet: Horror of philosophy* (Vol. 1). Zero Books.
Thacker, E. (2018). *Infinite resignation*. Repeater Books.

Tyler, R. (1949). *Basic principles of curriculum and instruction*. University of Chicago Press.
Wark, M. (2014). There is another world and it is this one. Retrieved January 20, 2014, from http://www.publicseminar.org/2014/01/there-is-another-world-and-it-is-this-one/#.U_zN4EuE6Ml
Weaver, J., & Snaza, N. (2016). Against methodocentrism in educational research. *Educational Philosophy and Theory, 49*(11), 1055–1065.

CHAPTER 8

Cosmology and the Anthropocene: Speculative-Educative-Artistic Practices for a Planetary Consciousness

jan jagodzinski

The event of the Anthropocene has already happened, but it seems 'we' are slow to recognize that the Earth's phase change, abetted through our species anthropogenic activity, has already set the parameters of what will be a possible outcome: our species extinction as well as our possible planetary remodification. The rhetorical 'we' addresses a globalized capitalist system of trade agreements, financial exchanges, government lobbying and corporate greed and corruption which ensure a thin layer of elites the 'good life' of health and prosperity regardless of the country named. The gap between rich and poor continuously widens, a claim that has, sadly, become a cliché. It is not an all-encompassing 'humanity' that bears the weight of responsibility for this state of affairs, more so the blame belongs to well-off nations (primarily OECD countries) who use up most the

j. jagodzinski (✉)
University of Alberta, Edmonton, AB, Canada
e-mail: jan.jagodzinski@ualberta.ca

M. Paulsen et al. (eds.), *Pedagogy in the Anthropocene*, Palgrave Studies in Educational Futures,
https://doi.org/10.1007/978-3-030-90980-2_8

world's resources, and then use other nations as dumping grounds for their e-waste, plastics and garbage (Angus, 2016, pp. 224–230).

The stock market remains the acephalic mechanism of neoliberalist fundamentalism, which divines and resolves the construction of social life by factoring in unforeseen contingencies in the name of algorithmic control (Ramey, 2016). The contemporary world of nation states is pervaded by a paranoia that has the rise of strong authoritarian figures, extremes of left and right politics, border walls and politically motivated fortifications that make immigration almost impossible; yet the dynamism of migration is not a 'problem' to be solved, it is indisputably part-and-parcel of globalism, not only shaped through military conflicts that leave cities in rubble, a form of urban 'clear-cutting,' but increasingly migration from no longer liveable climate conditions caused by fire, drought, and coastal devastations via rising sea levels (e.g., Maldives), hurricanes, and outright oceanic plastic pollution.

The Anthropocene is not a future disaster that must be prevented. This is the fantasy of extending the Holocene indefinitely. Rather the epoch presents already a changed ontology, a new geological and geopolitical era, a difference in kind and not degree, marked most prominently by the burning of fossil fuels. The string of climate conferences, highlighted by the Paris Agreement, COP21 in 2015, and then followed by COP24 in Katowice, Poland in 2018, which was to operationalize the treaty targets, has largely failed. COP25, in Madrid, 2019 marked the date Trump's Whitehouse officially pulled out of the Paris Agreement. The Biden administration has thankfully restored its initial commitment in 2021. Realistically, limiting global warming to 1.5–2.0 C by mid-century seems unlikely given the state of the Capitalocene to follow Jason Moore's (2016) characterization. Treating the Anthropocene as the Capitalocene should be thought on two levels; the first level is to recognize that we have indeed entered into a new geological epoch; the science here is indisputable. As such, the Anthropocene far outdates capitalism. Anthropocene science studies fundamental shifts in the Earth System, far beyond the Holocene and the human impact on this system by various measurable indicators: CO_2, artificial nitrogen, species extinctions, ocean acidification, sea level, holes in the ozone layer and (seldom mentioned) population growth, each of which has a critical limit in relation to the sustainability of our species. The second level is to recognize, and concur with Moore (2016) that: "The Capitalocene signifies capitalism as a way of organizing nature—as a multispecies, situated, capitalist world-ecology"(p. 6).

However, the Anthropocene cannot be completely equated with global capitalism. The Anthropocene exceeds the geology of our species. It should *not* be equated with the geology of the system of capitalist power as Moore's (2015) work claims. While these two senses of the Capitalocene overlap, they are *not* identical. As Adrian Parr (2017) maintains, there needs to be a reckoning with militarism, market fundamentalism, global inequality to find an alternative political vision. How the economy of capitalism hinders or intensifies the state of the Earth system is an overriding question in relation to what would be an educational imaginary that is adequate to this planetary condition. This is the pressing concern with no easy or foreseeable solutions.

Ecomodernist Humanism

The question of sustainability and survival looms large. Sustainability has already been hijacked by the corporate world through 'green capitalism' (Parr, 2009). This, however, is only the tip of the iceberg. The epitome of this development goes under the seductive name: *ecomodernism* (other signifiers such as eco-constructivists, eco-pragmatists and eco-environmentalists are often used). The hijacking of the Anthropocene takes place by forwarding an extreme form of humanism (hence the name eco*modernism*) where the environmental narrative is captured through pro-environmental policies grounded in new technologies and capitalist growth. It is not climate denial that is at issue; rather it is how the Earth is to be managed within the framework of entrepreneurship, venture capitalism and investment. The Anthropocene is given its literal and economic expression: the Age of Humans. It is a 'Good Anthropocene' providing humans an opportunity and the capacity of both controlling Nature and creating Life. Nature no longer exists 'in-itself' as separate entity, it is simply material to be researched, its properties understood and manipulated as a world-for-us. While there have been calls for an 'ecology without Nature,' especially by Timothy Morton (2007), and much earlier, a recognition of an 'end of Nature' by Bill McKibben (1989), these stark proclamations had completely other intentions in mind: the first was the recognition that nature|culture were constantly entangled and could not be separated, their entwinement required speculation given that 'reality' always had a unknown remainder. The second was a clarion call that our relationship and understanding of Nature had changed as global warming and ozone layer depletion became known forces of geo-change.

Ecomodernists, in their *Manifesto* (www.ecomodernism.org/) present a strict 'correlationalism,' as Quentin Meillassoux (2008) called it; that is, a separation between subject and object where the neoliberal subject prevails; Nature paradoxically no longer 'exists' as some mysterious unknown, given that it is entirely subjected to ever increasing epistemological know how through sophisticated AI and genetic technologies. In brief: Nature is reduced to complete transparency and representation as it becomes material to be constructed and manipulated for the benefit of 'humanity.'

Ecomodernists oppose the efforts to limit greenhouse gas emission. The U.S. panel on climate change at COP24 infuriated many as claims were made as to the progress of fossil fuel technologies (Russia, Saudi Arabia and Kuwait were the other three countries that defended oil production). Ecomodernists maintain that shale gas and improved nuclear reactors, with fission 'just' around the corner, will produce all the needed energy. Consumerism will then continue business as usual. Behind this 'Good Anthropocene' are the conservative forces of capitalist think tanks and investment: *The Breakthrough Institute, The Long Foundation, The Nature Conservancy, The Gordon and Betty Moore Foundation*. This helps to explain why Trump and his administration had no use to participate in any form of climate resolutions or emission targets, and why the US Environmental Protection Agency (EPA) was so heavily deregulated and made ineffective so that natural resources could be further exploited. The bottom line is the capitalist machine through technological innovation will indeed geo-form the planet. We have a bright future ahead of us. As the icon, Alfred E. Neuman of the US based *Mad Magazine* says, "What, Me Worry?" Given this scenario, it is not difficult to imagine the reorientation of education given the ecomodernist scenario. Variations on the STEM (science, technology, engineering, mathematics) curriculum are forwarded, and a 'Maker Culture' is marketed for education (Halverson & Sheridan, 2014).

A planetary consciousness establishing the fantasy of everyone pulling together to make fundamental economic changes and change life-style choices appears to be an ideological utopian impulse. This is generally a 'critical posthumanist' position that has become a veritable doxa in various circles of the humanities and social sciences. The appeal to the entanglement between the more-than-human-world, including in some cases a recognition of AI (or inhuman) contribution in such assemblages, results in an emphasis on the ethical and political consequences of the Anthropocene. The ethical demands of attunement and responsibility to 'materiality' are at the forefront of what are materialist (feminist) ontologies put forward by Karen Barad (2007), Jane Bennett (2010) and Rosi Braidotti (2013a, 2013b). Kantian and post-Kantian 'rational cosmopolitanism' has been presented as one answer: the positive power; that is affirmative power of ethical virtues, which will overcome and surpass conflict, personal interests and desires motivated by bodily material empiricism. By extending humanity's rational potential, forwarding the *affective* ties of immaterial labour of communication and passions of love, admiration, devotion, a new collective being can emerge through such a 'civilizing process' (Elias, 1978). Such a sentiment takes on various forms: Michael Hardt and Antonio Negri's (2000) creation of a new commonwealth via the multitude of collective and creative human spirit; Jürgen Habermas' (1986) proposal of communicative consensus; Etienne Balibar (2008) Spinozian inflected 'transindividuality,' Jacques Derrida's (2001) 'refuge city,' which extends infinite hospitality to the wholly other, a 'community without community' is striven for, but never reached. Such proposals are shaped by what Gilles Deleuze (1994, pp. 223–224) identifies as a fundamental contradiction of bourgeois 'thermodynamic' ideology: on the one hand, an intensification of the global world order is claimed free of calculating economic codes and differences, while on the other hand, the (democratic) extension of inclusivity (justice, rights, hospitality) never frees itself of global economic disparity and violence. This results in a schizophrenic-paranoid process, a process repeated by Kantian paradoxical aesthetics: on the one hand a universal transcendental aesthetic is possible via the *apriori* forms of space and time that condition the possibility of experience and cognition (*Kritik der Reinen Vernuft*, 1781), while on the other hand the aesthetics of subjective judgment based on feelings of pleasure and pain that holds no immediate universal validity is also forwarded (*Kritik der Urteilskraft*, 1790). '*Sensus communis aestheticus*' (aesthetic common sense) as the universal principle of aesthetic judgment was the Kantian solution to this dualism: what Deleuze referred to as the "dogmatic image of thought." The result is the problematic of a deferred form of action: the compromised actuality of the 'not yet' (Jacques Derrida) or the futurity of what is 'to come' (Deleuze & Guattari) (see also Colebrook, 2013).

Posthumanist and Posthuman Tensions

At base, these are ***posthumanist*** or (for some) metamodernist (van de Akker et al., 2017) as opposed to ***posthuman*** and ***ahuman*** developments (jagodzinski, 2017; MacCormack, 2020). The former posthumanist developments both modify and include the developments of poststructuralism in the 1990s where linguistic discursivity continues to decenter the human by recognizing the more-than-human via interspecies relationships. Karen Barad's (2007) 'agential realism' and Judith Butler's (1993) theorizations on the performative body, while divergent directions, both stress the linguistic discursivity of matter as an onto-epistemology couplet supplemented by an ethics. In posthumanist developments, the alien Other consists fundamentally in difference from the animal (in relation to the autonomous difference of a 'human self'). It does so to enrich the 'self,' to expand its horizons, so to speak, with the danger of generating yet another form of exceptionalism. Alien and exotic otherness is approached through a variety of tropes of being more open, ethical, and welcoming to the Other's 'face,' as in Levinasian ethical developments, strongly supported by Karen Barad (2007, pp. 408–413). The 'becoming of difference' as a relational difference is celebrated and achieved at the expense of the animal by considering its welfare and attunement to the parameters of its existence (its niche ecology, its 'world'). Anthropocentrism is certainly mitigated, replaced by a posthumanist nomadic subjectivity wherein the animal remains *for-us*, often tied to the animal industrial complex via 'free range' humane treatment and doing away with 'needless suffering' through humane slaughter (Petersen, 2019a, 2019b). Haraway's (2003) championing of pets, like her dog Cayenne, seems to overlook all the violence and cruelty inflicted by pet owners. As Deleuze and Guattari (1987) note, it is easier to love animals whose facial expressions are closer to us, and our dispositions, while abjection is to be found everywhere: worms, insects, spiders, viruses. These species are more difficult if not impossible to anthropomorphize (like the box jellyfish). But even here, Myra Hird (2010), drawing on Donna Haraway and Lynn Margulis, goes as far to claim that bacteria are super-agential having "invented symbiogenesis" and are responsible for "evolve[ing] all life…on Earth" (p. 38). The implication is that bacteria are people too! If bacteria, why not viruses then? The folly of such thinking is vividly illustrated by the Covid-19 pandemic. Most often biophilia is praised at the expensive of its more troubling side: biophobia.

It is the virtue of co-existence that is forwarded in these natureculture 'hybrids,' as a politics of 'dwelling' (Ingold, 2005). 'Life,' as it is theorized in the biopolitics of this millennium, continues to be characterized as kinship with bodies that enhance or reject recognition of human-becoming through sexual and animal difference, whereas indifference and non-identity haunt life that remains defined by autonomous, individual, self-determining and personal Man (Colebrook, 2015). Man, as MacCormack (2017) rightly argues, is defined by the effects of knowledge: discourse is a form of bio-power that is able to 'know' its nonhuman Other. There is always a 'difference' that makes a 'difference,' to incorporate the recognition of personhood and rights within the umbrella term of a 'democracy of things,' to reach (it is claimed) a fully inclusive planetary consciousness where equity rather than equality is established. Participation remains 'open' and 'pluralistic' in the sense that multiple voices are given a say as to human-more-than-human relationships (Lindgren & Öhman, 2018); the shift to 'more-than-human' as opposed to the nonhuman (tending to a binary) indicates that anthropocentrism is being addressed as a contested position that demands argumentation, reflection and debate.

The extensive literature in posthumanist environmental education, the so-called 'animal turn,' embraces the above perspective (Stevenson et al., 2013; Rice & Rud, 2016).The relationality with nature is recognized, the human 'is' decentered through human–nonhuman *encounters* where stress is placed on the exchange of feelings that take place, highlighting the ethics involved (Andrzejewski et al., 2009; Bonnett, 2020). Multispecies relationships become a variation of multiculturalism and the democratization of extension of identities. Rivers and trees take on personhood or subjecthood, their rights protected. Ecopoetic descriptions of the sensuousness of nature are exemplified (Abram, 1997; Bai, et al., 2010). It is a form of de-anthropocentrism, a 'default anthropocentrism' (i.e., being *with* the world, Martineli, 2008) and an 'interspecies articulation' according to Pauliina Rautio (2013), one that preserves the human as stewards and sustainers of the Earth, an education for environmental and sustainability education (ESE) and an education for sustainable development (ESD).

Given the entanglement between naturecuture, the emphasis remains on the ethical relationships that are generated with other species. It is not so much the manipulation of 'nature' as trying to accommodate and adjust to its diversity and to live-well with the ecological complexities. An emphasis on the technologies of nature plunge us more toward the posthuman where both the physiological and psychic modifications as to what is 'human' are modified. More often referred to transhuman developments, in contrast the posthumanist developments so far discussed, which care less about technological manipulations of body-mind such as genetic engineering, prosthetics and the like. The 'critical posthumanism' of new materialism continues to emphasize issues concerning sex, gender, race, ableism as related to broader ecological issues. The range of science and environmental educators who have embraced Donna Haraway's (2003, 2008) concerns 'when species meet' or her 'companion species' and Karen Barad's (2007) 'agential realist' with its 'diffractive methodology' have developed such posthumanist pedagogies for the Anthropocene under the proliferation of 'new materialist' rhetoric and affect theories (i.e., Hultman & Lenz Taguchi, 2010; Ceder, 2019) along with a variety of (what I would call) *neo*-qualitative educational research methods rather than the usual post-qualitative research moniker, given that phenomenological human experience remains overdetermined, yet recognizing the nonhuman ('more-than-human' in some circles). In this regard, child-animal environmental education is especially prevalent (Taylor & Pacini-Ketchabaw, 2015), with the term 'childnature' used as a way to rethink childhood itself in relation to the environment (Cutter-Mackenzie-Knowles et al., 2020).

This proliferation of natureculture ontologies becomes extended in yet other ways through the variety of indigenous pedagogies that push back on postcolonialization to reclaim a sense of place, what in the western throught was referred to as *genius loci*, (lococentric education, Garrard, 2010) where the land takes on spiritual, ancestral and animistic meaning (Hawke, 2012). The literature here, of course, is extensive and growing as indigenous peoples around the globe attempt to decolonize the established curriculum and turn it toward their own core values in relation to land and its sustainability (i.e., Smith et al., 2019). These movements remain 'posthumanist' in the sense that such indigenous pedagogies, in relation to the 'climate crisis,' begin to modify their traditional ways of living on the land to cope with the changed conditions of encroachment via postcolonialism, yet not lose their core beliefs that gives them meaning and identity through elder teachings, rituals and language. Narrating stories and 'walking' (Donald, 2021) have become central tropes for such a pedagogy as it provides the necessary centering for well-being as the land (Earth) becomes rhetorically, constitutionally, and lawfully the 'first' nation to be colonized.

The posthuman side of these tensions raise the more technoscience and cybernetic relations between humans and the more-than human, further complicating the implications for the Anthropocene in the way natureculture hybridity is taken up. In evolutionary terms, animals are both material and semiotic, especially dogs (Haraway), which for Derrida (1992) marks a carnophallogocentrism, a 'carnivorous virility' when it comes to a genealogy of their 'symbiogenesis' with species homo, placing Man, once more at the center when it comes to the technoscience of breeding (Freccero, 2011). This cynanthropic becoming, argues Carla Freccero (2011) is tried to the machine of transnational capital and the prison industrial complex. It also raises questions as to what the ontological choreography is when it comes to cats (descendants from *Felis silvestris lybica*/wildcats), which also have a historical complexity in the fertile crescent of the Near East and Egypt, and by extension, all animals. The conundrum is that there is no nature-nurture binary but a complex genealogy that takes into account the colonial and racist past of animal breeding, and the power structures that go along with it. The western 'great chain of being,' for instance, assigns an animal to every saint, raising complex questions as to those symbiotic prosthetic relationships that are theologically, spiritually and psychically at play. St. Francis of Assis is considered the apotheosis of this way of thinking. The hierarchy of lions, panthers, and tigers. when it comes to royalty is more evident of 'carnivorous virility.' Perhaps the term for such analysis (cf. Freccero) is 'anthrocynic becoming' revealing the complexity of human-nonhuman (animal) relationships when it comes to power differentials of race, class, sex, gender, age, and so forth.

More recent posthuman developments present a further dynamic that faces the fourth industrial revolution based on NBIC technologies (nano-bio-info technologies and cognitive science), so strongly promoted by ecomodernist developments within capitalist markets. The trajectory of these developments have become disturbed and questioned as the awareness of 'human' extinction grows. Nowhere is such anxiety most profound and heightened given gene editing technologies (CRISPR-Cas9) that can modify the human germ line as to what is and is not inheritable, and mRNA vaccines (messenger ribonucleic acid) that can potentially dampen the effects of parasitic viruses in the future. Viruses are 'particular' particles; they are singularities; reproductive forms that do not fit 'life' as it is defined in the humanist paradigm. Which came first, one asks? the virus or its host? If they come together, what sort of human–nonhuman relationship is formed? (Claverie, 2006; Moreira & López-García, 2009). Viral movement is a replication of difference, requiring a host to reproduce more copies of the original. A paradox emerges: usually, the original only becomes original after a copy, but the copy now becomes a variant. The original 'disappears.' Only particular characteristics are retained (the spike (S) protein of the coronavirus for instance). Its mutation, which is its sexual-genetic 'expression' effects each body differently. The virus, as Covid-19 has shown, cannot be anthropomorphized, nor can it be 'controlled.' In terms of 'personification,' they are said to govern us, rule us; they reign, are fickle, unreasonable, whimsical and inconstant; shape shifting in their mutations. Viral ontology presents the paradox that viruses are not able to reproduce themselves by themselves; they are the products of living organisms that have exceeded the very vital contexts of their existence, freed of them by (in the case of humans) killing them to 'live' if they are pathogenic. Yet, most viruses are harmless. The virus is '*trans*boundary in nature' belonging to both biology and biochemistry—alchemical in their action. In some circles the portmanteau term 'tranimal' (Kelley, 2014, p. 226) has been used to include not only such viruses, but any animal that has been genetically modified. The virus is self-genetically modifying in this sense. Viruses are tranimal as technically they are described as a small quantity of genetic matter, either DNA or RNA, which appear in single or double strands.

The paradox of immunology now shows itself in the Anthropocene, especially the Covid-19 pandemic which is usually a direct result of zoonosis (unless one accepts the Wuhan 'lab accident' narrative). As Roberto Esposito (2008) has shown, an 'immunological mind-set' governs globalization, each nation protects its own population within the impossible larger whole where resources and sustainability of the Earth are ultimately limited. The relevance of viruses that move throughout the world, and contribute to the global genetic pool, shows that life localized within bounded membranes, pitted 'against' the world is not possible. The host-parasite dialectic shows the primacy of the body: we are the jealous proprietors who covet our vital property. The 'scale' of the body represents the essential psychological-political-biological metonym for 'human' life (Campbell, 2006; Cohen, 2009, 2011). For Michel Serres (2007), the parasite confounds the hierarchy of life-non-life binary, as well as the host-parasite (friend-enemy) couplet: "man (sic) is the universal parasite … we parasite each other. The parasite is a being and nonbeing at the same time" (pp. 8, 10). Relationality, in general, is both necessary and disturbing. Viral warfare (or, germ warfare) is, by definition, a losing proposition, which may inadvertently be the case with Covid-19 pandemic as its 'origins' remain obscure, politically, socially and economically. Ed Cohen (2009) suggests that the 'viral factories' that become established in our bodies introduce a new productive relationship. "At the cellular and molecular levels, they [viruses] confuse the categories of properties, of ownership, of propriety, of self and other" (25). The virus, in Deleuze and Guattari terms, would be *anorganic life* that is *indifferent* to our existence, upsetting the strong biophilic sensibilities of so much environmental education.

Added to this difficulty is the growing power of AI, what I refer to as the 'inhuman' (to add to the 'non-human' [or more-than-humans] that posthumanist developments have recognized). This further puts into question what is 'human' (aside from the usual socio-historical exclusions from this categorization based on a long list of descriptors: sex, race, age, mental ability, physical, and so on). In this realm we are no longer dealing with naturecultures, but a future shift in ***kind*** and not *degree* toward a 'speculative posthuman' (Roden, 2014). A point may be reached where a 'disconnection' from the human occurs (Roden, 2013), and the 'truly' alien posthuman is invented by machines making machines. 'Ecology without ecology' doesn't seem so strange to stage as the direction is science-fiction with a strong dose of aesthetics and ethics in the service of epistemologies and projected ontologies ('worlds,' for some). Pushed to the extreme would be transhuman developments as promoted by Nicholas Bostrom (2014) but *without* the extreme 'superhuman' implications, but more questions of the elimination of certain diseases (cancer especially), and the ability of paraplegics to walk, the growing of bones and organs via stem-cell research and so on.

Posthumanism and posthuman directions are propelled by the need to question anthropocentrism and, in the indigenous pedagogies a renewed anthropomorphism and neo-animism. The idea that a panpsychism pervades matter underlies these directions in various degrees and forms. Tyson Lewis (2010), for instance, attempts to 'rethink' Hardt and Negri's (2000) overly speciesist and decisively human multitude by recognizing the discourse of critical animal studies (CAS). The human remains overdetermined, even when it comes to Derrida who came late to the zoological problematic. There is no escape from such posthumanist frameworks for this requires a cosmology that recognizes both nonhuman (organic and inorganic as the blurring of the non-living with the living) as well as inhuman (smart and artificial technologies) contributions as necessary agential energetic forces to be recognized and worked with in response to the Anthropocene problematic. Science, Technology, Engineering, Art, Mathematics (STEM) pedagogies of science have been questioned by a variety of art(s) based pedagogies that play with the sensuality of materiality, promoting STE*A*M to interject posthumanities discourses to these sciences, but this is only part of the difficulties that pedagogy faces in the Anthropocene.

Cosmology

Against this backdrop, there is need to understand the cosmic artisan in contemporary times, a 'conceptual persona' first introduced and scattered throughout Deleuze and Guattari's *A Thousand Plateaus* (1987, *TP*) where the mention of a New Earth, a 'people to come,' talk of the alchemy of metallurgy and the machinic phylum, provide inspirational ways to think the (recent) turn toward the 'new materialism,' Object Orientated Ontologies and future of the Earth. The Cosmos for Deleuze and Guattari should not be confused with any harmonious notions of a universe as classically developed by the transcendent views of the Old Testament, Plato and Aristotle, where super-earthly narratives of harmony of the heavenly spheres are equated to heights of spiritualized ideality; or, with modernist scientific cosmology (Galileo and Copernicus) where there is silence on ethical and moral issues, until perhaps now with the geodynamics of the Earth looming in scientists faces. Astrophysics becomes more and more pressing as talk of escaping the Earth, colonizing Mars, offer flights of sci-fi fantasy. Cosmic evolution becomes a binding concept across many disciples where *energy* (and not matter), or energy '*as*' matter prevails, portending a future, as Eric Chaisson (2013) the distinguished astrophysicist says, "be it [one of] complex survival or simple termination" (p. 436).

Following Nietzsche, Deleuze and Guattari turn their attention to the unfamiliarity of this world, to an Earth in-itself, to explore the strangeness of a deterritorializing Earth, an Earth in perpetual becoming. Their cosmology has more to do with quantum physics, which presents an unknowable cosmos (dark energy and dark matter and a puzzle concerning the weak force of gravitation), as well as a knowable order of forces known as the Standard Model of Quantum Particles (SQM), which is able to describe three of the four known fundamental forces: electromagnetic, weak, and strong interaction in the Universe, while the weak force of gravity continues to allude it. This enables a classification of known elementary particles, including the recent 'empirical' discovery of the missing Boson 'particle.' The Standard Quantum Model captures only certain dimensions of the world-in-itself. The rest, like viruses, remain allusive, mysterious and even magical. The recent muon g-2 experiment, for instance, has shown that its magnetism is of a greater magnitude than was once measured, suggesting that there are other undiscovered particles in the vacuum tube, which would upset SQM cosmology.

Deleuze, along with Guattari present a conceptual model of this quantum physical reality that they name chaosmos, or rhythm-chaos: the symbiotic continuum of order and disorder, chaos and cosmos via an enteral return of difference, dissimilarity and divergence through repetition: the perpetual deterriorialization, territorialization and reterritorialization of becoming. The 'given as given' of this virtual dimension is often referred to as the Outside ("an outside which is farther away than any exteriority" Deleuze, 1994, p. 259). This is the realm of complete deterritorialization, involuntary encounters, vital intensities, and desiring machines. François Laruelle (2013) uses the term 'radical immanence' for the same concept of the Real (or One). For Deleuze and Guattari this realm is like a 'cosmic egg': "nonstratified, unformed intense matter" before its 'hatching,' which then leads to stratification (geological, biological, cultural). These are the virtual forces of chaos (dark matter, dark energy) that are non-representational, have no substance, 'pure' chaos. Deleuze and Guattari conceptualize this through the term differenciation (with a 'c') to get at the virtual forces of chaos in-themselves that appear and disappear. The plane of immanence becomes a conceptual term that acts like a membrane for forces at the virtual level that becomes actualized into particles: energy becomes matter ($E = MC^2$). As the Standard Quantum Model (SQM) shows, the entanglement between energy-matter is a wave and particle (corpuscular) phenomena. Deleuze calls this differentiation (with a 't) where order appears—cosmos. Particles are virtual energy, while waves are actualized matter. Differenciation|differentiation, or virtual-actual are in constant entanglement mediated by and on the plane of immanence, which is 'simply' the processes of becoming, of transversal change. While it is outside the scope of this chapter, such a quantum theory is contrary to Karen Barad's (2007) 'agential realism' that relies on a peculiar interpretation of Niels Bohr retaining a Derridean poststructuralism concerning discursive meaning. Her apparatus theory cannot be equated with Deleuze and Guattari's assemblage theory (*agencement*); put another way: her 'intra-relationality' is not the same as their 'trans-versaity' (jagodzinski, 2022, in press).

The meaning of an event (as an actualization) can be counter-effected, or counter-actualized. It can be empirically revisited, explored and rethought. Such an counter-actualization is a refrain (*ritournelle*), which captures the forces of the chaosmos. A *ritournelle* 'grounds' the territory of the event, establishing a milieu, an interval. Refrains assemble milieus and rhythms. Thinking (as in true creativity) is grappling with the unthought of the Outside through which glimpses of the world-as-it-is become 'visible' via philosophy (concepts), art (affects/percepts) and science (mathematical formulations). Thinking is a form of deterritorialization, with paradigm shifts of knowledge being discourses of an event (territorialization), which have been counter-actualized or counter-effectuated (reterritorialized) in specific domains (i.e., The parameters set by Newtonian physics still enables a spacecraft to reach the Moon.). Cutting edge developments in philosophy, art and science "force us to think." An avant-garde in this sense is a 'future present,' (first synthesis of time), while the future of the past (second synthesis of time) develops with the counter-actualizations of an emergent discourse. The pure form of time (third synthesis of time) is the cosmic event itself, a caesura (or cut) in time.

Deleuzian synthesis of time provides a complex structure to grasp various possibilities of becoming of assemblages that are being actualized. For the Anthropocene, an understanding of Earth time (or deep time) is extended to cosmological evolution itself. Cosmology, one speculates, presents a *hologram* of the universe; in other words, the virtual forces of chaos that are harnessed in particular domains (territories, milieus) as tentatively grasped paradigmatically by way of philosophical concepts, aesthetic blocs of percepts and affects and through speculative mathematical systems are, in effect, only 'parts' (like in a holograph) that are said to contain the whole universe. Hints of the whole elusive universe are speculated via various Grand Unification Theories and Supersymmetry. Here caution should be noted. It is highly doubtful that a unified theory of 'everything' will ever be achieved, which maintains that the laws of the universe are immutable. I would follow the work of Lee Smolin (2019) who (convincingly) makes the claim that physical laws evolve (theory of 'fecund universes'). In quantum physics, 'fields' are synonymous with milieus, although milieus have the added feature in the French of denoting 'surrounding,' 'medium' (as in science), and 'middle' (Massumi, 1987, *TP*, xvii). Deleuze and Guattari point out that milieus (the 'fields' of quanta) are not ever *fixed* territories; rather a territory 'organizes' a multiplicity of milieus that is temporarily *marked* by limits. Milieus therefore are actualized as territories; they are not territories until they are marked as such. Territories have to become enacted and remain in a metastable state. When this happens, the territory expresses itself. There is an emergence of qualities of expression that define that particular territory. "It is the mark that makes the territory" (*TP*, 315). "The notion of the milieu is not unitary: not only does the living thing continually pass from one milieu to another, but the milieus pass into one another; they are essentially communicating" (*TP*, 313). Milieus are more like vibratory 'middles' or 'dimensions,' or directions in motion, a 'block of space-time' that moves according to varying speeds and slownesses within a physical system. They are metastable, subject to change when new elements enter its assemblages (cf. Gilbert Simondon, 2017). Milieus transcode and transduce each other, as such they are thoroughly relational. There is no 'set' code, only arhythical codings in the exchange between milieus. The result is rhythm: "What chaos and rhythm have in common is the in-between—between two milieus, rhythm-chaos or the *chaosmos*" (*TP*, 313, original emphasis). "Rhythm is the milieus' answer to chaos" (313). Rhythm turns out to be a figure that is involved in an event of deterritorializing transformation between heterogeneous bodied; it is non-metrical, unbounded and unfixed, escaping the impositions of regular pulsations.

Deleuze and Guattari's cosmology is primarily ethical; it's aim at self-transformation, new ways of perceiving the world and experimenting with the deterritorializing earthly forces. Artists have the privilege and cosmological capacity to create a "new Earth" (*TP*, Chap. 11). Deleuze and Guattari call on Gilbert Simondon's developments of ontogenesis to overcome Aristotelian forms of hylomorphism: moving matter as material is quite unlike the matter-form model. Simondon's paradigmatic technological exemplar of making a 'clay brick' as a symbiotic coming together of heterogeneous substances is perhaps equally iconic of Deleuze and Guattari's wasp-orchid exemplar. Matter is self-organized via the immanent power of morphogenesis. In *TP* they call on the alchemical tradition and the smithery of metallurgy where form is never separated from matter as metal flows and is forged. Artisans seek to know a material's condition from within, and intuit how to pass from one mode of its existence to another; much like the ceramist or the baker works where the right amount and consistency of ingredients and process to transform clay into ceramics, flour and water into bread. Both processes need energy from the kiln to make this happen. In Simondon's (2009) terms, at the 'preindividual' level of ontogenesis, "*form, matter, and energy preexist in the system*" (7, original emphasis). Energetic materiality in movement carries singularities or haecceities that have varying intensive affects and are topological in their forms. It is a perspective that reorientates thought from the perspective of the organism and organic life to the perspective of the cosmos and nonorganic life, life that is *indifferent* like the viruses mentioned earlier. Life is not a 'form' but a complex relation between differential velocities. Material has "three principal characteristics" (*TP*, 408): matter is molecularized and understood for its properties; what forces matter can harness, and what processes can be consistently applied to it. The artisan 'surrenders' to the material and follows where it leads for matter has *nomos*. It has a free distribution of properties that are subject to change. As Deleuze was to say latter, there are 'micro-brains' everywhere you look.

The artist-artisan taps into a cosmological continuum of material-forces, extracting 'chaoid sensation' (Deleuze & Guattari, 1994, p. 206) that cuts across the usual dichotomies: nature-culture, nature-technology, human-animal, inanimate-animate. The machinic phylum (or 'technological lineage,' *TP*, 406) consists of the non-organic 'consciousness' of metallurgy; its matter-flow is "expressed in panmetalism." Metal is "coextensive to the whole of matter, and the whole of matter to metallurgy. … Metal is the conductor of all matter" (*TP*, 411). The significance of this is laid out in today's high-tech media as Jussi Parikka (2014) has usefully shown the link between 'deep time' of the planet and the technological developments, where mining (and pillaging) of (scarce) minerals becomes crucial in order to ensure constant production and upgrading of digitalized media. He calls this development the 'Anthrobscene,' which make visible the ubiquitous media that are the result of ecological violence: e-waste and unsustainable geo-practices such as fracking that are necessary to maintain technological and corporate cultural networks where worker exploitation continues on as in bygone periods. An alternative media history that extends toward geophysics of media culture becomes possible. The 'geology of the media' (Parikka, 2015) requires that the 'temporality' of the tech-media products need to be rethought as the 'deep time' of fossilized e-detritus, the toxic residue whose decay cycles may last longer than our species survival, what Timothy Morton (2013) calls 'hyperobjects,' which raise more questions than they answer (Bradley, 2019). Such geophysics extends from interest in alchemists to contemporary mining practices, the discovery of new minerals, as well as the invention of new elements, and the affective materialities that present themselves for the 'new media.'

Parikka (2015) invests in the possibility of a 'psychophysics,' which supplements, if not replaces the psychogeography as the all too human study of cities by the Situationists updating Walter Benjamin's *flaneûr*. The shift is to an inhuman orientation: "a cartography of architecture of the technological that is embedded in the geophysical" (p. 79). Parrika pushes Anthropocene thinking in the direction of the posthuman as there is a strong recognition that technology transforms our species both physically, psychically, and spiritually (i.e., affectively). In this sense he supplements work that both Michel Serres (2001) and Bernard Stiegler (2018) have developed; the former with his ideas of exo-Darwinism in relation to technologies, and the latter attempting to rethink the thermodynamics of the Anthropocene through negentropy where new technologies might be able to harness new energy sources (e.g., fusion).

Contemporary Cosmic Artisans of the Anthropocene: Culling a Pedagogy ... of Sorts

Since the advent of ubiquitous computerization and digitalization (ca. mid-twentieth century), and with the beginning of the twenty-first century designated as the biotech society (Rifkin, 1999), bioengineering was accelerated by the Human Genome Project (launched in 1990 and ending in 2003). Virtually all 'new' technologies have by-and-large overcome the hylomorphic paradigm of industrial production, pithily stated as the 'heat, beat and treat' methods of harnessing Nature. Strongly stated, in relation to Deleuze and Guattari's 1980s explorations in *TP*, art and design has turned 'cosmological.' When it comes to technological DIY side of this development the so-called Makerspaces, broadly defined as "community-oriented places in which an ethos of do-it-yourself (DIY) experimentation with new technologies and materials coalesces with the goals of sharing knowledge and collaboration" (Reider & Elam-Handloff, 2018). Suffice to say that (perhaps), following Deleuze and Guattari, a zone of indiscernibility is indeed established in these makerspaces where human and inhuman 'smart' technologies come together forging a field of becomings through mutual use, modulation, incubation and mutation. Such an artisan fits their description of "a *homemade* atomic bomb" (*TP*, 345, added emphasis). The artisan is indeed the hero in this endeavor as labor is certainly not Taylorized; processes are grasped, shared, researched and experimented with. Skill and knowhow are forwarded as craft, art, science, and technology collapse together. For Sjoerd van Tuinen (2017) this presents a mannerist genealogy traceable to the sixteenth century. Artists and artisan so engaged are primarily digital designers engaged in processes that utilize the latest technologies where coded algorithms become the agential artist via the robot or via a 3-d printing machine. Quayola 's *Captives #B8-6-0* 'replicates Michelangelo's *Captives* (unfinished slaves) by enlisting a large, digitally controlled milling machine. The machine 'frees' the slaves from the polystyrene blocks as documented on video (there is no audience). Dominique Moulon (2018) offers a host of other examples in his *Art Beyond Digital*. The recycling of trash, garbage, and waste to decommodify (e.g., Germano Celant's *arte povera*) is also part of a 'new alchemy,' as in Wilhelm Mundt's 'packaged studio shit.' These are his sculptural stones, a sort of record of the 'crap' he leaves behind while working in his studio, which is re-packaged and (again) de-commodified, but only paradoxically as these 'stones' are exhibited and sold. Then there is 'reversed alchemy,' turning electronic e-waste (shit) into gold (Feigelfield & Parikka, 2015). Or, we might point to Wim Delvoye's *Cloaca* (2000–2007), a series of mechanical contraptions that utilize the latest technologies that do the 'job' of the human intestines: they shit. The cured 'turds' sell for 1000 dollars each. Shit, after all, is money according to Freud, and this is an alchemist's trick to change shit into gold like the previous examples. The springing up of makerspaces in communities, universities, libraries, museums, K-12 schools present a maker movement that fits nicely into the entrepreneurial spirit of commodification linked to capitalism, the state and to the private institutions they are founded and supported by. While there is a 'hacker ethic' as well that differentiates itself from market forces, its range and impact remains limited (Kostakis et al., 2015). The best 'hackers' are asked to join the social media industries, especially Google, Apple, Amazon and Facebook.

At the turn of the twenty-first century, a number of important books emerged: Donna Haraway's *Modest_Witness@Second_Millennium. FemaleMan©Meets_OncoMouse™*, N. Katherine Hayles' *How We Became Posthuman*, and Jeremy Rikin's *Biotech Century* to mention key authors who began to raise issues about bioengineering. The digital Makerspaces that were dominated by telematic art with its emphasis on 'dry medium' were supplemented, if not supplanted by 'life' itself. Life now became the 'play' thing via genetic manipulation, stem cell research and transgenic Genetically Modified Organisms (GMOs). Telematics was supplemented, supplanted and infused with bioengineering; a generation of 'exo-Darwinian technologies (Serres, 2001) began to develop using nanotechologies. These developments enabled cross-modular synesthetic sensations to take place such as the 'seeing' with your tongue, modifying what a body could 'do.' Deleuze and Guattari's concept of the Body without Organs (BwO) has been successfully appropriated by a third wave of cyberneticists, paving the way for a transhumanist future exemplifying the potential of the Anthropocene as the 'age of man' (More & Vita-More, 2013). The Body without Organs is a concept Deleuze and Guattari develop that helps comprehend how bodily organs are organized, whether they are productive or not, healthy or not, cancerous or not. The point is such an imaginary body is materialized through the technologies that actualize it, by drugs and 'medical herbs' of all kinds, or an entire host of 'smart' machines that now monitor the functioning of various organs and bodily systems, conscious and non-conscious alike: an endless list of cyborgian gadgets—heart monitors, respiratory aids for asthma, blood analysis, contact lenses, and so on.

In these developments, the cosmic artist-artisan now emerges in several directions where art, craft, technology and science come together under the broader umbrella of biomimesis (Benyus, 1997). There is a particular strain of bioart that is deeply steeped in the craft traditions where the 'play' with living materials takes precedence in the name of providing for a waste-free environment; the production cycle of the commodity ends up being completely biodegradable. The most common living materials used here are the various forms of fungi. Scientists and artists work together in various corporate sponsored and entrepreneurial start up labs where fungi are used to detoxify plastics, microorganisms such as yeast are used to produce bioethanol, leather products are being replaced with products made of mycelium, which is also being used as a building block material. Hemp and kelp are other living fibers that are explored for clothing as well as disposal shoe design. There is a whole industry around spider's silk. 3-d printing of a certain fungi has now become possible. It seems the possibilities here are limitless. All these initiatives are consonant with the idea of Gilbert Simondon's (2017) notions of individuation, transduction and 'disparation.' It has taken on the signifier 'endosymbiosis' with nature (jagodzinski, 2019a).

The bioart-craft referred to above is a subset of biomimicry that was spearheaded at the turn of the century. Janine Beuyus' *Biomimicry: Innovation Inspired by Nature* came out in 1997 promising a new revolution in industrial production. Biomimesis, sometimes called 'synthetic biology' and a 'synthetic aesthetic' (Ginsberg, 2014), calls on nature as its inspiration and innovation to bring together engineering, design and biology in new ways to grow physical devices and structures (cells, from the ground up, rather than from the top down). 'Life' rather than the 'gene' is its focus, although both are obviously entangled. It is the careful examination of nature's design that can be expanded. For instance, the study of how birds conserve energy by changing their wing shapes in various conditions of flight so that such knowledge can be transferred over into the aircraft industry; or, new material synthesized from spider's silk, mycelium bricks, termite-mound inspired buildings are now possible. Perhaps Rachel Armstrong's (2018) 'living architecture' where proto cells are developed, manufactured and grown to replace rotting wood in cities such as Venice is a remarkable example of the power of such technologies.

The corporate world, especially venture capitalism, is highly invested in biomimetic design. In brief, biomimesis, while so full of promises, is thoroughly harnessed for its utilitarian usefulness for industry and the military (Johnson, 2010). Nonhuman life is harnessed to a capitalist imaginary. Nonhuman life becomes intellectual property (as opposed to raw material) to be harnessed as an active product of knowledge (and not for passive consumption). It is what nonhuman entities can 'do' through their *physiological* capacities to guide and 'teach' us how to do it (nature's 'wisdom'), which is what becomes identified as 'enclosed' intellectual property (the patent), and not the nonhuman entity 'in-itself' (Goldstein & Johnson, 2015). Nature becomes productive in its processes that can be mimicked for specific utilitarian needs and effects as animal and machine boundary breaks down. In effect, they are the nonhuman extension of the 'Maker entrepreneurial culture' that has spread throughout universities and industry. Each organic *creature* is now carefully dismantled, dissected and explored for its capacity to perform, evaluated for its *bio-value*. It's 'vitality index.' Such knowledge also generates the building blocks to construct 'wetware machines' as neurobiology meets robotics so that bio-tech-science can begin to mimic nature's own creatures (e.g., stigmergy navigation in ants and geese, bat sonar, lizard limb regeneration and so on). The proviso being, only those nonhumans whose productivity can be exploited are chosen. The sleight of hand to stave off accusations of enslavement, which would be ineffective anyway given that The Universal Declaration of Human Rights does not apply the nonhuman other, is to maintain that Nature is now honored and respected for its capacity to 'show us the way' in collaboration. Nonhuman life now becomes '*terra economica*,' a repository of limitless potential to become capital (Goldstein & Johnson, 2015, p. 76). The economic blueprint for such a world order is presented by "natural capitalism" (Hawken et al., 1999) with claims to "intentional biology," meaning more and more knowledge leading to complete control of nature.

A byproduct of biomimesis has been the recent development in the field of biosensing (Johnson, 2017). LimCo International has developed a LimCo BioSensor System (LBS). Based in Germany, this corporation uses multiple species of whole-body organisms to monitor fresh and marine water sources for pollutants. Between eight and 96 sensor chambers house an array of animal species (fish, worms, crustaceans, mollusks, microorganisms), which are carefully monitored for their 'expressions of life.' Their 'behavioural fingerprints' are exposed to different ecological system from which their reactions indicate the severity of hazardous anthropogenic chemicals, biologically produced toxins, and blue-green algae. Pollution is thus made 'measurable.' Bioanalytical Microsystems and Biosensors Laboratory (BMB), located at Cornell University, has synthesized liposomes that are used in small-scale technological devices that can signal the presence of pathogenetic organisms: toxins in foods, drinking water and in the generalized environment. The Wyss Institute at Harvard reengineers plants with Green Florescent Protein (GFP) to identify pathogens. Microorganisms are now used in biomining and bioleaching to extract minerals from ores beyond the mine.

While it is true that biosensing enables a way of seeing with nonhuman life, expanding the human sensorium and has the potential to shift perspectives in environmental health, it is also true that this industry is deeply tied to the profit industry. It selects what trait is useful in a plant, animal or microorganism. This applies also to 'brain' research with "artificial animals" as they are called—hybrid robotic creatures modified from existing animals, mostly rats and fish. These hybrids become 'tranimals' (Kelley, 2014). Thomas DeMarse develops neural interfaces by growing neurons (rat nerve cells) that integrate as a network on chips creating an organic computer or 'living computer' that is able to autopilot a jet plane in a stimulator. Sanjiv Talwar's ratbot (robo-rat) has electrodes implanted in its medial forebrain bundle (MFB) and sensorimotor cortex of the brain so it can be 'controlled' at remote distances via a computer terminal. Then there is Mussa-Ivaldi's neurobiotic explorations of the lamprey to understand how its olfactory systems and motor orientations work. It was possible to remove the entire eel brain and keep it functioning in a nutrient medium with electrodes implanted and then installed in a robot-cart. This 'brain,' as the sole processor, was able to move this cart forward through photophilic means simply by having a light shine on it. While these are all bioengineering wonders meant to improve our living standards, in effect they exemplify the way all organic life is there to be used only to service 'us.'

Given this sweep of technological openings that have such a strong appeal to technological innovations with almost 'magical' properties of control, pedagogy becomes a rather 'dull' endeavor if the latest machine learning and digitalized resources are not harnessed for the classroom. The Covid-19 pandemic has shown that online learning, while limited, is possible. It seems that this residual experience will remain as an incentive by many institutions as a cost-saving strategy. Internet technologies have clearly demonstrated that knowledge as information is not where education can find its *raison d'être* any longer. Not only do designers have to think 'otherwise' (Marenko & Brassett, 2015), but the biopolitics of life and surveillance become crucial issues to be discussed in classes if there is to be a rethink of science, technology, and the humanities.

Sequel to Come?

To end, the cosmic artisan has been essentially captured by capitalism. These recent developments of craft-bioart, biomimesis and biosensing seems to play so well into the 'Good Anthropocene.' Some have called these developments a 'radical Enlightenment' (Gare, 2014) in the way art, science, technology have come together that may provide a turn to an 'ecological civilization.' As it stands, however the underlying philosophy of biomimesis, biosemiotics and ecosemiotics is anthropocentric as it is ultimately, a world-for-us that is promoted, more and more control of nature as its 'workings' are symbiotically uncovered. There is a whole other side to this story by artists who actively push back these developments, but with 'climate change' looming in our face, as it were, all bets seem to be on these new technologies. Philosophers such as Eugene Thacker (2008) have tried to rethink the philosophy of biology as biophilosophy, which looks at life as a multiplicity and brings in the fundamental aspects of death that are very much part of life. There are bioartists, significantly like Natalie Jeremijenko, who attempt to reverse the picture so that we learn from the ecologies of various animals, and there are bioartists who, as 'hackers,' attempt to provide a minoritarian position that questions the bioengineering of new organisms (here TC&A, CAE, Kac remain strong critics) (jagodzinski, 2019b).

In relation to the question of a planetary consciousness, which I raised at the start of this chapter, it seems that explorations such as Rosi Braidotti's (2013b) nomadic form of reflexive cosmopolitanism, or Ron Bogue's (2012) 'chaosmopolitanism' imaginaries should not be lost but entertained and worked in relation to the other proposals such as Hardt and Negri's conception of the 'commons' despite their limitations. We need more imaginary projections, perhaps like that of Arran Gare's call for a radical Enlightenment, which does not lose site of the inhuman and non-human assemblages that are in play, yet exemplify the need not to lose sight of 'species-integrity.' Perhaps most radically, as Patricia MacCormack (2020) insists in her manifesto for the end of the Anthropocene, there is the urgent need to initiate a strict form of animal abolitionism. Yet. if our species extinction is on the table, meaning that death needs to be continuously rethought in relation to life, which is what consumes the biopolitics of capitalism, then Ray Brassier's (2007) controversial 'transcendental realism,' where nihilism as a crisis of meaning needs to be faced, should not be dismissed outright either. This plunges us into the direction of *dark pedagogy* that has been profoundly explored by Jason Wallin (2014, 2015). If there is a 'New Earth' to come, vigilance needs to be maintained toward the unthought. A 'new (grand) narrative' is badly needed, but one that circumvents capitalist machinations and staves off the coming dystopia. Pedagogy finds only singular pockets of 'smooth' space to raise difficulties with teachers and students who sense the urgency of the 'now' of climate change. As an educator, one does not have the luxury to simply give up, at the same time, involvement of any kind enables determination to persist.

References

Abram, D. (1997). *The spell of the sensuous: Perceptions and language in the more than a human world*. Vintage.

Andrzejewski, J., Baltodano, M., & Symcox, L. (Eds.). (2009). Social justice, peace, and environmental education: Transformative standards. New York and London: Routledge.

Angus, I. (2016). *Facing the Anthropocene*. Monthly Review Press.

Armstrong, R. (2018). *Soft living architecture: An alternative view of bio-informed practices*. Bloomsbury Visual Arts.

Bai, H., Elza, D., Kovacs, P., & Romanycia, S. (2010). Re-searching and re-storying he complex and complicated relationship of biophilia and bibliophilia. *Environmental Education Research, 16*(3–4), 351–365.

Balibar, É. (2008). *Spinoza and politics.* (P. Snowdon, Trans.). Verso.

Barad, K. (2007). *Meeting the universe halfway: Quantum physics and the entanglement of matter and meaning.* Duke University Press.

Bennett, J. (2010). *Vibrant matter: A political ecology of things.* Duke University Press.

Benyus, J. (1997). *Biomimicry: Innovation inspired by nature.* Harper Collins.

Bogue, R. (2012). Nature, law and chaosmopolitanism. In R. Braidotti & P. Pisters (Eds.), *Revisiting normality with Deleuze* (pp. 98–112). Bloomsbury.

Bonnett, M. (2020). *Environmental consciousness, nature and the philosophy of education: Ecologizing education.* Routledge.

Bostrom, N. (2014). *Superintelligence: Paths, dangers, strategies.* Oxford University Press.

Bradley, J. (2019). Transcendental monsters, animism and the critique of hyperobjects. *Australian Journal of Environmental Education, 35*(3), 163–172.

Braidotti, R. (2013a). *The posthuman.* Cambridge Polity Press.

Braidotti, R. (2013b). Becoming-world. In P. Hanafin & B. Blaagaard (Eds.), *After cosmopolitanism* (pp. 8–27). Routledge.

Brassier, R. (2007). *Nihil unbound: Enlightenment and extinction.* Palgrave Macmillan.

Butler, J. (1993). *Bodies that matter: On the discursive limits of sex.* Routledge.

Campbell, T. (2006). "Bios," immunity, life: The thought of Roberto Esposito. *Diacritics, 36*(2), 2–22. Retrieved March 31, 2021, from http://www.jstor.org/stable/20204123

Ceder, S. (2019). *Towards a posthuman theory of educational relationality.* Routledge.

Chaisson, E. J. (2013). A singular universe of many singularities: Cultural evolution in a cosmic context. In A. H. Eden, J. H. Moor, J. J. Søraker, & E. Steinhart (Eds.), *The singularity hypothesis: A scientific and philosophical assessment* (pp. 413–439). Spring-Verlag.

Claverie, J.-M. (2006). Viruses take center stage in cellular evolution. *Genome Biology, 7,* 110. https://doi.org/10.1186/gb-2006-7-6-110

Cohen, E. (2009). *A body worth defending: Immunity, biopolitics, and the apotheosis of the modern body.* Duke University Press.

Cohen, E. (2011). The paradoxical politics of viral containment; or, how scale undoes us one and all. *Social Text, 29*(1), 15–35.

Colebrook, C. (2013). Destroying cosmopolitanism for the sake of the cosmos. In R. Braidotti, P. Hanafin, & B. Blaagaard (Eds.), *After cosmopolitanism* (pp. 166–182). Routledge.

Colebrook, C. (2015). What is it like to be a human. *Transgenders Study Quarterly, 2*(2), 227–243.

Cutter-Mackenzie-Knowles, A., Malone, K., & Hacking, E. B. (Eds.). (2020). *Research handbook on childhoodnature: Assemblages of childhood and nature research.* Springer International Publishing.

Deleuze, G. (1994). *Difference and repetition.* (P. Patton, Trans.). Columbia University Press.

Deleuze, G., & Guattari, F. (1987). *Thousand plateaus: Capitalism and schizophrenia, vol. 2.* (B. Massumi, Trans.). University of Minnesota Press. *TP* in text.

Deleuze, G., & Guattari, F. (1994). *What is philosophy?* (H. Tomlinson & G. Burchell, Trans.). Columbia University Press.

Derrida, J. (1992). Eating well or the calculation of the subject. In E. Weber (Ed.), *Points: Interviews 1974–1994* (pp. 255–287). Stanford University Press.

Derrida, J. (2001). *Cosmopolitanism and forgiveness.* Routledge.

Donald, D. (2021). We need a new story: Walking and the wâhkôhtowin imagination. *Journal of the Canadian Association for Curriculum Studies, 18*(2), 53–63.

Elias, N. (1978). *The civilizing process.* Urizen Book. original 1939.

Esposito, R. (2008). *Bios: Biopolitics and philosophy.* (T. Campbell, Trans.). University of Minnesota Press.

Feigelfield, P., & Parikka, J. (2015). Media ecology out of nature: An interview with Jussi Parrika. *e-flux,* 62. https://www.e-flux.com/journal/62/60965/media-archaeology-out-of-nature-an-interview-with-jussi-parikka/

Freccero, C. (2011). Carnivorous virility; or, becoming-dog. *Social Text, 106, 29*(1), 177–195.

Gare, A. (2014). Deep ecology, the radical enlightenment, and ecological civilization. *The Trumpeter, 30*(2), 184–205.

Garrard, G. (2010). Problems and prospects in ecocritical pedagogy. *Environmental Education Research, 16*(2), 233–245.

Ginsberg, D. A. (2014). *Synthetic aesthetics: Investigating synthetic biology's designs on nature.* The MIT Press.

Goldstein, J., & Johnson, E. (2015). Biomimicry: New natures, new Enclosures. *Theory, Culture & Society, 32*(1), 61–81.

Habermas, J. (1986). *The theory of communicative action: Reason and rationalization of society.* Polity Press.

Halverson, E. R., & Sheridan, K. M. (2014). Maker movement in education. *Harvard Educational Review, 84*(4), 495–504.

Haraway, D. (2003). *The companion species manifesto: Dogs, people, and significant otherness.* Prickly Paradigm Press.

Haraway, D. (2008). *When species meet.* University of Minnesota Press.

Hardt, M., & Negri, T. (2000). *Empire.* MIT Press.

Hawke, S. M. (2012). Water literacy: An 'other wise,' active and cross-cultural approach to pedagogy, sustainability and human rights. *Continuum, 26*(2), 235–247.
Hawken, P., Lovins, A., & Lovins, H. (1999). *Natural capitalism: Creating the next industrial revolution*. Little, Brown and Co.
Hird, M. (2010). Meeting with the microcosmos. *Environment and Planning D. Space and Society, 28*(1), 36–39.
Hultman, K., & Lenz Taguchi, H. (2010). Challenging anthropocentric analysis of visual data: A relational materialist methodological approach to educational research. *International Journal of Qualitative Studies in Education, 23*(5), 525–542.
Ingold, T. (2005). Epilogue: Towards a politics of dwelling. *Conservative Society, 3*(2), 501–508.
jagodzinski, j. (2017). Betraying further. Arts-based education at the 'end of the world'. In j. jagodzinski (Ed.), *What is art education: After Deleuze and Guattari* (pp. 267–309). Palgrave Macmillan.
jagodzinski, j. (2019a). Death after life: The ethico-political response of artists towards the Anthropocene. In C.-P. Buschkuehle, D. Atkinson, J. Baldacchino, & R. Vella (Eds.), *Art—ethics—education*. Sense Publishers.
jagodzinski, j. (2019b). Artistic research in an expanding field: The case of bioArt. In P. de Assisi & L. D'Errico (Eds.), *Artistic research: Charting a field in expansion*. Rowman & Littlefield.
jagodzinski, j. (2022). Quantum creativity: Afracting new materialism in the Anthropocene. Qualitative inquiry. Special issue: Posthuman creativities. Daniel Harris & David Rousell (Eds.). (in press).
Johnson, E. R. (2010). Reinventing biological life: Reinventing 'the human'. *Ephemera: Theory & Politics in Organization, 10*(2), 177–193.
Johnson, E. R. (2017). At the limits of species being: Sensing the Anthropocene. *The South Atlantic Quarterly, 116*(2), 275–292.
Kelley, L. (2014). Tranimals. *TSQ: Transgenders Study Quarterly, 1*(1–2), 226–228.
Kostakis, V., Niaros, V., & Giotitsas, C. (2015). Production and governance in hackerspaces: A manifestation of commons-based peer production in the physical realm? *International Journal of Cultural Studies, 18*(5), 555–573.
Laruelle, F. (2013). *Principles of new philosophy*. Bloomsbury Academic.
Lewis, T. (2010). Swarm intelligence: Rethinking the multitude from within transversal commons. *Culture, Theory and Critique, 51*(3), 223–238.
Lindgren, N., & Öhman, J. (2018). A posthuman approach to human-animal relationships: Advocating critical pluralism. *Environmental Education Research, 25*(8), 1200–1215.
MacCormack, P. (2017). Ahuman abolition. In C. Garner & P. MacCormack (Eds.), *Deleuze and the animal* (pp. 25–46). Edinburgh University Press.

MacCormack, P. (2020). *The ahuman manifesto: Activism for the end of the Anthropocene*. Bloomsbury Academic.
Marenko, B., & Brassett, J. (2015). *Deleuze and design*. Edinburgh University Press.
Martineli, D. (2008). Anthropocentrism as a second phenomenon: Semiotic and ethical implications. *Social Semiotics, 18*, 79–99.
Massumi B. (1987). Notes on the translation and acknowledgements. In G. Deleuze & F. Guattari (Eds.) and B. Massumi (Trans.), *Thousand plateaus: Capitalism and schizophrenia, vol. 2.* (pp. xvi–xix). University of Minnesota Press.
McKibben, B. (1989). *The end of nature*. Random House.
Meillassoux, Q. (2008). *After finitude: An essay on the necessity of contingency*. (R. Brassier, Trans.). Continuum.
Moore, J. (2015). *Capitalism in the web of life: Ecology and the accumulation of capital*. Verso.
Moore, J. (Ed.). (2016). *Anthropocene or capitalocene: Nature, history and the crisis of capitalism*. PM Press.
More, M., & Vita-More, N. (Eds.). (2013). *The transhumanist reader*. John Wiley & Sons, Inc.
Moreira, D., & López-García, P. (2009). Ten reasons to exclude viruses from the tree of life. *Nature Reviews Microbiology, 7*(4), 306–311. https://doi.org/10.1038/nrmicro2108. Epub 2009 Mar 9.
Morton, T. (2007). *Ecology without nature: Rethinking environmental aesthetics*. MIT Press.
Morton, T. (2013). *Hyperobjects: Philosophy and ecology after the end of the world*. University of Minnesota Press.
Moulon, D. (2018). *Art beyond the digital*. LINK Editions.
Parikka, J. (2014). *The Anthrobscene*. University of Minnesota Press.
Parikka, J. (2015). *A geology of media*. University of Minnesota Press.
Parr, A. (2009). *Hijacking sustainability*. MIT.
Parr, A. (2017). *Birth of a new earth: The radical politics of environmentalism*. Columbia University Press.
Petersen, H. (2019a). *Schizoanalysis and animal science education*. Bloomsbury Academic.
Petersen, H. (2019b). The contested space of animals in education: A response to the 'animal turn' in education for sustainable development. *Education Science, 9*(3), 211. https://doi.org/10.3390/educsci9030211
Ramey, J. (2016). *Politics of divination: Neoliberal endgame and the religion of contingency*. Rowan & Littlefield.
Rautio, P. (2013). Being nature: Interspecies articulation as a species-specific practice of relating to environment. *Environmental Education Research, 19*(4), 445–457.

Reider, M. D., & Elam-Handloff, J. (2018). Makerspaces. *MLA commons: Digital pedagogy in the humanities, concepts, models and experiments*. https://digitalpedagogy.mla.hcommons.org/keywords/makerspaces/

Rice, S., & Rud, A. G. (Eds.). (2016). *The educational significance of human and non-human animal interactions: Blurring the species line*. Palgrave Macmillan.

Rifkin, J. (1999). *The biotech century: Harnessing the gene and remaking the world*. New York, N.Y.: Penguin Putnam Inc.

Roden, D. (2013). The disconnection thesis. In A. H. Eden, J. H. Moor, J. J. Søraker, & E. Steinhart (Eds.), *The singularity hypothesis: A scientific and philosophical assessment* (pp. 281–298). Spring-Verlag.

Roden, D. (2014). *Posthuman life: Philosophy at the edge of the human*. Routledge.

Serres, M. (2001). *Hominescence*. (R. Burks, Trans.). Bloombury Academic.

Serres, M. (2007). *Parasite*. (L. Schehr, Trans.). University of Minnesota Press.

Simondon, G. (2009). The position of the problem of ontogenesis. Trans. G. Flanders. *Parrhesia, 7*, 4–16.

Simondon, G. (2017). *On the mode of existence of technical objects*. (C. Malaspina & J. Rogove, Trans.). Univocal Publishing.

Smith, L. T., Tuck, E., & Yang, K. W. (Eds.). (2019). *Indigenous and decolonizing studies in education: Mapping the long view*. Routledge.

Smolin, L. (2019). *Einstein's unfinished revolution: The search for what lies beyond the quantum*. Alfred A.Knopf.

Stevenson, M., Brody, M., Dillon, J., & Wals, A. J. (Eds.). (2013). *The international hand-book of research on environmental education*. Routledge and American Educational Research Association.

Stiegler, B. (2018). *The neganthropocene*. (D. Ross, Trans.). Open Humanities Press.

Taylor, A., & Pacini-Ketchabaw, V. (2015). Learning with children, ants, and worms in the Anthropocene: Towards a common world pedagogy of multispecies vulnerability. *Pedagogy, Culture, Society, 23*(4), 507–529.

Thacker, E. (2008). Biophilosophy for the 21st century. In M. Kroker & A. Kroker (Eds.), *Critical digital studies: A reader* (pp. 132–142). University of Toronto Press.

van de Akker, R., Gibbons, A., & Vermeulen, T. (Eds.). (2017). *Metamodernism: Historicity, affect and depth after postmodernism*. Rowan & Littlefeld International.

van Tuinen, S. (2017). The cosmic artisan: Mannerist virtuosity and contemporary crafts. In I. R. van de Akker, A. Gibbons, & T. Vermeulen (Eds.), *Metamodernism: Historicity, affect and depth after postmodernism* (pp. 69–82). Rowan & Littlefeld International.

Wallin, J. J. (2014). Dark pedagogy. In P. MacCormack (Ed.), *The animal catalyst* (pp. 145–162). Bloomsbury.

Wallin, J. J. (2015). Dark posthumanism, unthinking education, and ecology at the end of the Anthropocene. In N. J. Snaza & J. A. Weaver (Eds.), *Posthumanism and educational research* (pp. 134–150). Routledge.

CHAPTER 9

Lying on the Ground: Aesthetic Learning Processes in the Anthropocene

Helene Illeris

Prologue

On Friday 6th of October 2020, Fridays for Future Denmark had a public action in front of the Danish parliament. After a week-long occupation of the square, they collectively performed the funeral of their own future, their hopes and their dreams. After the funeral rite, the young activists lay down on the cold granite setts under the gray autumn sky, silently listening to Greta Thunberg's voice from the loudspeakers: "Right here and right now is where we draw the line" she shouts, "You have stolen my dreams and my childhood with your empty words and yet I am one of the lucky ones".

Afterwards the microphone is open, and one after the other the activists stand up and bear witness about their hope and dreams, fear and anger. A young girl states: "Climate is a lot about graphs and politics, but it is important that we dare to be together around our grief and vulnerability". In pronouncing the last words her voice breaks. (Krogh, 2020; Skolelever i over 170 timers uafbrudt aktion, 2020)[1]

H. Illeris (✉)
Faculty of Fine Arts, Department of Visual art and Drama, University of Agder, Kristiansand, Norway
e-mail: helene.illeris@uia.no

M. Paulsen et al. (eds.), *Pedagogy in the Anthropocene*, Palgrave Studies in Educational Futures,
https://doi.org/10.1007/978-3-030-90980-2_9

Introduction

Today's young people live in a world of limit experiences. Many are haunted by doubts and fear for their future as part of an anthropocene planet. To *Generation Z*, born between the late 1990s and the early 2010s, life-perspectives seem precarious and they generally feel more anxious about their future than previous generations (McMaster, 2020). According to the *Future of Humanity*-survey (Amnesty International, 2019),[2] 41% of young people aged 18–25, consider climate change to be the most important issue facing the world, and an increasing number of them are demanding radical changes to stop the threatening consequences of global warming (Barbiroglio, 2019).

In order to confront the politicians who are responsible, week after week the activists of #Fridays for Future (FFF) stand, sit and lie in front of parliament buildings all over the world. Inspired by Thunberg's school strike, they repeatedly expose their bodies to the public, enacting what Judith Butler (2015, p. 18) has termed "an unforeseen form of political performativity that puts livable life at the forefront of politics". As can be seen from the prologue, the symbolic funeral in front of the Danish Parliament in October 2020 was a direct expression of grief and despair among the young participants. By lying down on the ground as "dead corpses" symbolizing "the millions of humans who have died from climate changes and all the millions who will die" (Krogh, 2020), the activists materially performed the connection of their living bodies to death and with that, as we shall see, to the ground that sustains their human existence.

Presentation of the Chapter

In this chapter I will use the example of the Danish FFF action as an entrance to an inquiry into which ways human bodies might be involved in the transition from a well-known human-centred way of relating to the world to an unknown anthropocene way. Theoretically, I will focus on *aesthetic learning processes* (ALPs), a notion that is widely employed in contemporary Scandinavian educational research, and one that I believe can gain new relevance in the light of the pedagogical challenges posed by the Anthropocene. Empirically I will zoom in on the moment where the young activists are lying down on the parliament square, and on the meeting between their bodies and the granite pavement. Following object-oriented conceptions of aesthetic experience as *attunement*, I will argue that, although the attention of the activists was directed at the voice of Greta Thunberg, the lying bodies and the granite pavement were secretly involved in an aesthetic process of tuning to each other. I thus deliberately change the focus of the text from the conscious staging of the symbolic form by the activists to the imagined aesthetic experiences of the bodies and the granite setts, the paving stones. In order to relate these speculative ideas to concrete educational practices, I end the chapter by offering an educational proposition which aims to connect the explicit environmental political intentions of the activists with the implicit political intentions of reconceptualizing ALPs for the Anthropocene.

The inquiry is divided into three separate sections:

- Section I presents ALPs as conceptualized in Scandinavian educational discourses since the early 1990s. Through a discussion of the two partly overlapping approaches of production and reception, it shows how ALPs sustain modernist conceptions of learning as a human-centred activity.
- Section II proposes an object-oriented reconceptualization of ALPs in the Anthropocene in the light of the imagined aesthetic experiences of the bodies of the young activists lying on the ground.
- Section III uses the insights from sections I and II to introduce *proposition* as a creative and experiential way to enact ALPs as part of a pedagogy for the Anthropocene.

I. Aesthetic Learning Processes in Scandinavian Education

In Scandinavian[3] education, ALPs are considered a key approach to education (cf. e.g. Hohr & Pedersen, 1996; Austring & Sørensen, 2006; Lindstrand & Selander, 2009; Fink-Jensen & Nielsen, 2009; Johansen, 2018). While, according to Tavin (2007), Anglo-American researchers are often familiar with more general terms such as “aesthetic consciousness”, “aesthetic processes” and “aesthetic modes of knowing”, in Scandinavia ALP has been the favourite term to denote processes where learning occurs through sense-based and emotional approaches (Illeris, 2012). Recalling Friedrich Schiller’s early romantic work *On the Aesthetic Education of Man* (1795/2004), ALPs are basically understood as a playful form of mediation between the form-drive of the self-conscious enlightenment subject and the matter-drive of man as part of nature, living "in the here-and-now, carried along by the stream of time" (Hohr, 2002, p. 64).

Regarding educational practices, the concept of ALPs is used in two different ways: it is used in relation to the pedagogy of specific "aesthetic subjects", mainly the arts (visual art, music, theatre etc.), and it is used to indicate a dimension of all learning that can thus be connected to all themes and subjects (Lindstrand & Selander, 2009, p. 13; Fink-Jensen & Nielsen, 2009, pp. 192–220). The basic understandings are also twofold. In relation to arts education, ALPs have mainly been understood as a *mode of production* where learners are taught how to communicate their impressions of the world through symbolic form (Austring & Sørensen, 2006, p. 107). As a dimension of all subjects, ALPs are mainly considered a *mode of perception* where learners relate to the existing world through their senses to increase the emotional depth and inner motivation of the learning experience (Wickman & Jakobson, 2009, p. 130). As we shall see, this is especially important for environmental education where ALPs are considered a way of giving learners a deeper understanding of *nature*.

ALPs as a Mode of Production

Historically speaking, the concept of ALPs was developed by critical Scandinavian educational researchers in the early 1990s (cf. e.g. Løvlie, 1990; Schou, 1990). In arts education, ALPs became part of a paradigm shift where the child-centred pedagogy of the 1960s and 1970s, was substituted by research-based approaches inspired by critical theories of cognition and learning (Illeris, 2012). Of particular importance was the redefinition of the aesthetic experience in education from an open and playful approach to creativity, to a specific form of cognition that could be expressed through "form-producing activities" (Hohr & Pedersen, 1996). By combining Schiller's ideas of aesthetic education with Alfred Lorenzer's (1972) socialization theory and Susanne Langer (1960) symbol theory, the Norwegian educational researcher Hans Jörg Hohr explained ALP as experiential learning through the production of "created and intentional form", for example, design, play, ritual, dance, song, music, painting, film, theatre and literature (Hohr & Pedersen, 1996, p. 23).

In later publications on ALPs, such as Austring and Sørensen's *Aesthetics and Learning* (2006) the "aesthetic mode of learning" is translated into a model of didactic progression "where learners through aesthetic mediation transform their impressions of the world to aesthetic expressions of form allowing them to reflect and communicate about themselves and the world" (Austring & Sørensen, 2006, p. 107). Through three distinct phases, *impression* provided by direct sensation through contact with inner and outer realities, *expression* focused on the production of own pictures or other expressive forms, and *reflection* where learners communicate about their products, the subject learns how to communicate symbolically about his or her personal way of perceiving the world. In this way, ALPs contribute actively to education as a human-centred activity aiming at elevating personal sensuous impressions to interpersonal communication through form-giving activities where "the world" is understood as a surrounding that provides the subject with "impressions", or as inert matter, providing the subject with "materials" to be formed 2021.

ALPs as a Mode of Reception

When ALPs are understood as a dimension of all learning, sensuous experiences are more likely to be understood as having an educational value in themselves. In environmental and sustainability education, ALPs are often understood phenomenologically as related to sensuous experiences in, of, and with nature. While most programmes of environmental and sustainability education employ models of understanding from the natural sciences, ALPs emphasizes sensuous and emotional experiences which are able to loosen the separation between the human self and nature as a nonhuman form of existence. According to the Danish researcher and nature guide Lasse Thomas Edlev, in ALPs the learner

> has to be able to connect sensuously and to be emotionally open to new sides both of nature and of oneself, maybe even to be able to experience the oceanic feeling it can be when the separation between the I and nature loosens up. (Edlev, 2009, p. 20)

With reference to the concept of *mimesis* as developed by Danish existential phenomenologist Mogens Pahuus (1988), Edlev explains that:

> To sense natural phenomena [...] includes a kind of creative activity where the forms and rhythms of nature are bodily experienced, where the senses and the emotional register develops and where the ability to contact and

> modes of expression are trained. In this sense an experience of nature can be perceived as an 'aesthetic learning process'. (Edlev, 2009, p. 16)

The "ability to contact" should be trained to develop the senses and the emotional register as a value in itself. Following environmental phenomenologists like David Abram (1997), we here see a sensibility towards nature not only as surrounding, material or matter, but as a more-than-human form of being that learners can contact.

While in arts education ALPs as a mode of production tend to echo the enlightenment idea of using individual expression to elevate pupils from a state of passive, sense-based reception to a state of active communication (Bengtsson, 2019), the approach to ALPs as a mode of perception employed by environmental education is more open-ended and directed at the value of sensuous experience as related to nature. Inspired by existential phenomenology, the concept of mimesis is used to indicate a reciprocal process where the human body tunes to nature by assimilating its forms and rhythms (Pahuus, 1988). In environmental education, the form-giving imperative of arts education, where the scope of ALPs is to make learners become *more* human, is loosened in favour of ALPs as a way to let go of oneself, becoming more nature and *less* human.

II. Aesthetic Learning Processes as Attunement: Lying on the Ground

When a lies on the it is as close to as it might physically be, drawn down by the force of gravity. Lying down minimizes the use of physical strength and maximizes the use of other senses. It helps the to feel itself, it allows it to listen, to smell, to see and to touch and it gives room for dreams and , for relaxation and sleep. By being open to sensations and dreams, and by its passivity, defenselessness and , the lying on the ground connects to a realm of unspoken forces and intensities that operate beyond the symbolic messages of acting as "dead corpses".

In the first chapter of the anthology *Art in the Anthropocene*, the editors Heather Davis and Etienne Turpin (2015, p. 11) argue that in the Anthropocene, "Our sensorial and perceptive systems are being refashioned at rates that we can barely keep up with, as the world around us changes so rapidly", and that "our current climate demands a different kind of aesthetic and sensorial attention". According to them, we need art and aesthetics in order to learn to develop "techniques to begin to think through the limits of our temporal frameworks, and then thinking beyond them" (pp. 12–13).

In pedagogy, this means that we have to think of new ways to support the learner with the refashioning of their human perceptive systems, which will allow them to think beyond their own species, and to connect with other forms of existence on their own terms. In the new material reality of the Anthropocene, we need to find paths to re-embrace the matter drive that aesthetic education, according to Schiller (1795/2004), should help humans to control by elevating it to "play". What Kant (1798/1991, p. 54, quoted in Bengtsson, 2019, p. 68) saw as the innate "laziness and cowardice" of man, might now be praised as a human strength, forcing us to slow down our ever accelerating level of activity and instead become more receiving, more accepting, more earth-bound. In the words of the three Scandinavian authors of the book *Dark Pedagogy: Education, Horror and the Anthropocene* (2019), the self of Anthropocene *Bildung* needs to be "ecologized":

> The self at the core of Bildung is to be ecologized and turned into a primarily natural phenomenon. Or, to put it in another way, the cultural formation of self is to be rethought as a process of natural becoming within the nonhuman parameters of the environmental conditions and foundations of modern societies. (Lysgaard et al., 2019, p. 16)

The Bodies

I will now return to the bodies of the activists lying on the square in front of the parliament after the funeral of "the future as we know it", and discuss how to relate this action to a reconceptualization of ALPs.

If we follow the arts education approach, where ALPs are understood as a mode of production, the funeral can be seen as a playful way to give form to the young activists' feelings of anger and despair. By staging a ritual, they manage to communicate not only their political position, but their personal emotions related to climate changes. By lying on the square as "corpses", they symbolically show their solidarity with the dying planet and with people whose lives are already seriously affected by the changed human living-conditions of the Anthropocene era (Krogh, 2020; Skolelever i over 170 timers uafbrudt aktion, 2020).

If, in contrast, we follow the environmental education approach, where ALPs are seen as a mode of perception, the action becomes less about political communication and more about getting in touch with personal feelings. Here, the moment of lying down gains significance as a moment where the activists sensuously connect to the earth on which their future depends. From this position the less spectacular moment of just lying on the ground, showing and accepting bodily vulnerability, can be more important than the funeral itself. Instead of a strong, active and autonomous self, the body on the ground represents the greatest fear of many educators: the fear of doing nothing, of passivity, inertia, of pupils acting as objects without the individuality and autonomy so cherished in the *Didaktik*-tradition of European pedagogical thought (Lysgaard et al., 2019, p. 15).

The Little Death of the Self

Stefan Bengtsson, one of the authors of *Dark Pedagogy*, compares the "little death" of the passive subject to human fear in front of our own vulnerability:

> As the subject has to think in order to be a subject, the act of not-thinking might remind the subject of its *vulnerability*. We might re-encounter this fear of passivity and the little death in education with the pre-eminent demand of action and focus on activity. (Bengtsson, 2019, p. 69)

Bengtsson relates vulnerability to the (re-)opening of the porous and dependent kind of self that modern education has so eagerly tried to close by understanding education as the formation of a strong, coherent and enclosed self. Instead, the uncontrollable forces of the Anthropocene make us aware that the human self is not transparent, that it is out of our grasp, and that it cannot be given any authentic or new form. Realizing this causes an anxiety that, from a speculative point of view, might also be a reason for the grief of the young activists: What they have lost is not only the future of life as they know it, they are also losing the idea of self as they imagined it. Instead of the promised development of a coherent and capable self, able to design its own future by working hard and making reasonable choices, what they are actually dealing with when lying on the parliament square is what Bengtsson calls a twilight self:

> [...] Vulnerability relates to that feeling of dark, shimmering self not apparent to myself. When the self is reflecting on itself there is not mere nothingness, but rather a *twilight self*, a self barely visible in darkness. What the self finds is not an absolute nothingness, but rather a *presence* of absence. (Bengtsson, 2019, pp. 78–79)

Attunement as an Aesthetic Mode of Connecting

Theoretically *Dark Pedagogy* is driven by Object-Oriented Ontology (OOO, pronounced 'triple O'), a speculative realist approach[4] coined by Graham Harman (2011, 2018a) and Timothy Morton (2013, 2016). At the core of OOO is the recognition that *being* is not reserved for humans, or that the being of humans is not different from the existence of everything else: plants, bacteria, rocks, monsters or memories. In OOO all forms of being, even ideas and feelings, are understood equally as *objects*, and each object exists in two forms: the *real object* which is per definition withdrawn and unknown even to the object itself, and the *sensual object*, the object as it appears, and which will thus always be a correlate of the experience of another object (Harman, 2018a, pp. 78–80; Illeris, 2020, pp. 156, 162). An important thing to notice is that even if all things, including immaterial phenomena are objects, it does not mean that they are objects in the same way, or that all objects are real. On the contrary, each object has its own form of existence, which it does not share with any other object and that is withdrawn even from the object itself. In addition, there is no privileged interaction between objects, meaning that the interaction between two things is as important as the relationship between a human and a nonhuman form of existence (Harman, 2011, p. 6).[5]

From a background in literary theory, Morton has written extensively about the aesthetic experience in the light of OOO (Morton, 2013, 2018). To him, aesthetics is what happens when two objects unintentionally reach out towards each other, causing a third object, "the relationship", to come into being (Morton, 2013, p. 23). The reciprocal attention between objects happens through *attunement*, meaning that what Harman terms the *sensual object*s reach out to each other and link to each other's phenomenological qualities. Morton thus reinterprets Harman's description of the interaction between nonhuman forms of existence, seeing these interactions as aesthetic events:

> Aesthetic events are not limited to interactions between humans or between humans and painted canvases or between humans and sentences in dramas. They happen when a saw bites into a fresh piece of plywood. They happen when a worm oozes out of some wet soil. They happen when a massive object emits gravity waves. (Morton, 2013, s. 19–20)

From this quote, what was going on the square could be understood as an aesthetic event characterized by the attunement among objects of sensuous qualities such as temperature, consistence and surface. Attunement also incorporates what Morton (2018, p. 128) calls temporality formats: the long, slow, planetary time of the granite against the fast pulsing, short-lived human-biology time of the bodies. The history of excavation and elaboration of stone for human purposes against the stone's resistance, its hardness, its unwillingness to adapt but also its humble acceptance of lying there, paving the streets and squares, firstly in the old baroque city, and later on as part of rapidly changing urban trends, most recently as a part of terror protection measures (Danske Landskabsarkitekter, 2019). Compared to the human bodies that lie down, rise and proceed to live their short lives, the setts are *stayers*. Most humans would not expect them to act differently, to be anything apart from inert matter, but still one could argue that fighting against climate change is also a fight to give something back to the stones, to let them exist in their own right.

Thus, when I imagine the relationship between the bodies of the activists and the granite setts, I imagine the interaction between two forms of existence independently of the mind, attention or consciousness usually thought to be a prerequisite of such a connection. In Morton's vocabulary, the body and granite tune to each other independently of the fact that the thoughts of the activists are directed at something else, for example, at Greta's voice or the consequences of global warming.

ALPs as a Mode of Attunement

In the anthropocene era, human bodies are an inseparable part of geological and biological changes that now are happening so fast that the human time of the historical period intersects with the geological time of the Earth (Colebrook, 2017, p. 1). To grasp the significance of the

Anthropocene it is thus crucial that we as humans get in touch with the planet that we are a part of, not only as a resource for human expansion, but as a world of objects, each with their own mysterious form of existence. The right to be, to live and to die with dignity is not only a human right; it is planetary.

Following Morton's theory of attunement, ALPs can become possible modes of connection among humans and other forms of planetary existence and among nonhuman objects. By revitalizing the idea of the aesthetic experience as attunement between objects, we can find ways for pedagogy to reconceptualize learning as truly situated and practice-bound (Illeris, 2016).

As Davis and Turpin indicate, our modes of perception are changing. Almost without noticing, we have begun to *see* and to sense with whom we coexist and many of us are experimenting with possible reterritorializations of our ways of living. Like the young people on the square, we sense our aching bodies, exhausted from a life of contradictions and dilemmas between what we sense and how we live. In this situation, ALPs are not a mediation between human "drives", but rather a way to learn how to exist as indeterminate, vulnerable beings living in landscapes of multiple densities, temporalities and rhythms. If we stop ignoring everything that does not fit the time line of human progress, we can adopt ALPs as ways to revitalize possible forms of relationship between forms of existence though sensuous attunement and imagination.

Thus, when, for example, I try to understand a granite sett, I search for information from geology and urban history, but what I learn is not so much a series of facts as it is a possibility of attunement to another object. When I relate to the impossibility of imagining the movements that formed the earth and with it the granite, it seems almost impossible that right here and now I can strike the setts with my hand. I can perceive its inaccessibility and I can tune to the melancholia and joy of both of us being here, having sympathy, reaching out, although being apart.

On the one hand, the setts and my human body share some objects, certain minerals, for example, or our presence in the city of Copenhagen, but obviously, we do not share these relationships in the same way. Being a human, I have no chance to feel the minerals inside me, but knowing that my body hosts iron, magnesium, calcium and a number of other minerals, I can feel a kind of chemical solidarity with rocks:

> The aesthetic experience is about *solidarity* with what is given. It is a solidarity, a feeling of alreadiness, for no reason in particular, with no agenda in particular—like evolution, like the biosphere… (Morton, 2018, p. 121)

Following this thinking, one could imagine that human bodies and granite setts tuning to each other could indicate a way for ALPs, not as a mode of either production or reception, but as a *mode of attunement* to/with other forms of existence.

III. Proposition for Aesthetic Learning Processes in the Anthropocene

> I am lying on the grass. It is a cloudy day. A light rain touches the skin of my face. I can feel the cold of the through my clothes. Gravity is keeping my limbs down. In my ears I hear past voices of teachers telling me to relax. I follow the clouds with my eyes. My tunes to the wind. To the being of wind. Then to the humid ground. My hearing tunes to traffic, wind, birds. I imagine planet traveling through space at an incredible speed. I tune to this speed. The ground tunes to my.

Propositions

The above text is the result of a proposition that I offered to myself while writing this chapter. The proposition was formulated in these words: *Go outside and lay your body flat down on the ground. When lying there, let the objects around you tune to your body and let your body tune to them.*

During the last year, inspired by a/r/tography (Leblanc & Irwin, 2019) and by the SenseLab collective founded by Erin Manning (Manning & Massumi, 2014), I have worked with propositions as a didactic tool for ALPs in Sustainability Art Education (Illeris et al., 2022). Together with colleagues and students, I have explored how to make and enact propositions as "a theoretical lure or provocation that combines virtual potentials of the speculative imagination with the empirical dimensions of embodied experience in the actual world" (Roussel et al., 2018, p. 25). Inspired by the use of the term by Alfred North Whitehead (1929/1978), a proposition is a situated event, an opening that makes human and nonhuman experiences with and of the world take new directions, or as formulated by Manning in a chapter written together with Brian Massumi:

> The proposition is so much more than a statement. It is nothing less than a worlding. It is a serial iteration of the world's complexing, and re-complexing, of its own relational potential. The proposition is the force of thought gone worlding. (Manning & Massumi, 2020, s. 8)

In philosophy Morton's OOO-approach and the process-philosophy adopted by Manning and Massumi might be seen as difficult to bring together, but as a pedagogical approach, my experience is that propositions are able to draw attention to attunement in a very simple way without reducing the complexity of the event.

In my educational practice, a proposition is a simple proposal aimed at exploratory practices, in order for humans as well as nonhumans to tune to the complexity of the world as it exposes itself through innumerable objects of all kinds. To my students I have described propositions in the following way:

- A proposition is as an open invitation that someone/something offers to you.
- It is not an assignment but an occasion to open your worldviews and let them develop in new and unexpected directions.
- Instead of explain and simplify, a proposition maintains and explores complexity
- A proposition works from a premise of equality instead of hierarchy
- A proposition is an occasion to experience sensuous knowledge in the making
- A proposition is a practice, meaning that you can only create propositions by practising them yourself before you offer them to others

In the proposition *Go outside and lay your body flat down on the ground...*, the important thing is that your body can tune to the ground and vice versa, and that you allow your mind and self to become less while your body attunes to all the objects inside, outside and around you—including sensations, memories and wishes. This implies a suspension of action-oriented forms of being in favour of passiveness, something that we are used to seeing as embarrassing and painful. According to enlightenment philosophy "to be an object" is the worst you can be, but following the thoughts of OOO this proposition entails exactly this: the presence of the absence of being some*one*, of being a coherent self (Bengtsson, 2019, p. 79).

To me, practising this proposition has given something that is difficult to translate into words. On the scale of human mind, lying down has so many connotations (e.g. of sleep, death, sunbathing), and, of course, many of them appear in my mind when I do it. On the scale of my human body, lying down makes me feel heavy, my breath gets deeper, my senses open. On the scale of the objects tuning to me, yes, we connect or rather as sensuous objects we form a new form of existence, a new object, which I cannot capture at all. I become something like a me-body-grass-wind-air-ground-lips-birdsong-breath-cold-childhood-memories-and-much-more-object. But I also somehow become a lazy-shameful-uncomfortable-stupid-embarrassing-object in need of movement and escape.

Why this discomfort? In the Anthropocene, pedagogy is about opening a vulnerable rift between appearance and essence, and thus of rethinking and re-experiencing the human form of being a part of the world (Morton, 2013, 188; Bengtsson, 2019, p. 80). By enacting a funeral, the young FFF activists have yelled out their grief and desperation regarding the future of the planet. By lying down, they have shown their vulnerability and symbolic connectedness to death. Inspired by this action and by my own experiments with propositions, I am convinced that the Anthropocene calls for a pedagogy of death and of transformation. Pedagogy is the funeral, the first steps into the deep unknown of anthropocene forms of perception where human bodies and granite setts begin to sense each other, explore each other, pay tribute to each other. Although propositions might sound like doing very little, it might be a place to start with in order to re-calibrate human perception to the aesthetic reality of the Anthropocene.

Conclusion

Like the rest of society, the school of today is haunted by a constant and output-oriented demand of activity. Teachers have to make detailed programming of each lesson to "teach to the test" (Biesta, 2009). In such a regime, there is no time for students to delve mindlessly and await other forms of existence to tune to them in reciprocal exchanges.

In school, aesthetic learning processes, initially conceived of as a playful and sensuous form of human experience, are often reduced to stereotyped form-producing activities, at least when it comes to arts education. In environmental education ALPs risk being reduced to an add on to the prevailing science-based approach to nature. However, within Scandinavian environmental education, we also see the contours of a phenomenological approach where attunement to nature can be considered an ALP in itself, outside the artistic demand of expression and production.

Thinking with the action of FFF Denmark and with the young bodies lying on the ground after the symbolic funeral, the phenomenological conception of aesthetic learning processes as "mindless" attunement can be taken further in a reconceptualization of the relationship between all objects, human as well as nonhuman. If we bury the enlightenment idea of ALPs as processes where humans become *more* human by expressing themselves through material production, and substitute it with aesthetic learning processes as processes where humans learn to become *less* human and more attuned to all the objects with whom they co-exist, there is hope for a less anthropocentric (post-)anthropocene era. If we manage to operate with learning as a negative outcome, trying to do less, produce less, consume less, we might actually begin to live the way the young activists wish for: living our lives closer to the ground, to granite, to Earth.

Notes

1. All translations from texts in Scandinavian languages are by the author.
2. In the *Future of Humanity*-survey on the most important issues facing the world, conducted in 2019 by Ipsos MORI on behalf of Amnesty International with 10.000 informants aged 18–25 from 27 countries, 41% respondents selected climate change, followed by pollution (36%) and terrorism (31%). Among environmental issues, global warming ranked highest, at 57%. (Amnesty International, 2019).
3. By Scandinavia, I intend Denmark, Norway and Sweden.
4. Speculative realism is a contemporary philosophical current that aims to think reality in itself, independent of the habitual 'correlationism' of most modern philosophical thought (see also Harman, 2018b).

5. It should be noted that Harman (2011, Chap. 8) avoids the binary of withdrawn and sensual object through the construction of his own quadruple object diagram with twelve sets of relations between real and sensuous objects and real and sensuous qualities.

References

Abram, D. (1997). *The spell of the sensuous.* Vintage Books.

Amnesty International. (2019, December 10). Climate change ranks highest as vital issue of our time—Generation Z survey. https://www.amnesty.org/en/latest/news/2019/12/climate-change-ranks-highest-as-vital-issue-of-our-time/

Austring, B. D., & Sørensen, M. (2006). *Æstetik og læring. Grundbog om æstetiske læreprocesser* [Aesthetics and learning: Manual on aesthetic learning processes]. Hans Reitzels Forlag.

Barbiroglio, E. (2019, December 9). Generation Z fears climate change more than anything else. *Forbes.* https://www.forbes.com/sites/emanuelabarbiroglio/2019/12/09/generation-z-fears-climate-change-more-than-anything-else/?sh=70685b2a501b

Bengtsson, S. (2019). Death. In J. A. Lysgaard, S. Bengtsson, & M. H.-L. Laugesen (Eds.), *Dark pedagogy. Education, horror and the anthropocene* (pp. 63–83). Palgrave Macmillan.

Biesta, G. J. J. (2009). *Good education in an age of measurement: Ethics, politics, democracy.* Routledge.

Butler, J. (2015). *Precarious life. The powers of mourning and violence.* Verso.

Colebrook, C. (2017). We have always been post-anthropocene. The anthropocene counterfactual. In R. Grusin (Ed.), *Anthropocene feminism.* University of Minnesota Press.

Danske Landskabsarkitekter. (2019, October 15). *Cirkulær Økonomi på Christiansborg Slotsplads* [Circular economy at the square of Christiansborg castle]. https://www.landskabsarkitekter.dk/Aktuelt/cirkulaer-okonomi-pa-christiansborg-slotsplads/

Davis, H., & Turpin, E. (2015). Art & death: Lives between the fifth assessment & the sixth extinction. In H. Davis & E. Turpin (Eds.), *Art in the Anthropocene. Encounters among aesthetics, politics, environments and epistemologies* (pp. 3–30). Open Humanities Press.

Edlev, L. T. (2009). Naturoplevelse og naturfaglig interesse. Hvad er forholdet mellem æstetiske læreprocesser og naturfaglig undervisning? [Nature experience and environmental interest. How is the relation between aesthetic learning processes and environmental education?]. In K. Fink-Jensen & A. M. Nielsen (Eds.), *Æstetiske læreprocesser i teori og praksis [Aesthetic learning processes in theory and practice]* (pp. 13–18). Billesø og Balzer.

Fink-Jensen, K., & Nielsen, A. M. (2009). *Æstetiske læreprocesser i teori og praksis. [Aesthetic learning processes in theory and practice].* Billesø og Balzer.

Harman, G. (2011). *The quadruple object.* Zero Books.

Harman, G. (2018a). *Object-oriented ontology. A new theory of everything.* Pelican Books.

Harman, G. (2018b). *Speculative realism. An introduction.* Polity Press.

Hohr, H. (2002). Does beauty matter in education? Friedrich Schiller's neo-humanistic approach. *Journal of Curriculum Studies, 34*(1), 59–75.

Hohr, H., & Pedersen, K. (1996). *Perspektiver på æstetiske læreprocesser* [Perspectives on aesthetic learning processes]. Dansklærerforeningen.

Illeris, H. (2012). Aesthetic learning processes for the 21st century: Epistemology, didactics, performance. *Journal of the International Society for Teacher Education, 16*(1), 10–19.

Illeris, H. (2016). Adult education in art galleries: Inhabiting social criticism and change through transformative artistic practices. In D. E. Clover, K. Sanford, L. Bell, & K. Johnson (Eds.), *Museums and adult education: Animating social and cultural change* (pp. 229–242). Sense Publishers.

Illeris, H. (2020). Intimacy, solidarity, fragility. Everyday objects and ecological awareness in arts and crafts education. In K. Bergaust, R. Smite, & D. Silina (Eds.), *Oslofjord ecologies. Artistic research on environmental and social sustainability* (pp. 153–168). RIXC Center for New Media Culture.

Illeris, H., Knudsen, K. N., & Skregelid, S. (2022). *A/r/tografi som tilgang til udvikling af en sanselig bæredygtighedsdidaktik i kunstfagene* [A/r/tography as approach to the development of a sensory sustainability education in the arts]. Manuscript submitted for publication.

Johansen, M. B. (Ed.). (2018). *Æstetik og pædagogik* [Aesthetics and pedagogy]. Akademisk forlag.

Kant, I. (1991). What is enlightenment? In H. S. Reiss (Ed.), *Kant: Political writing.* Cambridge University Press. (Originally published 1798).

Krogh, M. A. (2020, October 6). Unge klimaaktivister: Vi er nødt til at se sorgen i øjnene [Young climate activists: we have to face our grief]. *Information.* https://www.information.dk/indland/2020/10/unge-klimaaktivister-noedt-se-sorgen-oejnene?lst_tag

Langer, S. (1960). *Philosophy in a new key. A study in the symbolism of reason, rite and art* (3rd ed.). Harvard University Press.

LeBlanc, N. & Irwin, R. L. (2019). A/r/tography. In Noblit, G. (Ed.). Oxford Research Encyclopedia of Education (n.p.). Oxford University Press. https://doi.org/10.1093/acrefore/9780190264093.013.393

Lindstrand, F., & Selander, S. (Eds.) (2009). *Estetiska lärprocesser: Upplevelser, praktiker och kunskapsformer* [Aesthetic learning processes: experiences, practices and forms of knowing]. Studentlitteratur.

Lorenzer, A. (1972). *Zur Begründung einer materialistischen Sozialisationstheorie* [On the necessity of a materialist socialisation theory]. Suhrkamp.

Løvlie, L. (1990). Den estetiske erfaring [The Aesthetic Experience]. *Nordisk Pedagogik, 1990*(1–2), 1–18.

Lysgaard, J. A., Bengtsson, S., & Laugesen, M. H.-L. (2019). *Dark pedagogy. Education, horror and the Anthropocene*. Palgrave Macmillan.

Manning, E. & Massumi, B. (2014). *Thought in the Act: Passages in the Ecology of Experience*. University of Minnesota Press

Manning, E., & Massumi, B. (2020). For a whiteheadian laboratory. How do you make yourself a proposition? In R. Faber, M. Halewood, & A. M. Davies (Eds.), *Propositions in the making: Experiments in a whiteheadian laboratory*. Lexington Books.

McMaster, G. (2020, January 28). Millennials and Gen Z are more anxious than previous generations: here's why. *Folio*. University of Alberta. https://www.ualberta.ca/folio/2020/01/millennials-and-gen-z-are-more-anxious-than-previous-generations-heres-why.html

Morton, T. (2013). *Realist magic: Objects, ontology, causality*. Open Humanities Press.

Morton, T. (2016). *Dark ecology. For a logic of future coexistence*. Columbia University Press.

Morton, T. (2018). *Being ecological*. Pelican Books.

Pahuus, M. (1988). *Naturen og den menneskelige natur* [Nature and human nature]. Philosophia.

Rousell, D., Cutcher, A. L., Cook, P. J., & Irwin, R. (2018). Propositions for an environmental arts pedagogy: A/r/tographic experimentations with movement and materiality. In A. Cutter-Mackenzie et al. (Eds.), *Research handbook on childhoodnature*. Springer.

Schou, L. R. (1990). Undervisningens æstetiske dimension [The aesthetic dimension of teaching]. *Nordisk Pedagogik, 1990*(1–2), 19–29.

Skoleelever i over 170 timers uafbrudt aktion for klimaet [Students in more than 170 hours of demonstration for the climate]. (2020, October 6). *Arbejderen*. https://arkiv.arbejderen.dk/indland/skoleelever-i-over-170-timers-uafbrudt-aktion-klimaet

Tavin, K. (2007). Eyes wide shut. The use and uselessness of the discourse of aesthetics in art education. *Art Education, 60*(2), 40–45.

von Schiller, J. C. F. (2004): *On the aesthetic education of man* (R. Snell, Trans.). Dover Publications. (Original work published 1795).

Whitehead, A. N. (1978). *Process and reality* (corrected ed.). The Free Press. (Originally published 1929)

Wickman, P.-O., & Jakobson, B. (2009). Estetiska lärprocesser i naturvetenskap—att behandla en förgiftning [Aesthetic learning processes in science education: To treat a poisoning]. In F. Lindstrand & S. Selander (Eds.), *Estetiska lärprocesser: upplevelser, praktiker och kunskapsformer [Aesthetic learning processes: Experiences, practices and forms of knowing]*. Studentlitteratur.

PART III

Interspecies Inclusion and Environmental Literacy

CHAPTER 10

Embodying the Earth: Environmental Pedagogy, Re-wilding Waterscapes and Human Consciousness

Shé M. Hawke and Reingard Spannring

Embodying Knowledge: An Introduction

In this chapter we argue that humans have a duty of care towards nature that must be more effectively and inclusively realised before further willful damage is unleashed on the planet. Our focus is specifically waterscapes in relation to how they interact with other natural, earth and cultural systems. As Steffen, Crutzen and McNeil declared in 2007 when naming the Anthropocene Epoch and its fallout, "the future of Earth's environment and its ability to provide the services required to maintain viable human

S. M. Hawke (✉)
Mediterranean Institute for Environmental Studies, Science and Research Centre of Koper, Koper, Slovenia
e-mail: she.m.hawke@zrs-kp.si

R. Spannring
Institute for Educational Sciences, University of Innsbruck, Innsbruck, Austria

M. Paulsen et al. (eds.), *Pedagogy in the Anthropocene*, Palgrave Studies in Educational Futures,
https://doi.org/10.1007/978-3-030-90980-2_10

civilizations" (614) and their inter-species and elemental interconnections, is compromised now more than in any "other comparable period" (617). We intentionally position nature—and in the case of waterscapes presented here—as both sentient and intelligent creators, and conveyors of knowledge—beyond most human conceptions of what constitutes sentience and intelligence—and complete with their own systems of signification and pedagogy, for which many humans hold only a rudimentary grasp. This is not to say that nature does not cause chaos and suffering itself, or have the "ability to act" (Bannon, 2020); the issue is more how human centric approaches view what nature is doing in and of its own cycles, and how the Anthropocene epoch has interfered dangerously with some of those cycles and inter-relationships, as the complex adaptive systems they are. In the context of this chapter, nature is positioned as the physical elemental, biotic, abiotic, floric, faunal and environmental aspects of the spectrum of all life, independent of and historically located in opposition to human culture. We aim to further dissolve this dangerous binary through a methodology that incorporates all life as equal parts of a web of inter-relationships.

Western management systems have until recently, overwhelmingly evolved to a place of working against nature not with it, or controlling and managing it according to primarily human economic needs. For example, the damming of rivers to support economic needs that are often out of balance with the capacity of the natural environment. This has resulted in the planet's diminished capacity to continue to provide the service of *all* life, that asks us to re-vision how and what we think we know. For this reason, we re-visit and engage with invitations from First Nations Indigenous pedagogy and ontology. Through this collective knowledge that pre-dates dominant Eurocentric approaches by over 40,000 years, different insights, values and practices are now more available for sustainable cross-cultural collaboration. However, as Stephen Muecke (2011) suggests, Indigenous representations of knowledge and measurements of value, do not fit neatly into "western phenomenological orthodoxy" (2) or disciplinary systems that have self-created a legitimised system of knowledge that has often been reductionist in terms of cultural diversity, and elemental respect. A new and inclusive lexicon and collaborative approach is currently needed so that attempts to learn from the broad spectrum of nature story and non-western cultures can expand, and isn't misread as soft anthropomorphism or cultural tokenism.

Maori scholar, Lewis Williams explains that, "The compulsions and ravages of modernity, Euro-centrism and colonial capital expansion on our ecology is a well-worn story and one that needs to be repeated here" (2012, p. 400), because such approaches have dictated a sustained disembodiment from elemental sources of life and culture. Our positioning, which also reflects the post-humanities turn in environmental education research, disrupts the human-centric western/industrialised belief that humans are the only species engaged in intelligent dissemination of knowledge. In this chapter, we offer the broad methodology of re-wilding in the specific context of "water literacy" (Hawke, 2012) via examples of three re-wilded waterscapes, and finally integration of that knowledge into participatory engagement with university students as citizen scientists.

We propose a re-wilding that invites all actors/agents into inter-species and elemental dialogue through: (a) affiliation, (b) being with more than human worlds and (c) deep listening. This methodology simultaneously supports the concept of embodying knowledge through conscious experiences with nature and biosocial entanglements, and dare we say it, through love of nature and desire to be more holistically affiliated with it and educated by it, through what Edward O Wilson calls "biophilia … an urge to affiliate with other life forms" (1984, p. 85) with love or affection. Being with, is embodied through conscious contact and connection with more-than-human life and worlds, through practices such as deep listening which is about being still and tuning in to the changing tones, murmurs and sounds of waterscapes and *their* companion species, such as croaking frogs and bird song.

This chapter responds to what we see as an Anthropocene urgency to affiliate, be with and listen to, what nature and particularly water are trying to teach us. We are motivated by the impoverishment of the planet and its waters that have been tamed, contained and hyper produced beyond sensible limits. Equally, while the commercialized and touristic lens encourages people to 'look at' and visit natural environments, this often increases pollutant risk. This can paradoxically limit capacity to spend more conscious and intimate time with waterscapes.

Fundamentally, we define re-wilding as low impact, environmentally appropriate and conscientious human intervention and restoration of pre–existing wild areas that have suffered under sustained 'out-of-balance' human activity. This approach is also described by the European Commission's (EUC) "Biodiversity strategy for 2030 Section 2.2.7", where it holds out the prospect of restoring at least 25,000 kilometres of freshwater ecosystems to a 'free flowing state'.[1] So too, in Australia the World Wildlife Fund (WWF) has partnered with science and Indigenous Ecological Knowledge "to restore Australia's ecosystems and missing faunal links" by re-introducing keystone species through the Rewilding Australia initiative.[2] For the re-wilding component of our chapter, we call not only for a re-wilding of the environment and water specifically, but also for re-wilding the environment and "ecology of the mind" (Bateson, 2000), or human consciousness. In this way, we open up the space for more-than-human knowledge to inform how we re-educate humans for sustainable futures, by embedding the components of water literacy, alongside respectful engagement with Other knowledge. We will illustrate the water literacy/re-wilding approach with examples of waterscapes that are understood bilingually as embodying elemental, economic and socio-cultural value and knowledge, and then adapt the re-wilding concept and water literacy method to educational settings.

By connecting these streams of human consciousness and embodying them in a modern context, we also extend the notion of 'being with', drawn from our engagement with the lived experience and scholarly work of First Nations Indigenous Peoples. Concomitantly, we attempt to dismantle the over instructional ways that past and present Western pedagogy has taught and 'tamed' young minds, through a primarily Eurocentric, industrialized/commodification framework, and through related shortcomings in educational and knowledge practices. As Deborah Bird Rose suggests, a "border zone in which Indigenous ecological knowledge, Western scientific knowledge and Western philosophical and poetic inquiry converge" (2007, p. 9) is needed for a new ethos to develop. We explore these issues in our Pilot Case Study presented below.

Locating Time, Place, Water Literacy and Posthumanism

In 2012 after meeting with First Nations Indigenous scholar Professor Uncle Roy C. Gordon from the Far North Coast of NSW Australia, Widjabul Country, the term "water literacy" emerged (Hawke, 2012). Uncle Roy is an elder and water guardian of the area, with whom his ancestors have had kinship and guardianship responsibilities for over 40,000 years. He shared his people's experience and the 'look, listen and learn' (Bragg et al., 2007, p.4) approach that we apply in this chapter through the 'language of water', of 'being with' water, and embodying love and respect for it. Uncle Roy shared about the importance of knowing and sharing the water's being, and tangible and intangible values to a more general settler-descended audience bi-lingually (ibid.), through his collaboration with the "Sustainable Futures Project" (Bragg et al., 2007).

As we shall discover, this was not the only or first iteration/variation of the term 'water literacy'. As water issues and conflicts around water become more salient, 'water literacy' has become a widely adopted term. It now appears across diverse fields of scholarly inquiry and is in use in governments, municipalities, funding bodies and NGOs, although it is applied variously across these contexts and it often lacks a clear definition or description (McCarroll & Hamann, 2020). The most general and unspecified definitions (in their review of fifty-five articles), refer to water literacy as "water-related knowledge" (ibid., p. 9). More specific definitions refer to different kinds of cognitive and scientific knowledge sets, attitudes and values, cross-cultural meanings, and behaviour, both individual and collective. Overall, there is a strong bias towards anthropocentric water needs (primarily economic), and an overemphasis on the cognitive dimension and Western (or industrialized) scientific approaches to knowledge and individualized forms of learning rather than collective learning and transformation. They therefore suggest strengthening place-based learning approaches to foster local and hydro-social knowledge, interdisciplinarity and "deep cultural and eco-cosmological literacies" (ibid., p. 22) that convey historical hydrologies, cultural traditions as well as spiritual and ethics-based knowledges (Hawke, 2012). The latter is imbricated in the method of the "Sustainable Futures Project" (2007) described here, and the method we employ in our research and in our student Pilot Case Study. For our purposes, we divide the water theme and its values more simply into three basic spheres or flows that have particular under currents: ecological (biospheric), economic (egoistic/human centred) and socio-cultural (traditional) (Hawke & Palsson, 2017, p. 235; Jackson et al., 2019; Gratani et al., 2016; Jackson, 2006). This tripartite view enables us to examine more clearly the complex interactions between water values and understanding, rather than to privilege one aspect over another.

We further extend this view to include neglected or under-represented aspects of water literacy as echoed by other research. While framing concerns differently, such research foregrounds the role of outdoor experiences (Cachelin et al., 2009; Liefländer et al., 2013), which we further reference as embodied experiences of 'being with'. We also recognise the overwhelming and problematic dominance of the anthropocentric and utilitarian belief system, as others have done (Corral-Verdugo et al., 2003), which complicates methods of recognising water as its "own self" (Bird Rose, 2007, p. 18), as part of a "sacred geography", that connects organic and non-organic entities in a more-than-human world (Twance, 2019).[3]

Part of our scholarly, pedagogical and participatory intention is to enact a shift in consciousness and praxis in everyday life by partnering with nature, by engaging youth and communities in re-wilding consciousness and activities, and by developing eco-pedagogical capacity through water and environmental literacy. The concept of re-wilding is drawn originally from Deep Ecology and living wilderness knowledge in nature and in human consciousness (Seed et al., 1988; Bateson, 2000/1972; Carson, 1962; Leopold, 1949; Plumwood, 1993). We argue for a confluence of these ways of knowing through water examples that also connect with traditional First Nations Indigenous ecological knowledge. Each example emphasizes a different aspect of re-wilding and cross-cultural water literacy. The pedagogical service and wisdom that this "natural library" (Hawke, 2012, p. 239) provides, recognizes and exemplifies the movement from wild, to tame to re-wilding of thinking, along with the interconnectedness and affiliations of diverse communities in and with their waterscapes through this process.

So, let us take a closer look at what human vernacular might call a "natural library" in which all elements and species could represent specific "canons of knowledge" (Hawke, 2012, p. 239) and in which water might be considered a primary text for learning and sharing. Some of the 'books' in this body of natural knowledge are rare texts indeed. As the Anthropocene Epoch demonstrates, we have increasingly become disconnected from our elemental sources, so much so, that we now need expert and diverse translatory help to decipher what it is nature is trying to teach us, and how more holistically we might read nature. As Makere Stewart-Harawira says, "increasingly scholarship is re-turning to First Nations knowledge, not as a panacea for all the environmental ills of the world, but as an alternative way" (2012, p. 74). The students in our cohort clearly identified the gaps in learning and some of what we/they do not know, or have forgotten.

This further situates our approach in the literature on water literacy and within posthumanist environmental education research. Posthumanism is implied in the environmental and animal rights movement, which along with other progressive movements such as anti-colonialist, antiracist, antisexist, and LGBTQ inclusion, strives to give a voice to humanism's Others. It seeks to disable the dualisms and hierarchies that enable the marginalization and exploitation of what the humanist tradition deems only partially human, subhuman or nonhuman. Posthumanism is not anti-humanist since it retains the modernist notions of rights, justice, equity and freedom. However, it does open up a new perspective that de-centres the human, focuses on inter-relationships and connectivity and thereby disrupts the dominant discourse of educational humanism that focuses on the production of the human over and against what it perceives as the Other (Snaza & Weaver, 2015). This raises critical questions such as: How can we conceive of education if the learning outcome is not predetermined; and how can we foster learning that is not reduced to object appropriation but acknowledged as an open process of co-evolution with other knowing human and nonhuman subjects and agents (Spannring, 2019) that are part of a complex adaptive system of inter-relation (Spannring & Hawke, 2021)?

These questions are based on our understanding that nature emerges *with* humans and the world (cf. Taylor, 2017), thereby de-centering the human as the sole learning subject. This further advances the notion that "water [and by extension nature] is an agent in its own becoming" (Hawke & Palsson, 2017, p. 240), and has the capacity to exchange *its own* version of knowledge. The aim of such an approach is to re-discover what it means to live with nonhuman others (Crinall & Somerville, 2020), attending to the inextricably entangled life-worlds and vulnerabilities (Taylor, 2017), in specific moments of embodied, curious and creative mutual "becomings" (Haraway, 2008). This requires a method that goes beyond Cartesian rationalist modes of being and knowing (Hawke & Palsson, 2017, pp. 235–236), and is unconstrained by fixed methods and forestalled results and reduction to only a human knowing subject (Snaza & Weaver, 2015). Such methods have punctuated the world since the beginning of the Industrial Revolution. The more-than-human paradigmatic shift, leads us to an open-ended research/learning process that emerges through "exchanges between bodies and places in everyday lives" and experiences of intensities and sensations (Crinall & Somerville, 2020, p. 10), in which flourishing for all life may be more possible.

Adopting a humble and curious stance towards waterscapes in a slow pedagogy of place affords researchers and young people the space to "think with and listen to water", to "become worldly with water", its fluidity, creativity, connectivity, gestationality, unknowability and destructiveness (Pacini-Ketchabaw & Clark, 2016). To understand best how to invite students and youth, and indeed our own colleagues into this slow pedagogy and sense of 'being with', we engage in an applied cartography of shifts in consciousness. Yet as we shall see by the student commentary at the end of this paper, invitation into such a multi-species, elemental and cross-cultural dialogue disrupts conventional ways of knowing. While this was our aim, the Pilot Study demonstrates that rigid instruction and measurable outcomes have been so concretised in teaching and learning that introducing some freedom in co-created instructional design was confronting, and further that the measuring of that freedom was difficult to quantify.

From Wild, to Tame to Re-wilding

The Anthropocene epoch and its hyper production (Steffen et al., 2007), has cemented some critically shocking behaviours, that have de-wilded, un-balanced and over-exploited many natural ecosystems. How we break redundant patterns of knowledge that have tamed natural or pre-existent wild systems to serve this epoch, and how we travel forward taking the best and the worst knowledge from the past as our teacher, is imperative. How do we now re-new our perception and re-wild our thinking and waterscapes towards an inclusive and boldly un-tame 'best practice' in the twenty-first century? As well as looking at the recent flourishing of re-wilding, we also look to re-wilding consciousness from the past.

Re-membering great teachers like Copernicus from the late 1400s can remind us how opening up our thinking about the universe and the planet could sustainably change things. Copernicus asserted that the earth was not actually the centre of the universe, and therefore neither was 'man'. A century later Descartes sought to reduce everything to the neat tidiness of mathematical equations and binary oppositions, which turned out not to be a good fit for nature. Further along, Rachel Carson's *Silent Spring* (1962), arguably signifies the birth of the deep ecology movement which later inspired Arne Næss (1989) in his ecosophy approach. Carson was writing at a time when women in science were rare and easily dismissed. Joanna Macey (1983), author of *Despair and Personal Power in the Nuclear Age*, further provoked environmental communities around the world to speak up about injustices to nature through reckless industrialisation and its toxic outflow of effects, and the simultaneous injustice to human beings. She also co-authored with Næss in the revolutionary book,

Thinking Like a Mountain: Towards a Council of All beings (1989). She co-designed inter-connected approaches of how to cope with "planetary perils", and voiced a broadly held experience that "Feelings of pain for our own world are natural and healthy" (Macey, 1983, pp. 22–23) if dealt with effectively. We take our lead from the solid epistemological and physical ground of these past innovations (that recognized a de-naturing of the environmental/ecological world), in partnerships with First Nations ecological knowledge. In so doing, we aim to further grow and transform new ways of thinking, being, and doing critical and new environmental pedagogy, and the healing to the environment that accompanies such a collaborative pathway.

How we empower/inspire students/colleagues to stay awake to new possibilities and to imagine new ways of being, absolutely involves going out with nature, dis-integrating the finitude of the classroom but complementing its ample technological resources. Change, transformation and empowerment depend on creating new meaning and taking a risk with our thinking that must be intelligent and inclusive of different world views and lived experience. We now turn to recent examples where re-wilding human consciousness and waterscapes has created a shift in which water literacy, ecological, economic and cultural values are evident.

The "Sustainable Futures Project", (Bragg et al., 2007) which elder Prof Uncle Roy Gordon was a part of, "[c]onstitutes a multi-literate and collaborative project about water sustainability involving the local water authority of Lismore, Rous Water, NSW, Australia, and The Widjabul/Bundjalung peoples and their multi-generational Knowledge that includes resource management as well as sacred ontology and cosmogeny" (Hawke, 2012, p. 238). This initiative combines the discursive and classically western pedagogical material with earlier Traditional Knowledge made accessible in "The Water Walk: A Users Guide to the Far North Coast Water Cycle and the Rous Water Walk at Emigrant Creek" (Bragg et al., 2007). In addition, walks at the local water place (Emigrant Creek Dam and Rocky Creek Dam), guided by a representative from the local water authority (Rous Water) and a Widjabul elder were made available to participants.[4] This enabled the hearing of knowledge and science from two world views now working together to sustain not only humans, but all the life in and around that river/dam system, and also to listen to the water. Widjabul elder Auntie June Gordon suggests of this coalition in which water is the meeting place between two world views; it is also an opportunity for reconciliation with nature and with each other, in other words:

"'social sustainability' cannot be separated from 'environmental sustainability'"(2007, p. 10). It may not be possible to undam this waterscape, however, this example demonstrates, an undamming of 'tamer' thinking, and re-wilding it from the rich depository of knowledge from a culture over 40,000 years old, and in confluence with innovative modern scientific approaches, towards shifts in consciousness that shine broader light on perceptions of water.

We see another example of this re-wilded, cross-cultural reconciliation method in Aotearoa/New Zealand in the case of *Te Awa Tupua*/ Whanganui River (in Pakeha/English). This river held special cultural significance for the local *Iwi* (tribe, clan), and economic significance for settler-descendants. After a decades long process for recognition of its elemental and cultural value, this body of water was eventually accorded "legal personage", sovereign status on 5 August 2014, later ratified in 2017 (Ruru, 2018; Strang, 2020) because both settler-descended people (white/migrant) and Maori people recognised the river as a living being with power and agency of its own, and its own system of knowledge transfer. This ground breaking case makes clear that while water can be sacred, it is also involved in the physical life and sustenance that it provides for both neighbouring ecosystems and people; hence, it has ecological value, economic value and cultural and spiritual value. The success of the river to attain its sovereignty is an example of cross-cultural collaboration embedded with a re-wilding consciousness, and embodied engagement in and with rivers. These two examples provide hope for a wilder consciousness about waterscapes to further evolve along natural lines.

Our last example is the Elwha River waterscape of the Lower Elwha Klallam Tribe, USA, which had been dammed by 'settler descended' humans to serve the perceived needs of colonial/capitalist industry for a hundred years. The fish ladders for the salmon that had been promised when the dams were built, never materialized. The subsequent prevention of native salmon jumping upstream to their spawning ground had a devastating effect on their population and flow-on effects on the whole ecosystem and the economic and cultural life of the First Nations people of the area. However, "*some* salmon still kept on jumping against that concrete wall … year after year. They simply didn't give up trying" (Müller, 2020, p. 62, orig. emphasis), using their own communicative strategies to enact their desired result. Between 2007 and 2014 local environmental activists and First Nations Indigenous peoples came together to continue the decades-long quest for the removal of the two dams, for ecological as well as cultural reasons, including the sustainable and healthy supply of salmon. The eventual demolition of the dams demonstrates not only a re-wilding of human consciousness, but in practicality, the re-wilding of the river through community engagement in which cross-cultural knowledge and pedagogy was effectively shared. This further illustrates water literacy, affiliation and the deep listening capacity and actions of the local communities, "who finally understood what the salmon had been saying all along" (ibid., p. 72) that the river needed to flow. As a consequence, it became possible for natural flows, eddies, rapids and pools to resume a dynamic co-evolvement, and for humans to become more "water literate" in the process (Hawke, 2012; Spannring & Hawke, 2021).

Student Participatory Engagement: Concepts, Context and Practice

To further discover ways of knowing around water, we co-designed a seminar "Education for Sustainable Development" in winter term 2020/2021 at University of Innsbruck that focused on the hydrological cycle and waterscapes. The seminar was part of a supplementary study programme "Sustainability," which is open to students from diverse faculties. The cohort comprised mainly bachelor students with some master students participating (See Table 10.1). Due to timetabling conflicts, 8 of the

Table 10.1 Demographic data on student participatory engagement

Gender	*Initial*	*Water body*	*Discipline*	*Education level*
Male	D	Kranebitter Gorge, Innsbruck, Austria	Biology	MA
Male	P	Tauern National Park, Carinthia, Austria	Biology	BA
Female	V	Lake Lutten Bavaria, Germany	Geography	BA
Female	Ju	Isar Rive & Gleirsch Gorge, Bavaria, Germany	Economics	MA
Female	C	Ice waterfall near Innsbruck, Austria	Education Science	BA
Male	Jo	Pragser Lake, Dolomites, Italy	Accounting	MA
Female	M	Rhine, Switzerland	Economics	BA
Female	S	No walk, Austria	Teacher training (biology)	BA

original 12 completed the course, which was taught online due to COVID. We provided material from First Nations Indigenous scholars, such as Lewis Williams who suggests "every person on this planet has the innate human capacity to be Indigenous: that is to be in intimate relationship or resonance with the world of spirit, the earth and other human beings" (Williams, 2012, p. 93). In addition, we presented re-wilded thinking and river and landscape examples of *Te Awa Tupua*, Sustainable Futures and Elwha River and scaffolded the information into the course design and delivery. Some of the students were exposed to Indigenous cultural, elemental and water knowledge for the first time, and learned of how it might benefit the re-wilding and sustainability practices of the future as our analysis demonstrates.

The students were introduced to the topic with an overview of water-related problems and hydro-social issues with a strong emphasis on culture, power and ethics. They were given additional reading (Simmons et al., 2007; Müller, 2020; King, 2020; Harding, 2020; Bannon, 2020; Strang, 2020; Hawke, 2012), which they discussed in small groups. They also received lectures that introduced them to water literacy and cross-cultural water engagement through the thematics, and examples we use in this chapter. Their task had three phases:

- Phase 1: To meet with and explore a familiar waterscape where they were encouraged to 'be with' and 'listen to' water and to develop an understanding of 'affiliation'
- Phase 2: To write a scholarly but self-reflexive essay based on their experience and the literature they were provided.
- Phase 3: To share their embodied experiences to the cohort through a multi-media presentation, and explained connections with scholarly material and their developing understanding

The observations and analysis we offer in this chapter are based on the assignment and the group discussions, recorded, transcribed and condensed here to connect with our aim of discovering student capacity for embodying affiliation, being with, deep listening and water literacy.

A critical finding was that the majority of students found this new participatory pedagogical approach both challenging (75%) and beneficial (81%). For example, **D** says: "With the task I was a little overwhelmed … because I am a dry natural scientist after all." **P** expressed similar sentiment by stating: "We are already so designed that we get guidelines (so and so has to be done). When you have to create your own reference to it, it's really challenging."

A vast majority of students (81%), reported that the approach was beneficial. For example, **S** says, "This purely scientific approach to date does not cover topics for the average citizen." And for **D**: "Since one of my main areas of studies in biology is limnology and the ecology of flowing waters, I didn't have to learn much from a technical point of view, but once again I was able to take a closer look at the perspectives of non-scientists." And **Jo**, who says: "The seminar has strengthened my resolve to continue to hold on to what I believe, for example, not to see water as a commodity … Yes, also the access to water …. To connect water with home." **M** also noted the benefits of relating water to home: "That the knowledge of Indigenous people in dealing with nature and water can be incorporated into future renaturation projects … in our home environment."

A majority of students demonstrated understanding and/or action of affiliation with water (69%). For example, **D** reported that more can be learned from nature than at a desk and quoted Lewis Carroll to express his sense of affiliation: "I wonder if the snow loves the trees and fields, that kisses them so gently." The experience of **S** on affiliation is reflected in the recent birth of her baby. In her presentation she showed a clip of the baby bathing and reported on the "evolutionary biological journey" and the womb as a primordial connection to water.

The vast majority of students demonstrated sophisticated understanding and/or action of 'being with' water (81%). For example, **V** who went with her dog, reported being "emotionally charged" and acknowledged, "emotion and water are always connected and … can have different meanings for everyone." **C** describes her phenomenological and embodied response through the journey of icicles: "I was out for a walk … nothing special grows along this forest path, but because it had rained, a real icicle landscape formed … Icicles are somehow always in process … I can totally switch off there " And **S** in relation to her new born and what it brought up for her in relation to the unit, says: "the likeness of fins and fingers in different historical species, the likeness of shark and human embryos", which combines 'being with' and 'water literacy'.

'Deep Listening' (45%), or being still with the water and leaning in to its sounds, was only noted in a few comments. **P**, who took his dog with him to the Liesen River, recalls the legend: "that gold seekers had not been prepared to listen to the water [Liesen means listen in the regional dialect] and were killed and carried away by a gush of water" And **Ju**: "Even when I go [to Isar River and Gleirsch Gorge] over and over again, I am speechless at the movements in nature."

All students also demonstrated a new grasp of (and inspiration by) cross-cultural approaches to water within the context of water literacy and teaching. For example, **Jo** says: "I was particularly impressed by the Maori understanding of rivers, lakes, mountains and forests. The indigenous people of New Zealand see themselves as a small part of a delicate cycle … If one part is disturbed, this … brings it into imbalance." **V** also reported: "I particularly liked the approach of getting to know … the meaning of water in other cultures," but added that the " 'indigenous' or ethical approach would be great" but she was not sure that would be accepted by everyone.

Indeed, **Jo** and **V** specifically problematized over-tourism as examples for a Western lifestyle that violates and prevents a holistic water-literate approach. Both point to the particular role of the social media/influencers and the selfie-culture in jeopardizing nature reserves: "Social media turn this and other lakes into places of pilgrimage for the perfect selfie" (**V**). "For the perfect shot—although they have nothing to do with nature, and actually abuse it a bit… sometimes the site sees 15,000 people a day." (**Jo**)

For **M**, water literacy was demonstrated through understanding processes and cycles: "Re-wilding rivers means—giving nature a space to unfold again, that is then backwards but somehow also forward again [to return] nature back [to] its original form … at least halfway". "Within the literature … there was much about Indigenous knowledge … and I just asked myself: has this knowledge been lost somewhere?" **S** adds to this dimension with a metaphor: "We work like a river enclosed on a concrete bed, and perhaps we should let ourselves be guided back to working like a natural river that somehow meanders along, bumps into something, spreads out, narrows, which is actually how thoughts should flow."

In relation to the more than human approach, two respondents were accompanied by their dogs as "companion species" (Haraway, 2003). Some respondents also noted the interconnections between different organic and inorganic life forms, for example, **D**: "Abiotic and biotic cannot exist without each other", and **P**: "I was able to record some beautiful moments, the symbiosis that exists between animals and plants." Conversely, **M** noted the disembodied connections created by the Anthropocene: "'The straightening of the river …shows that the anthropocentric influence on nature is well in the human consciousness … it is primarily about setting priorities and that we humans are definitely not in balance with." **Ju** also notes the impact of the Anthropocene particularly and says "In both places [Isar River and Gleirsch Gorge] you are in the middle of nature, the mountains, the rivers, the flora … The water has immense power and shapes the river bottom … you can see the traces of the water in the sand." In conjunction with that observation of the "lively space … draws attention … to the sensitive structure… It would be gratifying if many more people were more sensitive to the impact of our actions."

What we can ascertain from this Pilot Group is that there is a keen willingness to engage with nature differently, given the opportunity, and to think about the interconnections between not only water, but flora and fauna and ourselves. As **S** aptly says, demonstrating all facets of the participatory pedagogical approach: "I liked the emotional approach to this sustainability topic, especially water … it simply brings these different disciplines, which the students now bring with them, a very multi-faceted picture". However, despite their enthusiasm they also hinted at the difficulty of the dominance of anthropocentric logic. This pertains not only to economics, politics and culture, but also and above all to the educational system that needs to dismantle its concrete bed and allow students' learning to flow like a natural river, as **S**' metaphor intimates.

Conclusion

The student cohort, while small, provides a critical snapshot of what is possible in how we transform educational practices and grow an affiliation with the elemental world that sustains itself and us. In addition to transformed thinking, and a re-newed 'ecology of mind' (Bateson, 2000), a new lexicon and open-ended approach is emerging. Combined with the three water examples of re-wilding that we shared with the students, and with the broader introduction to water literacy through affiliation, being with and deep listening, we can establish a clearer eco-pedagogical future that builds on the architecture of what we have discovered here. It is clear, that lessons from the past and the present are part of a more sustainable future.

Based on the remarkable response from our student cohort, in conjunction with our prior inter-disciplinary research, we have now begun to develop an AquaMooc, which is a massive open online course. It serves as a learning and research platform open to the public and that can be used in upper secondary school and university undergraduate programmes.[5] The creative industry of the students will be embedded into this learning platform, based on their experiences with re-meeting and learning with their local waterscape. Further, we will enhance the AquaMooc with the literature that supports this innovative way of learning and teaching. We believe the research would benefit from a longitudinal study with a larger sample and possibly different learning settings to augment our cross-cultural and more than human approach, and understand its broader transformative potential.

Re-visioning education, while putting it into a more critically aware framework can sustainably alter how communities and students can converge in support of all life on planet earth through education reform and perhaps towards an 'ecodemocracy' (Kopnina et al., 2021). Through this re-meeting with waterscapes that we have described, we offer possibilities of advancing the pedagogical method to include all of nature, broadly understood. If humans can present more as learners of the natural environment rather than as owners, then true stewardship and intelligent pathways will become more evident, as we lean in to know better what that natural library is teaching us. How humans serve the 75% of the planet that is water, along with its elemental and multi-species interconnections, forms part of the necessary expanded human consciousness that we call for here.

Acknowledgements This chapter was funded by the ARRS (Slovenian Research Agency, project J7-1824) and the FWF (Austrian Science Foundation; project I 4342-G) on the joint project "Surviving the Anthropocene through Inventing New Ecological Justice and Biosocial Philosophical Literacy". The authors wish to thank the students in the Education for Sustainable Development cohort, 2020/2021., from the University of Innsbruck, Austria

Notes

1. https://ec.europa.eu/info/index_en, accessed 20 November 2020. Along with shifts in policy and funding organisations towards rewilding, there are now numerous papers and examples available that talk about re-wilding in a modern context (e.g. 'Rewilding Scotland' https://www.rewild.scot/explore-rewinding; and https://www.euronews.com/living/2021/04/26/scotland-could-become-the-world-s-first-rewilding-nation-how-did-they-get-here), accessed 28 April 2021, including chapters within this volume.

2. https://www.wwf.org.au/what-we-do/rewilding-australia, accessed 22 February 2021.
3. See also Washington et al. (2017) for a broader discussion on the necessity of ecocentrism as the key pathway to sustainability.
4. https://rous.nsw.gov.au/the-water-walk-rocky-creek-dam, accessed 23 March 2021.
5. Please also see our project website: https://www.uibk.ac.at/bgl/surviving-the-anthropocene/.

References

Bannon, B. (2020). The bonding properties of water: Community, urban river restoration, and non-human agency. In I. Leman Stefanovic (Ed.), *The wonder of water. Lived experience, policy, and practice* (pp. 151–170). Toronto University Press.

Bateson, G. (2000 [1972]). *Steps towards an ecology of mind*. Chicago University Press.

Bird Rose, D. (2007). Justice and longing. In E. Potter, A. Mackinnon, S. McKenzie, & J. McKay (Eds.), *Fresh water: New perspectives on water in Australia* (pp. 8–20). Melbourne University Publishing.

Bragg, E., Acret, A., & Gordon, R. C. (Eds.). (2007). *The water walk at Emigrant Creek Dam: A users guide to the far North Coast Water Cycle*. Rous Water. https://rous.nsw.gov.au/the-water-walk-rocky-creek-dam

Cachelin, A., Paisley, K., & Blanchard, A. (2009). Using the significant life experience framework to inform program evaluation: The nature conservancy's wings & water wetlands education program. *The Journal of Environmental Education, 40*(2), 2–14. https://doi.org/10.3200/JOEE.40.2.2-14

Carson, R. (1962). *Silent spring*. Penguin.

Corral-Verdugo, V., Bechtel, R. B., & Fraijo-Sing, B. (2003). Environmental beliefs and water conservation: An empirical study. *Journal of Environmental Psychology, 23*(3), 247–257. https://doi.org/10.1016/S0272-4944(02)00086-5

Crinall, S., & Somerville, M. (2020). Informal environmental learning: The sustaining nature of daily child/water/dirt relations. *Environmental Education Research, 26*(9–10), 1313–1324. https://doi.org/10.1080/13504622.2019.1577953

Gratani, M., Sutton, S., & G., Butler, J. R. A., Bohensky, E. L. & Foale, S. (2016). Indigenous environmental values as human values. *Cogent Social Sciences, 2*(1), 1185811.

Haraway, D. (2003). *The companion species manifesto: Dogs, people, and significant otherness*. Prickly-Paradigm Press.

Haraway, D. (2008). *When species meet*. University of Minnesota Press.

Harding, S. (2020). Water Gaia: Towards a scientific phenomenology of water. In I. L. Stefanovic (Ed.), *The wonder of water. Lived experience, policy, and practice* (pp. 13–26). Toronto University Press.

Hawke, S. (2012). Water literacy: An 'other wise', active and cross-cultural approach to pedagogy, sustainability and human rights. *Continuum, 26*, 235–247. https://doi.org/10.1080/10304312.2012.664120

Hawke, S., & Palsson, G. (2017). Water futures, biosociality and otherwise agency. *ANUAC, 6*(1), 233–252. https://doi.org/10.7340/anuac2239-625X-2838

Jackson, M., Stewart, R. A., & Beal, C. D. (2019). Identifying and overcoming barriers to collaborative sustainable water governance in remote Australian Indigenous Communities. *Water, 11*(11), 2410. https://doi.org/10.3390/w11112410

Jackson, S. (2006). Compartmentalising culture: The articulation and consideration of Indigenous values in water resource management. *Australian Geographer, 37*, 19–31. https://doi.org/10.1080/00049180500511947

King, S. J. (2020). What we are talking about when we are talking about water: race, imperial politics and ruination in Flint, Michigan. In I. Leman Stefanovic (Ed.), *The wonder of water. Lived experience, policy, and practice* (pp. 133–146). Toronto University Press.

Kopnina, H., et al. (2021). Ecodemocracy in practice: Exploration of debates on limits and possibilities of addressing environmental challenges within democratic systems. *Visions for Sustainability, 15*, 9–23.

Leopold, A. (2000 [1949]). *A sand county almanac: And sketches here and there*. Oxford University Press.

Liefländer, A. K., Fröhlich, G., Bogner, F. X., & Schultz, P. W. (2013). Promoting connectedness with nature through environmental education. *Environmental Education Research, 19*(3), 370–384. https://doi.org/10.1080/13504622.2012.697545

Macey, J. (1983). *Despair and personal power in the nuclear age*. New Society Publishers.

Meghan, McCarroll Hillary, Hamann (2020). What We Know about Water: A Water Literacy Review. Water 12(10) 2803–10.3390/w12102803

Muecke, S. (2011). Australian indigenous philosophy. *CLCWeb: Comparative Literature and Culture, 13*(2), 1–8. https://doi.org/10.7771/1481-4374.1741

Müller, M. (2020). When salmon are deemed superfluous: Reflecting on a struggle of stories. In I. L. Stefanovic (Ed.), *The wonder of water. Lived experience, policy, and practice* (pp. 57–74). Toronto University Press.

Næss, A. (1989). *Ecology, community and lifestyle*. Cambridge University Press.

Pacini-Ketchabaw, V., & Clark, V. (2016). Following watery relations in early childhood pedagogies. *Journal of Early Childhood Research, 14*(1), 98–111. https://doi.org/10.1177/2F1476718X14529281

Plumwood, V. (1993). *Feminism and the mastery of nature*. Routledge.
Ruru, J. (2018). Listening to Papatūānuku: A call to reform water law. *Journal of the Royal Society of New Zealand, 48*(2–3), 215–224. https://doi.org/10.1080/03036758.2018.1442358
Seed, J., Macy, J., Fleming, P., & Naess, A. (1988). *Thinking like a mountain: Toward a council of all beings*. New Society Publishers.
Simmons, B., Woog, R., & Dimitrov, V. (2007). Living on the edge: A complexity-informed exploration of the human–water relationship. *World Futures, 63*(3-4), 275–285. https://doi.org/10.1080/02604020601174927
Snaza, N., & Weaver, J. A. (Eds.). (2015). *Posthumanism and educational research*. Routledge.
Spannring, R (2019). Ecological Citizenship Education and the Consumption of Animal Subjectivity. *Education Sciences 9*(1) 41-10.3390/educsci9010041
Spannring, R. (2021). Youth in the anthropocene. Questions of intergenerational justice and learning in a more-than-human world. In R. Winter & G. Knapp (Eds.), *Globalization and youth: Developments, analyses and perspectives* (pp. 113–133). Springer VS.
Spannring, R., & Hawke, S. (2021). Anthropocene challenges for youth research: Understanding agency and change through complex, adaptive systems. *Journal of Youth Studies*. https://doi.org/10.1080/13676261.2021.1929886
Steffen, W., Crutzen, P. J., & McNeill, J. R. (2007). The anthropocene: Are humans now overwhelming the great forces of nature. *AMBIO: A Journal of the Human Environment, 36*(8), 614–621. https://doi.org/10.1579/0044-7447(2007)36[614:TAAHNO]2.0.CO;2
Stewart-Harawira, M. (2012). Returning the sacred: Indigenous ontologies in perilous times. In A. McIntosh, R. A. Roberts, & L. Williams (Eds.), *Radical human ecology: Intercultural and indigenous approaches* (pp. 76–89). Ashgate.
Strang, V. (2020). The rights of the river: Water, culture and ecological justice. In H. Kopnina & H. Washington (Eds.), *Conservation* (pp. 105–119). Springer.
Taylor, A. (2017). Beyond stewardship: Common world pedagogies for the Anthropocene. *Environmental Education Research, 23*(10), 1448–1461. https://doi.org/10.1080/13504622.2017.1325452
Twance, M. (2019). Learning from land and water: Exploring mazinaabikiniganan as indigenous epistemology. *Environmental Education Research, 25*(9), 1319–1333. https://doi.org/10.1080/13504622.2019.1630802
Washington, H., Taylor, B., Kopnina, H., Cryer, P., & Piccolo, J. (2017). Why ecocentrism is the key pathway to sustainability. *The Ecological Citizen, 1*(1), 35–41. https://www.ecologicalcitizen.net/pdfs/v01n1-08.pdf
Williams, L. (2012). He Whanaungatērā: The politics and practice of an indigenous and intercultural approach to ecological well-being. In L. Williams, R. Roberts, & A. McIntosh (Eds.), *Radical human ecology: Intercultural and indigenous approaches* (pp. 397–420). Ashgate.
Wilson, E. O. (1984). Biophilia. Cambridge, MA: Harvard University Press

CHAPTER 11

To Love and Be Loved in Return: Toward a Post-Anthropocene Pedagogy and Humanity

Lennart Nørreklit and Michael Paulsen

Introduction

This chapter reflects upon how the Anthropocene calls us to rethink humanity and pedagogy. It has three parts. Part I forwards a theoretical understanding of the Anthropocene, showing that what is most needed today is the development of dialogical relationships and life communities between humans and more-than-humans, especially those based on mutual love. Part II refines our conceptualization of these relationships and communities—what they are, how they are possible, and their value. Through this, we outline a concept of humanity that is not centered only

L. Nørreklit
Aalborg University, Aalborg, Denmark

M. Paulsen (✉)
Department of the Studies of Culture, University of Southern Denmark, Odense, Denmark
e-mail: mpaulsen@sdu.dk

M. Paulsen et al. (eds.), *Pedagogy in the Anthropocene*, Palgrave Studies in Educational Futures,
https://doi.org/10.1007/978-3-030-90980-2_11

on humans but also reaches out to our life fellows. Part III projects how the school—the main pedagogical institution—could be changed to support coming generations in fostering and maintaining dialogical relationships and life communities involving both humans and more-than-humans.

Part I: What Is Most Lacking in the Anthropocene

The Anthropocene may, for one, be understood as denoting **a new epoch** in the history of the Earth (Crutzen & Steffen, 2003; Ellis, 2018; Sørlin, 2017). In this epoch, human activities mega-affect all strata of the life-critical zone of the Earth (Latour, 2017). In addition, the Anthropocene may alternatively denote **a world view** that emerges today because we recognize that we *are* in an Anthropocene age, which gives us a new view of our past, present, future, ourselves, and the world (Hedin, 2018; Paulsen, 2019). While the understanding of the Anthropocene as a new epoch is the starting point for the understanding of the Anthropocene as a new world view, some may remain at the first level of understanding, never reaching the second. This means that one responds to planetary problems without changing their basic world understanding. Such a response will, to our mind, concern optimizing our consumption of resources, still putting only humans at the center (Paulsen, 2021). In contrast, reflection on the way we behaved in the preceding epoch, the Holocene, points to the need for a different world view—that is, a new way of understanding being in the world, where life and dialogue are the center and not humans in an instrumental relationship to the environment. In the new understanding, the response is not only to optimize resource consumption but to learn to create dialogical relationships and life communities together with more-than-human beings—to create love relationships between ourselves, life, others, and the world as such. The elaboration of this leads to our thesis that what is most lacking today is the development of vibrant relationships and life communities between humans and more-than-humans.

Two Understandings of the Anthropocene

First, the geological term "Anthropocene" denotes an epoch in Earth's history (Crutzen & Stoermer, 2000). According to geology, a climatically stable epoch in Earth's history started about 11,500 years ago, called the

Holocene. During this epoch, humans spread across the globe, invented writing, and developed cities, philosophy, democracies, science, and technically mediated activities that changed all strata of the life-critical zone slightly above and below the surface of the Earth. Because of this, we do not live in a climatically stable epoch any longer (Ellis, 2018). Thus, the technical term "Anthropocene" denotes a new epoch in which the Earth is changed by human activities.[1] Part of this technical concept is that today, through science, we acknowledge that the effects are caused, although not intended, by humans. Nevertheless, they destroy ecosystems and other species, and they weaken the living conditions of the Earth. The Anthropocene does not signify a human victory march. It is open to interpretations as to which activities have been most damaging, who is to blame, and what deep structures of society are the causes behind the effects (Tønder, 2020).

Second, the Anthropocene signifies a change in the dominating world view (Hedin, 2018). When we acknowledge that the development of human institutions (e.g., science, economy, law, economy, politics, and education) in the Holocene caused the environmental problems of the Anthropocene epoch, we begin to look at ourselves and our past, present, and future—and the world—from a new perspective. For instance, from a "modernization" and Hegelian perspective, history has come to an end because we are now living in late-modern societies in which all further progress only means an adjustment to existing institutions (Paulsen, 2022). Yet from an Anthropocene perspective, this cannot be true because our institutional nexus of the Holocene has led to a situation where we are on the verge of destroying—externally self-negating—the earthly basis of life. This tells us that something is wrong with our way of being-in-the-world as developed during the Holocene. Our Holocene life-form has seemingly been based on *Earth-forgetfulness*—meaning that we have been (1) human-centered and (2) treating everything else as mere background, scenery, and resources for human development—that is, as manipulable objects that can be generalized, calculated, ordered, and planned to concord with human needs (Heidegger, 1977). Thus, we have mistreated life, animals, and plants alike, forgetting that we are singular living creatures, coexisting with other singular living creatures, as part of a shared world.

Now we begin to acknowledge that our life-form, as developed in the Holocene, has been based on *a scenic and human-centered world understanding* (Paulsen, 2019). The Anthropocene, understood as a shift in world view, therefore implies (1) the problematization of the Holocene world understanding and (2) an opening up toward an alternative ***dialogical and zoë[2]-centered world understanding***. This involves looking for the possibility of dialogue and co-creation with more-than-humans, and new interspecies institutionalizations. Here, other beings are not approached as manipulable objects but as dialogue partners—irreplaceable co-fellows with whom we can co-generate life communities. This nascent shift in world understanding is seen in many fields: within philosophy (e.g., political ecology and speculative realism), aesthetics and art (e.g., Anthropocene expeditions, eco land-art experiments, and aesthetic laboratories for ecology), education (e.g., wild, eco, and dark pedagogies), and architecture and design (rewilding architecture, bio-designs, regenerative designs, and nature now conceived as a dialogue partner).

Two Responses

This emerging world view—this way of being-in-the-world—does not at once replace the Holocene understanding.[3] We see two responses to the Anthropocene:

1. A "Holocene response" in which humans try to solve environmental problems with the same institutions, logics, and world understanding that caused the problems. This implies that we continue to see the Earth as a pile of resources. Yet, we try to optimize our resource management, making it greener and more sustainable, putting restraints on our activities for the sake of (some) humans for their survival and material welfare. Thus, we turn the Earth into one large spreadsheet, build protective technologies (against, e.g., rising sea levels), and educate new generations to manage and use resources (recycle more, pollute less). With this response, humans try to improve calculations and control everything, including their own behavior. They try to negate the self-produced external-self-negation by adjusting and improving the apparatus for controlling Earth as a planetary spaceship. If this strategy works, it implies that the whole life-critical zone becomes even more a human-made world. The Anthropocene will be the end of history, the takeover of Earth as the teleological endpoint of humanity.
2. An "Anthropocene response" in which we alter our world understanding away from being *scenic* (see the world as a stage) and *human-centered* (only accept human beings as actors) toward being

> *dialogically open* and *zoë-centered* (see the world as populated by living beings, potentially able to interact with each other). Thus, such a response calls for dialogue with more-than-humans as co-fellows with whom humans can co-create life communities (Morton, 2017)—not exclusively for the sake of humans, but for other beings as well, recognizing their intrinsic value. This implies an alteration of our institutions, away from being only for humans and consisting only of interactions between humans, toward frameworks for interspecies dialogue and the development of valuable life communities in which both humans and more-than-humans participate. In this response, we cancel our former Earth-forgetfulness, relate differently to the Earth, and re-situate ourselves in a world teeming with life. Inherent to this response is listening to what these co-fellows want, allowing them to take initiatives and decide how the shared world shall be. This response points toward an open post-Anthropocene world. We withdraw and stop acting exploitative on Earth. Instead, we let more-than-humans appear, develop, and express themselves to us, and see what we can do together with them, for them—or just simply let them be.

Human power in the Holocene to transform the geosphere is generated through cooperation enabled by communication, especially through the development of monological/hierarchical command-and-control systems and the use of *general concepts* to produce scientific analysis, categories, and advanced technologies (propelled through the invention of writing; cf. Innis, 1986). During the Holocene, instrumental power steadily increased. These strategies, monologue communication and the extended use of general concepts, have had negative effects. Monologue communication produces power inequality, while general concepts abstract from subtle variations that are important to handling life and the way people relate. By typifying the world, the biosphere is transformed into monocultural production units. Yet every living being is an individual being that unfolds its life in a specific interaction with its environment. This interaction produces vibrant ecosystems that are replaced by monocultural production, eventually depleting the quality of all elements of the geosphere.

The increasingly dominant use of *general concepts* as cognitive mediators between people and their world detaches people from the world. Things as types are, in principle, replaceable. They cannot constitute intrinsic irreplaceable values that bond people to their lives, giving them meaning, joy, and happiness. Personal bonds develop through interaction that sets its mark in both individuals so that they reflect each other and thus belong together. A component in the bond cannot be replaced by a replica, not only because there is no exact replica, which there cannot be, but also because the person would feel deeply deceived. The communication that situates people in their world is dialogue. Within dialogue, people develop *singular concepts* (Nørreklit, 2008), a specific knowledge about the other.

An explanation: It is important to distinguish between *general concepts* and *singular concepts*. When, for instance, we identify something as a tree, we do it by applying a general tree-concept. Through this, we can identify and recognize trees. Yet we can also identify a specific tree—for example, "our lovely tree in the middle of the garden." To do this, we use a singular concept, of *this* tree, almost as we do for a special person. With this concept, we identify its uniqueness and create a special relationship to it. Sometimes—not always—we also create proper names to form a singular concept of something. Also, more-than-humans may develop both general and singular concepts. If a bird recognizes a threat, it alarms others in a general way. Yet it also identifies something as a unique singular being that it connects to and recognizes in its singularity—for instance, its mate or its own nest.

The importance of singular concepts was forgotten during the Holocene. They are linked to and developed through dialogue. They are the basis of bonding and relationships. They link people to intrinsic values of life—that which they love and live for. To reduce human cognition to the development of theories based on general concepts is a mistake. It alienates knowledge from values, leaving values to be associated with emotions that are subject to manipulation, making people live in a mentally non-situated condition. Their individuality is reduced to a vanishing subject, forming a person as an object for social control. The principle of generality—equipped with the power of science and social control—replaces singular relationships. The relations to all other living beings in the biosphere, to family, to friends—all relationships fall victim to the growth of control. What is missing most in the Anthropocene is therefore dialogue between unique beings and, with it, the singular concepts making it possible to create life relations of love and being loved.

Assessment of the Two Responses

The problem with the Holocene response is that even though it may be possible to improve resource management on spaceship Earth, it increases Earth-forgetfulness and misses bringing meaning to life. The technical control of resources has only instrumental value. It instrumentalizes the world, including ourselves. That which has ultimate value is a good life, which is not the same as the management and consumption—and thus destruction—of resources (Nørreklit, 2017). A good life involves enjoying existence with others. Even if the Holocene response would secure survival *and* give some space for a surplus of activities, it would narrow the good life by limiting it to a man-made/controlled world, marginalizing the capacities of all other than humans, including the parts of ourselves that transcend resource management. As we will show, the possibilities for generating a good life stem from being-together with more-than-humans. Also, in the Holocene response, humans would be seen as resource managers (and/or resources), rewarded for their instrumental capacities and punished for a lack thereof. Moreover, even if the negative side effects could be eliminated, it would be selfish with regard to more-than-humans to treat them as pure resources. Thus, even if the Holocene response might turn out to be effective, it would narrow the good life on Earth.

Conversely, an Anthropocene response creates a life in which there are more exiting life-creatures to interact with and care for, who may care for us, that we can develop valuable life communities with, which have value in themselves, and thus make up a good life for us and for them. It is possible to integrate resource problem solving into interspecies life communities. Such interspecies partnerships that we can learn from do exist outside the human kingdom—for example, bees and flowers, who mutually help each other.

What Has Been Forgotten in Holocene Earth-Forgetfulness

During the Holocene, humans evolved from a species among others to the dominant species that changed and recklessly damaged the Earth's systems, creating a deteriorating biosphere. This development was possible through the power generated by increasingly displacing dialogue with communication-based objectification. Humans became an unprecedented force in conquering and destroying the biosphere. This displacement caused the Earth-forgetfulness. Humans forgot what it is to be a living being in a rich biosphere. They even started forgetting what it is to be a living human. What people forgot was that life *is* a dialogue. To be alive is to interact with other beings. Even if people are alone and have no one to interact with, dialogues take place, mostly in their heads. Even when people use no language, but try to fall asleep, their imagination remains engaged until they sleep. Even without language, when mother and infant smile and laugh with each other, there is dialogue and understanding—a fusion of horizons and a thrill of being alive. Being alive is not to be able to objectify everything from the outer position of nowhere (Nagel, 1986). It is to be in dialogue, to be happy and enriched through the experience of the other. It is a relation of love and being loved.

Love gives people a reason to live. It gives them motives and makes them experience meaning. Why does a person live—is there a reason to live? Some consider this question with suspicion: Is that scientific? Does it make sense? The answer is simple. If people love life, if they love the world, or their fellow beings, or what they do and what they see, then they have a reason to live and intrinsic motives to do things in their life; then there are things that make them feel happy; and then they experience life as filled with meaning. But if they do not love anything, then they have no reason to live and no intrinsic motives for doing anything.

In the Holocene process of developing control and power, humans objectified the environment and even themselves. Nature changed from being something people lived in to a source of resources. Institutions meant to help people objectified them as clients, patients, voters, and so on. In the process of increasing control, the reasons for humans to live were sidelined by demands for correctness and control. The intuitive, loving way of relating to others was scrutinized and considered invalid.

Analyzing problems of life as solely problems of fulfilling needs strengthened the objectifying model. When asking whether love is a need, love itself is objectified. Children need love in the form of care; otherwise, love is basically the need for sex, which concerns the propagation of the species and hormonal issues of the individual. The question of why people would care about these things is not addressed. It is not scientific. But it is the meaning of it all, including the meaning of science.[4] Love is a commitment to the value it concerns. It makes people do their utmost and their best. Love of knowledge drives the researcher and makes the scientific enterprise an adventure. If science is driven by other motives, such as utility for industry, capital, or power, then ulterior motives control the show.

However, as the endeavors to fulfill needs and obtain control and power increasingly dominate, they replace the role of love. Even nature is no more an object for love—only sightseeing, relaxation, and the provision of resources.

No relation survives as a basic value against the Holocene surge of power. Everything is replaceable. Replaceability is ingrained in the control systems. It is the principle of economy, and it is the principle of scientific generalization—operation and calculating with types. And it is the principle of technology: One unit can be replaced by another.

However, replaceability does not apply to intrinsic values or to love. The relations that give meaning are unique relations. The other is a unique other. Maybe the other is a typical person, but for those who have a relationship to the person, they are a unique person who cannot be replaced. Maybe the woman or the man is an average person—but to a person who loves him or her, that individual is unique and irreplaceable. If the one dies, the other mourns and suffers. The same with the child: The parents may be average people, but to the child they are unique and of irreplaceable value, and so is the child to the parents. When the child loses or is removed from its parents, both receive an irreparable wound.

The real force to take care of the environment is based on love. If the environment is only a resource to be exploited, then the conceptual framework will continue to focus on exploiting the resources, disregarding the needs of the environment itself—the living creatures as well as those who love the environment. Only love for the environment can change that. By recognizing love as the fundamental value to be protected and cherished through our life and organizations, a replacement for the typifying and instrumentalizing control forms can be developed. A replacement in which leaders are chosen because they love this world, its life and beauty, and love to protect and take care of the assets to which they have been entrusted, not destroy them to make money. Love takes care of the fulfillment of needs, but the fulfillment of needs does not take care of love.

Still, one may differentiate between the aspects of love: *existential*, *subjective*, and *universal* love.

Existential love is the love of being alive to do and enjoy things. It is of utmost importance. Work should be things workers love to do; then it makes them happy. Many managerial control mechanisms do the opposite. Existential love wants to do good things: to offer a smile, encourage others, tell a joke, help solve a problem, and more. To exist is to do things; loving to exist is loving to do things. The dead do nothing. Existential love is a blessing—to the ones who love to exist and to all others as well. It makes everybody feel well through its inspiring and joyful attitude. It automatically spreads among people.

Subjective love is the love for the specific other—that is, for specific people (including more-than-humans) and other phenomena one relates to. These are the loved ones—family and friends, as well as, for instance, places or art. Subjective love enables people to structure their lives. Life is about doing things with those and that which one loves. It makes the existential love projects unfold. It is the basis for intrinsic values and creates deep relationships. It creates interactive relations, the dialogue of life. It gives activities direction and meaning. For people to lose their subjective love—because a loved one dies, leaves them, or is removed by force—is devastating. The bonds that furnished the energy and happiness of their activities, making it an engaging dialogue, have snapped. Life loses direction and meaning. Losing the love inflicts a critical wound. In such situations, people need to love themselves, appreciate their qualities to maintain the energy to care for themselves, and eventually find new meaning in life.

Universal love is an open, loving approach to the world. It makes the world a source of wonderful experiences. It makes life enjoyable in general by making the world an inspiring place. It opens possibilities to develop subjective love. Universal love contrasts with a paranoiac "we–they" thinking, which penetrates modern culture and politics and makes people live in fear of the other, seek control, and become ready to engage in conflict and war. The idea that "we are good and they are bad" is contradictory to a knowledge-oriented, loving, and scientific approach. It is not based on genuine evidence of danger, but *a priori* considers anything about the other as bad and hostile—even when they help and are friendly. This Holocene mindset is the real destroyer, unable to create dialogue and loving relations.

Thus, our argument is that what is most needed in the Anthropocene is not better resource management, but that we change our world understanding from a scenic, monological, objectifying, and human-centered view toward a dialogical, loving, and zoë-centered view in which we pay heed to and try to develop dialogue with more-than-humans and thus try to join in life communities in which both humans and more-than-humans can participate and enjoy.

Part II: Conceptualization of Interspecies Life Communities and Dialogical Relations Between Humans and More-Than-Humans

To come closer to what interspecies life communities and dialogical relationships are, we will apply an approach based on Wittgenstein (2010), who invites us to clarify concepts based on the construction of language games (Nørreklit, 2017, 2021). Yet the language games that Wittgenstein (2010) applies are based on a scenic and human-centered world understanding. Only humans are constructed as actors; other beings are treated as background, scenery, or resources to the interacting humans in the language games. Thus, language is understood as taking place in technical interchanges between human actors (Paulsen, 2019).

An example is the language game Wittgenstein (2010) presents in §2: Builder A communicates with Assistant B, who "has to pass the stones, in the order in which A needs them. For this purpose, they use a language consisting of the words: 'block,' 'pillar,' 'slab,' 'beam.' A calls them out;—B brings the stone that he has learnt to bring at such-and-such a call."

This language game incarnates a scenic and human-centered world understanding: (1) Only humans are actors in the communication—only humans do something; (2) everything else—the stones—are subsumed to human purposes; and (3) all elements are replaceable and abstract. We get no knowledge of the singular builders; they remain *societal types*. Also, we get no knowledge of singular stones; they remain instances of *general concepts*. In addition, (4) what is done is resource management: moving around with objects; and (5) the Earth, world, or place where it happens is taken for granted as an implicit background or scene.

Wittgenstein's game is not a good dialogue. The activity is not about love and care. It has the power of objectified communication. It consists of short monologues of commands. It is a master–slave relationship. Yet it is possible to alter Wittgenstein's language-game approach in the direction of a dialogical and zoëlogical world understanding, which also leads to a change from technical to poetical language games (in Wittgenstein's terminology, this implies a shift from one *lifeform* to another). In Wittgenstein's example, we can introduce signals between A and B with which they try to understand each other and adjust to the situation of the other. This changes the situation. Now the participants are themselves part of the process, and they are no longer acting as if from a view of nowhere. The dialogue of existence, of love and care, takes place. However, to capture the dialogical expression of life more deeply encompassing care and love for the biosphere, we need examples involving humans meeting other species to clarify our concepts of *life communities* and *dialogical relations* between humans and more-than-humans. Our examples take place in an urban and high-tech setting, which causes problems but also creates new possibilities to develop interspecies dialogues and communities. Although there are limitations for humans to communicate with other species, modern technology might enable contact and translations to overcome such limitations. Our focus is, however, not on the technology but what it means to human and more-than-human life.

Example 1: With no real motive, Neil went out on the south-facing balcony and looked at the herbs he had planted in the balcony boxes. There were three long boxes with nine different herbs. They were not beautiful flowers with many colors, as on other balconies, whose owners showed off the glorious creatures to the outside world. They were modest plants, although Neil was not interested in their useful and health-related properties. He was interested in their well-being. *How are they?* he thought. The plants were living beings, and they were now in his care. Neil had felt a certain desire for more life; now that something could grow, it was, in a way, wrong that nothing grew. For Neil, it could have been some of the showy flowers, but his wife Hilde was not interested in that. She was not interested in having plants, because she thought she did not have a green thumb. She did not like the idea of having plants under her responsibility. But she had bowed to Neil's initiative: *We need some plants on the balcony.*

After that, Neil started going out on the balcony to see what had happened to the plants. To his surprise, he could see that they were growing. He knew they needed water to grow, so he took care of that. It pleased him and caused him to go out on the balcony every day. Now, he had lived in the apartment for 10 years without having plants; many things happened in the apartment, but things did not grow by themselves. It was he and, recently, also Hilde who made things change in the apartment. The world was, in and of itself, a dead world in which he and Hilde walked around and expressed themselves. Things that happened were an echo of him and Hilde. That was life. It was not life in its entirety; there was a world outside the apartment that also contained life. There were plants and animals in the outside world, but Neil never came into contact with them.

Example 1: (continued)

In good weather, Hilde and Neil ate on the small balcony. After the plants arrived, Hilde had begun suggesting they should eat there, even though she never mentioned the plants. Several of them flourished. They were tiny flowers, but there were many. Not something one would notice from the neighboring balconies—but the insects noticed them. Although theirs was a balcony on the fourth floor, the bees were diligent. They came every day. Neil wondered if there could be nectar in the little flowers. Every little flower received many visits, so Neil watered them a little extra. The flowers obviously wanted to supply nectar to the bees, so they needed a little extra power. *How could the bees know there was nectar up here?* Neil thought. One day, he again observed a bumblebee flying from flower to flower collecting honey. Hilde had still not seen the bumblebees. "Hilde, come, come and see," Neil called, and he showed her the bumblebee. "Look at that, there, see how eager it is," he said with enthusiasm. Hilde felt that it was nice.

Several things are worth noticing in this language game. (1) The plants make a difference in the life of Neil and Hilde. Before the arrival of the plants, their world was without contact with more-than-humans. (2) Neil cares for the singular plants in themselves, for their well-being; he has no further purpose than to take care of them. (3) It makes Neil happy to see the plants grow, which he takes as a sign of their well-being. (4) The plants attract Neil and Hilde; they enjoy sitting on the balcony with the plants. (5) The plants also attract other living beings, especially bees, and Neil notices this. (6) Through this, Neil and Hilde come in contact with insects and care for their well-being; for example, Neil begins to give the plants more water, feeling that it might help the plants to grow and give more nectar to the bees. (7) Neil notices that the bees and the plants have contact with each other. (8) Thus, the plants open the world of Neil and Hilde to contact with more-than-humans, increase their curiosity and knowledge about more-than-humans, and bring life and happiness.

Does this mean that dialogue is taking place between Neil, Hilde, and the plants and insects? Hilde and Neil are most certainly developing singular concepts about their plants. These concepts invigorate their life by creating interest in what happens on their balcony. This makes them happy. Neil regularly engages in small internal dialogue when thinking about the plants. He had been longing for something and got it. And the plants made an impression on him, which echoed in his head and made him respond with care. The plants reacted to his care—they thrived.

There is a language game between Neil and Hilde: He tells her about the plants and insects. But is there a language game between the humans and the more-than-humans? There is semiotic interaction (Barbieri, 2008). Neil takes notice of the life of the plants and their relationships with the bees. He takes this as a sign of the plants' desire to give nectar to the bees. It is, as the plants tell him, what he should do if he cares for them, and when he does it, he sees that the plants are grateful. For Neil, there is dialogue resulting in understanding: the plants ask for water, and Neil answers by giving them water. His dialogue with the plants makes him curious and expands his knowledge: What do they need? How much water?

From this example, we make a first clarification of interspecies dialogue as a bodily mediated, sign-based language game in which humans take notice of what more-than-humans need to live well and take this as a set of questions, calling for answers in the form of helpful support. The telos of the language game is to help the singular plants and their friends.

Does this evolve into a life community in which both humans and more-than-humans participate in the co-creation of a good life? It seems so. Despite (and because of) differences, the plants, insects, Neil, and Hilde begin to form a shared world in which they are thrilled to do things that bring life value to all. Neil and Hilde support the plants, the plants give nectar to the bees, and the bees bring life and happiness to the balcony and pollinate the flowers. All creatures participate in their own way but support each other and co-create a life that would not be possible if one of them were missing. Life community does not imply a unity between the creatures; rather, they form a community of differences, a coherent plurality, or what Lingis (1994) calls a "community for those who have nothing in common"—except life (we add). Each creature is not fully accessible to the others, but withdraws; Neil infers what the plants ask for based on signs. Each creature has its own world besides the shared world. What makes Neil happier, and brings value into his life, is that the plants are something in themselves, not just an echo of Neil. It is the miracle of being able to do something good for the plants, although Neil himself is not a plant. To see and help other living beings grow and be well makes him happy.

Thus, interspecies life community is possible and evolves when humans and more-than-humans do and enjoy something together, despite, but also because of, being different beings.

Example 2:[5] A man, Foster, dives every day for a year, at the same spot, in a South African kelp forest. He comes across a young octopus that he is drawn to and comes to know over almost its entire lifespan (one year). In the beginning, Foster just pays a daily visit to the den of the octopus. But, as he explains, when you do the same thing again and again, and visit the same place, you begin to notice a lot of differences. You shift from being a visitor to being part of the environment; this makes a huge difference. After many days, suddenly the octopus makes *contact* with Foster, and a relationship of curiosity and care develops. Now the octopus also seeks out Foster, takes initiatives, touches him, plays with him, observes, hides, and seeks his help, too. A relationship of trust is built, and the octopus invites Foster into her world. Every day, Foster learns something new about the octopus. But also, he learns something about himself, his fragility and potential power to connect and be a joyful part of an environment, forcing him to connect with powers dwelling within himself.

When the octopus dies, after giving birth to her children, and is eaten by a shark, Foster is moved and mourns the loss of his companion.When the year with the octopus began, Foster was in a life crisis. He had worked too much and lost the meaning of life. But the experience with the octopus, the power of becoming a part of her world and doing what he loves to do—diving, being part of nature, and developing a fine, loving relationship—brought life value to Foster. He began to dive together with his son and felt that he became a better father who enjoys life with his son, too.

Several things are worth noticing in this example. (1) The difference between visiting and being part of a world together with others is crucial. *The visitor* is not involved; he is an outsider, and as such, he is nowhere. Foster is not a visitor, a tourist from another place. Foster is seeking peace from the pain that is tied to his world. Here is a fine world without his pain; in the beginning, he is not part of this world either. *Being part of* implies involvement and doing things together; it involves participating in dialogue in a shared world. The visitor looks at the objectified world, exploits its resources and pleasures, and leaves it, not caring for the traces he leaves behind. Foster is not a visitor, and he is not a part. What is he? He is homeless, seeking a way to be somewhere. But he is open to this world, unfolding a universal love for this world while seeking distance from his world of pain. (2) The repetition of daily activities at the *same* spot opens the possibility of delving deeper into *differences*. When Foster is on land, he is often thinking about the den—it is calling—so he dives each day to the world of the octopus, and her friendship is made possible by the creation of a specific space-time structure: diving at the same location every day for a long time. (3) Being in a kind of *wu wei*—that is, a state of active non-action[6]—Foster makes it possible for the octopus to take its time to begin to trust Foster and decide when and if contact and a relationship should be established. Foster allows the octopus to take the initiative. (4) The octopus surprises Foster. Its life is much richer than he imagined. It is intelligent, careful, curious, and inventive. When trust is built, she shows her activities to Foster, and she plays and does things together with him. He learns *from* and *with* her. (5) The life with her gives Foster more *power*, *knowledge*, and *life value*, and it comes as a whole. His existential love strengthens. This transfers into his relationship with his son. (6) The friendship between Foster and the octopus is a relationship between singularities, a subjective love emerged. The octopus is not just an octopus among other beings; it is not a *type*. Foster relates to the octopus as an irreplaceable and unique being and builds a singular love-based concept of her. Therefore, he mourns her death. He is grateful for having been part of her life and her being a part of his. (7) Foster experiences what questions the life of the octopus is based on—for instance, that her life is not just about hunting and thus an answer to the survival problem. The octopus also plays with fishes, which Foster thinks is an answer to her intelligence and quest for enjoyment. This makes him think about his own life and what questions *his doings* are answers to. From this, he recasts what questions his doings ought to be answers to and thinks more deeply about what a good life is.

Here, dialogue happens on two levels. First, the octopus takes the daily appearance of a peaceful Foster as a sign of trustworthiness. Then she decides to contact him and begins to touch him, letting him know that she trusts him. Next, further bodily dialogue takes place. They swim together, play together, and cooperate in an escape from a shark. This happens through bodily sensing, imagining, and guessing what the other wants and is able to do. Second, the creative doings of the octopus make Foster wonder what questions these doings are answers to. This makes him think about his life. Such processes Gadamer (2013) calls real dialogue, in which understanding on a deeper level is reached. There is a fusion of horizons, which enables Foster to re-engage in life. The fusion of horizons that takes place in and between the partners in the dialogue explains the sorrow of the loss. The expectation of the other in a dialogue is that he/she must come back because "I have something in my heart that ties me to life, and this part comes from you and depends on you." When the other is dead, this part of the person continues calling for the other.

Clearly, *interspecies dialogue* is possible and can take the form of a bodily mediated, sign-based language game in which one can sense what the other can do and wants to do, and respond to this, thus making it possible to do things together, learn *from* and *with* each other, and create an emotional relationship connecting singular beings who care for each other.

Through their daily encounters, Foster and the octopus develop a community. To begin with, Foster visits the den. When community is established, he is no longer a visitor but a part of their shared world. They live together, take part in each other's lives, and form a shared life. They are not together all the time—both have secrets and a life of their own—but the life community contributes to the values of their life. Thus, life communities between humans and more-than-humans are possible and evolve when humans and more-than-humans do something together that contributes to a shared good life, despite (and because of) their differences, making it possible to develop feelings for each other, trust and friendship, and knowledge about each other's singular modes of being and life problems.

To sum up: Our examples show that interspecies dialogical relations and life communities between humans and more-than-humans are possible. The essential element of dialogue and understanding is a fusion of horizons that produces understanding and the joy of living in community. It implies language games in which humans approach more-than-humans as singular creatures with whom they have dialogue despite (and because of) their differences, and with whom they share a life in which they help and enjoy each other. The preconditions are that humans approach the more-than-humans (1) as co-fellows, (2) in appropriate locations where both can live and be together, and (3) stay together for a timespan that allows both to build confidence. Dialogue in this life-giving form, which is the basis of *being* in the world and which is that which is most missing today, is still possible. It is dialogue not only as a language game among humans. Through its openness (Gadamer, 2013), it relates humans to the whole biosphere. One should add that the sensitive environmental awareness (openness, *wu wei*) of Neil and Foster not only also includes the non-living but displays a rhythm of impression and expression in which Neil and Foster sensitively scan the environment, seeking whether there is something living to attend to, disposing them to the dialogue and the joy and gratitude they feel when it is found.

Three sets of values are generated through establishing interspecies dialogical relations and life communities with more-than-humans. (1) *Ethical values*: Instead of treating more-than-humans as mere objects or resources, one takes responsibility for others and tries to contribute to their possibility of creating a good life. (2) *Life values*: A richer life is opened, with mutual enjoyment, excitement, encounters of differences, emotional bonds, even love and the happiness of being loved in return. Through creating interspecies life communities, the world becomes richer and more meaningful. The dialogue makes life valuable. The dialogue itself is a basic value—maybe even the basic expression of love. (3) *Epistemological values*: Understanding and learning can be reached both with and from other species. This includes increased self-knowledge, self-understanding, and deeper thoughts about the meaning of life. This is ideographic knowledge, meaning knowledge of singular beings, which increases our capacity for becoming Earth-caretakers in relation to unique and singular beings. In none of the examples is abstract and technical language constructed. Singular and sensing zoëlogical semiotic systems are developed, directed toward the singular and irreplicable beings one relates to in the life communities, where beings are approached as unique.

A humanity that moves toward a world understanding, in which more-than-humans are approached as co-fellows that we can create dialogical relationships and life communities with, will bring out a richer life, produce important life knowledge and teleologically point toward a post-Anthropocene humanity and Earth.

Part III: Toward a Dialogical and Zoëlogical Understanding of the School

This section addresses the field of education to demonstrate how it could be possible to alter a school, based on a Holocene world understanding, toward a school based on a dialogical and zoëlogical understanding, and thus educationally institutionalize interspecies dialogical relations and life communities. Let's start with an example of a Holocene school language game to see what is wrong, from an Anthropocene perspective, with the Holocene school.

Example 1: A teacher enters a classroom. Students sit in rows, behind desks. At the back of the room, animals are stuffed in showcases. The teacher begins to talk. He explains what they shall learn today and asks them to open their schoolbooks and look at the same page. He then explains how rivers work, the day's topic. He writes on the blackboard, and students take notes. Sometimes students raise their hands. He allows them to ask questions, and he answers. Then he asks the students to look at the test tubes they filled with water, animals, and plants from a nearby river during the previous lesson. Their task is to use the knowledge from the schoolbook to identify five things caught in the tubes and present the results to the class. Most students succeed in doing this. The teacher writes the results on the blackboard. At the end of the lesson, the students are asked to pour the contents of the test tubes into a sink at the side of the classroom.

In this example, there is the dominant voice of the teacher. Except for answering a few questions, he talks in monologues, socializing the children for a Holocene life. Everything else of the environment, the river and the living creatures of the river, is treated as resources, background, and instances of abstract knowledge and concepts, without intrinsic value. The language game between the students and teacher is abstract and disembodied from a specific locality. Students are asked to learn the same things, to turn to the same page, and the task selected by the teacher is only to make identifications and exemplify general knowledge of what species can be found in the river and what type of river it is. The living creatures are treated for human purposes and finally as garbage.

Let's take another example, now situated in the Anthropocene age, today, but still based on what we, in Part I, called a Holocene response.

Example 2: A teacher takes her class of students into a forest to collect waste that is brought back to the school, examined, recycled, and turned into artworks and other things the students create. The teacher teaches the students about the harmful effects of waste in a forest.

The intention is to teach new generations to take care of the planet by not throwing garbage into the woods, to pick up waste when they see it, and to recycle it. Yet this is only about resource optimization. The structure of the language game resembles Example 1: teaching primarily takes place in a classroom; the excursion to the forest is secondary. Communication is between humans, while everything else is treated as resources, background, and scenery for human expressions and development. There are no interspecies dialogues or communities. The children are trained to keep human life and nature separate—implying that human life in the woods is bad. This illustrates that a Holocene response to protect the biosphere is problematic, creating a cognitive and emotional gulf between humans and the geosphere.

Our third example moves in the direction of an Anthropocene response. Students are to unfold life in dialogue with their environment.

Example 3: A school class is to live for one year next to a river running through the Great Forest to become part of life processes where other creatures live. Every school day, they meet at the same place at the edge of the Great Forest. Sometimes they stay overnight, and other times they stay for several days. The students unfold a life in dialogue and interaction based on their observations, and care and fear of the life of the forest, guided by teachers. They encourage students to be cautious and to explore the environment, see if they observe something interesting, and learn from and with the living beings that live in the Great Forest. The students display initiatives and support each other, make common projects to their curiosity, and exchange stories about their experiences and fantasies, as well as getting information and help from the teachers at the campfire. Also, students and teachers seek additional knowledge—for instance, through the internet and mobile devices, when relevant. The students are encouraged to notice what would be good or harmful for the different creatures and try to support the lives and well-being of these beings.

Students divide into smaller groups that choose to live and care for different areas and creatures. They are trained in being patient and, like Foster, observe and wait for the animals and plants to make contact or only to slowly intervene, and only help when they sense it would do something good, as with Neil. The students are encouraged weekly to present new knowledge and outline their ideas and perspectives on how to proceed. The teachers support the children and participate in the life activities.

As time goes by, they learn about the specific spot and begin to notice differences, rhythms, and changes. Some creatures begin to acknowledge and get used to the students. One group builds trust and a relationship with two duck families. After a while, they can differentiate between their individual members. They follow and support their duck life, and the ducks are grateful and respond by being near the students and making joyful things. When the ducks have ducklings, these grow up with the students as a natural part of their life; they seek help and enjoy it when the students are around, and mutual enjoyment flows.

Yet, during the Christmas holidays, when it becomes cold, most of the ducks die. When the group comes back after Christmas, the students become sad and wonder if it would have been better not to support the ducks with food and other things. The children blame themselves, thinking the ducks might have developed their life-skills

Example 3: (continued)
better without depending on unreliable students. Some think it was wrong to betray the ducks during the holiday. Teachers support their reflections that have no definitive answer. Other groups also have experiences with animals and plants, life and death, as the year goes by. One group experiences how the old oak tree, which they became fond of, changes sex during the winter. They are excited about why their tree suddenly chooses another sex.

In this example, the students and teachers are not visitors. Teaching is situated in the Great Forest. The students experience new questions and become part of a new world. Interspecies dialogue and life communities develop, with emotional bonds between students and singular more-than-humans. Like the life community on the balcony attracted Neil and Hilde, life communities evolve between student groups and other beings in the forest. Their horizon assimilates the life of the forest. Different problems occur, but that is what teachers are for. They encourage, support, and co-create with the students and their life fellows. The knowledge developed is concrete and relates to unique, singular beings. The students learn something important about life, death, human intervention, and how to sustain life communities and life values. Like Foster, they create relationships with unique beings, developing their subjective love. They also experience death and mourning. They come into contact with life—including their own. Their love of life, their existential love, strengthens. Whatever forgetfulness of life and being there was, they readapt human nature.

This example illustrates what a school based on a zoëlogical and dialogical world understanding might look like. It may also integrate resource management. The Great Forest case might include learning to make edible things grow in ways that, like on the balcony between the bees and plants, do not harm but expand life possibilities and mutual enjoyment. Understanding life, they discover how abundantly productive nature is—not only the small flowers on the balcony but the trees, the bushes, and all things that live produce an abundance in biodiverse environments. Further, students might experiment with building zoë-friendly zones and buildings not only meant for humans, making compostable clothes and more, find inspiration in literature, invent narratives, thus making the students more self-sustaining while simultaneously considering other beings, but also applying and developing technologies to foster interaction and dialogue with other species that might express themselves in formats not directly perceivable by humans.

A realization of such ways inspires *a post-Anthropocene pedagogy* that develops educational institutions in educating humanity toward a *new Earth*, in which more-than-humans are giving space through dialogically based life communities. It involves humans who are not instruments under monological control, blind to the life of the world. Instead, they are (or become) responsible beings unfolding a life dialogue with human as well as more-than-human life, promoting a lush and healthy world.

Conclusions

Part I argued that current environmental problems can be responded to in two ways: a Holocene response, based on the same understanding and logic that has produced the problems, and an Anthropocene response based on a zoëlogical and dialogical world understanding, in which more-than-humans are approached as living creatures with intrinsic powers and values, with whom humans might interact and co-create a good life. We argued that the latter should have priority and that resource management should be subsumed and integrated into this. Part II used a Wittgensteinian language-game approach to clarify that interspecies dialogical relationships and life communities in which both humans and more-than-humans can participate are possible and have ethical, life, and epistemological value. Thus, it is possible and desirable to strive for it. Insofar as humanity is recreated in this direction, it will be a humanity that is not only turned toward itself but reaches out to, cares for, and tries to create a good life for and together with other living beings. Part III addressed the educational field and argued that a school that is based on a zoëlogical and dialogical world understanding is desirable and possible. It supports students in developing interspecies dialogical relationships and life communities and thereby contributes to the development of pedagogy for a post-Anthropocene Earth.

Many practical obstacles remain. Our aim has only been to show that *interspecies dialogical relationships* and *life communities* are *understandable*, *possible*, and *desirable* and should therefore be aimed for.

Notes

1. Proposed start dates for the Anthropocene vary from 12,000 to 15,000 years ago to the trinity test in 1945 or even later. See, for example, Moore (2016), Haraway (2016), Emmett and Nye (2017), and Laugesen (2018). For objections to *the word* and *concept* "Anthropocene" and discussions of alternative candidates, see Morton (2016, pp. 7–25; 2018, pp. 39–67), Latour (2017, pp. 111–145), Sørlin (2017), Moore (2016), and Haraway (2016, pp. 30–57).

2. *Zoë* means life in Greek; thus *zoë-centered* means life-centered, yet not necessarily in the biological sense developed through Holocene science, but in a more fundamental sense, entailing "subjectivity" or *being-togetherness as living creatures* (in Danish: as *væsner*). See also Chakrabarty (2015).
3. The Holocene is obviously not based on only one world understanding. We propose only that a scenic and human-centered world understanding has become *dominant*. What we call the Holocene world understanding obtained the power to transform the life-critical zone on Earth, marginalizing alternative understandings. This understanding may be of European origin; nevertheless, it has become dominant on a global scale in the late Holocene and early Anthropocene. One might object that we do not explain how the new alternative "Anthropocene" world understanding could increase in power and subordinate the scenic and human-centered world understanding. Our argument only intends to clarify the content of the new world understanding and pinpoint implications of making this the basis of education and humanity.
4. At the beginning of the development of science, the role of love was obvious. Scientists were the lovers of knowledge, *philo-sophia*. Even Newton considered himself to be a philosopher: His books on physics are titled as books on the philosophy of nature. The separation between science and philosophy as we know it today came later. The point is that if there is no love behind the knowledge claim, the individual may become a sophist.
5. This example is taken from the Netflix movie *My Octopus Teacher* (2020).
6. *Wu wei* is a contested concept within classical Chinese thought, especially within Daoism (see, e.g., Girardot et al., 2001). The concept has several meanings. Despite that it can be translated as "non-action," it is better understood as a certain situational way of spontaneous, open, free-acting being-in-the-world that concords with life and nature rather than with fixed social-human norms. As we apply the concept, we want to indicate that Foster does not act upon the octopus and her world; he does not try to alter and adapt the world to his own desires. Instead, he tries to be there, be a part of the environment, and become alive within this nexus. Thus, he performs a kind of active non-action wherein he opens up to life, including his own aliveness.

References

Barbieri, M. (2008). Biosemiotics: A new understanding of life. *The Science of Nature, 95*(7), 577–599.

Chakrabarty, D. (2015). *The human condition in the Anthropocene. The Tanner lectures in Human Values*. Yale University.

Crutzen, P., & Steffen, W. (2003). How long have we been in the Anthropocene era? *Climatic Change, 61*(3), 251.

Crutzen, P. J., & Stoermer, E. F. (2000). *The Anthropocene*. IGBP Newsletter, 41. Royal Swedish Academy of Sciences.

Ellis, E. (2018). *Anthropocene*. Oxford University Press.

Emmett, R., & Nye, D. (2017). *The environmental humanities*. The MIT Press.

Gadamer, H. G. (2013). *Truth and method*. Bloomsbury Academic.

Girardot, N., Miller, J., & Xiaogan, L. (2001). *Daoism and ecology*. Harvard University Press.

Haraway, D. (2016). *Staying with the trouble*. Duke University Press.

Hedin, G. (2018). *Det antropocæne landskab*. I Jordforbindelser. Udgivet i forbindelse med udstillingen Jordforbindelser: dansk maleri 1780–1920 og det antropocæne landskab. Aarhus Universitetsforlag.

Heidegger, M. (1977). *The question concerning technology – And other essays*. Garland Publishing.

Innis, H. (1986). *Empire and communications*. Press Porcépic.

Latour, B. (2017). *Facing Gaia*. Polity Press.

Laugesen, M. H.-L. (2018). Det antropocæne paradoks. *Turbulens: Forum for samtidsrefleksion*. http://turbulens.net/det-antropocaene-paradoks/

Lingis, A. (1994). *The community of those who have nothing in common*. Indiana University Press.

Moore, J. (2016). *Anthropocene or Capitalocene*. Kairos.

Morton, T. (2016). *Dark ecology*. Columbia University Press.

Morton, T. (2017). *Humankind – Solidarity with nonhuman people*. Verso.

Morton, T. (2018). *Being ecological*. Penguin Books.

Nagel, T. (1986). *The view from Nowhere*. Oxford UP.

Nørreklit, L. (2008). At tænke filosofisk [To think philosophically]. In M. Etemadi & J. Zeller (Eds.), *At tænke filosofisk* (pp. 7–35). Aalborg Publisher.

Nørreklit, L. (2017). Actor reality construction. In H. Nørreklit (Ed.), *A philosophy of management accounting: A pragmatic constructivist approach*. Routledge.

Nørreklit, L. (2021). The language game of goodness. *Journal of Pragmatic Constructivism, 11*(1), 4–18.

Paulsen, M. (2019). Understanding the Anthropocene world: Contemporary difficulties. *Journal of Pragmatic Constructivism, 9*(2), 16–21.

Paulsen, M. (2021). Cautiousness as a new pedagogical ideal in the Anthropocene. In K. Petersen, K. Brömssen, G. Jacobsen, J. Garsdal, M. Paulsen, & O. Koefoed (Eds.), *Rethinking education in light of global challenges: Scandinavian perspectives on culture, society and the Anthropocene*. Routledge.

Paulsen, M. (2022). The plateau of Learning in the Anthropocene: How to relate differently to the Earth. In D. Kergel (red.), Learning in the Digital Age: An Transdiciplinary Approach for Theory and Practice Springer VS.

Sørlin, S. (2017). *Antropocen*. Weyler.

Tønder, L. (2020). *Om magt I den antropocæne tidsalder*. Djøf forlag.

Wittgenstein, L. (2010). *Philosophical investigations*. John Wiley & Sons.

CHAPTER 12

Planetarianism Now: On Anticipatory Imagination, Young People's Literature, and Hope for the Planet

Marek Oziewicz

One of the greatest challenges facing education in the Anthropocene is to empower young people everywhere *to believe* that we are able to transition to an ecological civilization. The nourishment of this capacity is a different task than helping them grasp—in numbers and data—the scope of climate change, biodiversity loss, and other forms of ecocide currently unfolding under the business-as-usual operations of neoliberal petrocapitalism. The latter challenge is "merely" about honing young people's climate science literacy. And while the quantifiable nature of scientific understanding is not without its challenges, unless one is willing to contest the scientific consensus about climate change, this factual knowledge is neither hard to find nor controversial. Helping young people *believe* that we can transition from an ecocidal to an ecological civilization is a whole different matter.

M. Oziewicz (✉)
Department of Curriculum and Instruction, University of Minnesota, Twin Cities, Minneapolis, MN, USA
e-mail: mco@umn.edu

M. Paulsen et al. (eds.), *Pedagogy in the Anthropocene*, Palgrave Studies in Educational Futures,
https://doi.org/10.1007/978-3-030-90980-2_12

While it also requires factual knowledge, such belief is a form of anticipatory imagination that looks beyond the horizon of what is currently possible to envision what is desired and necessary. One name for this belief is hope and the concern of this chapter is with how literature for young people can be tapped to nourish hope for the planet.

The argument is framed by a larger premise, according to which appeals to scientific evidence and data are not sufficient to "communicate—and hence trigger—the social and political changes needed to address climate change" (Szeman & Boyer, 2017, p. 5). The challenges facing us require a radical transformation of our conceptual structures of perception and a new critical awareness of how to use stories for that purpose. The proposition explored here is that while the battle for our planet's future is fought on many fronts, it is first and foremost the battle of imagination: of whether we can imagine a biocentric future. My contention is that hope-oriented anticipatory imagination is a necessary precondition for disrupting ecocide and enabling meaningful change. The term I propose for this anticipatory imagination focused on the planet's future is ***planetarianism***. This chapter offers its programmatic exposition, theorizing planetarianism as a mode of engagement with the issues of climate change in and through literature for the young reader.

Dreaming Despair: The Present Moment and the Rise of Dystopia

The leading question for this section was whether it is more responsible for educators to respond to the Anthropocene by preparing young people for the inevitable collapse of a petrocapitalist world or by empowering them to collective action that has the potential to re-orient the course of our civilization toward regenerative futures built on respect for planetary boundaries. So framed, the question represents a false dichotomy, but it gestures at a pedagogical dilemma: how to teach about climate change without leaving the audience feeling hopeless. The false dichotomy part is that preparing for the demise of our unsustainable, profit-driven, extractive, and ecocidal civilization is not antithetical to forms of activism necessary for the emergence of an ecological civilization. In fact, a broad consensus exists that petrocapitalism must be abandoned as soon as possible (Klein, 2014; Ghosh, 2016; Szeman & Boyer, 2017; Read & Alexander, 2019). As Rupert Read puts it, we are facing three broad possible futures. One, we will somehow manage to rapidly transform our ecocidal civilization into an ecological one. Two, our ecocidal civilization will collapse, but it will manage to seed a successor civilization as it falls. Or three, our ecocidal civilization will collapse utterly and terminally; victim of climate instability, resource depletion, and the conflicts these emergencies would engender (Read, 2020, pp. 159–160).

The main political and financial institutions of our petrocapitalist status quo have so far denied considering *any* of these three futures. They are locked instead in a magical thinking narrative about the future of accelerating expansion, including into space, with more of everything and infinite growth. For serious consideration of our possible futures, one needs to turn to climate movements like Extinction Rebellion and Fridays for the Future, or to UN reports and scientific literature. When the 2018 IPCC Special Report gives us only until 2030 for "rapid, far-reaching and unprecedented changes in all aspects of society" to contain global warming at 1.5° C by the end of the century (IPCC, 2018), or when the 2019 UN IPBES Global Assessment Report on Biodiversity concludes that we need "a fundamental, system-wide reorganization across technological, economic and social factors, including paradigms, goals and values" to protect nature and thus our own survival (IPBES, 2019), this is the language of science talking about the choices we have. Likewise, when Greta Thunberg calls out governments and corporations for "making it look like real action is happening when in fact almost nothing is being done, apart from clever accounting and creative PR" (Thunberg, 2020, p. 109), this is our time's equivalent of "The Emperor's New Clothes": of teenage activists recognizing how our socio-political order is caught up in self-serving illusions. In *Pedagogy of the Oppressed* (1970) Paulo Freire proposed the term *conscientization* for the process of "achieving critical consciousness" of oppression or injustice that allows one "to go a step beyond the deception of palliative solutions [and] to engage in authentic transformation of reality" (2018, p. 183). I want to suggest that our time is one of the accelerating conscientization of climate change. One in which we are entering a popular revolt against a dysfunctional system.

Knowing these facts does not make the pedagogical dilemma easier. The choice between the goals of "steeling for despair" versus "offering hope" is not just uncomfortable psychologically but daunting in practical ways. How do we teach to normalize despair or desensitize our students to a possible environmental collapse—and should we teach it, even if we knew how? Likewise, how do we teach hope for the planet without making it seem like solutions are easy or the situation is not as critical as the scientific data suggest? In more nuanced contexts, can despair and hope be taught together or balanced out without turning our pedagogy into collusion with the ecocidal status quo? Or perhaps hope and despair are each more complex notions than the binary opposition we often take them to be? If so, is there a way of enacting hope that embraces despair or acting on despair to carry forth the seeds of hope?

The trajectories of young people's literature and scholarship over the past four decades suggest some answers to these questions. In the realms of literary criticism and the humanities, two developments merit special attention. The first was the emergence of ecocriticism, with the term coined in 1978 by William Rueckert, the professional organization Association for the Study of Literature and Environment established in 1992, and the seminal collection that defined the field, Cheryll Glotfelty and Harold Fromm's *The Ecocriticism Reader: Landmarks in Literary Ecology* published in 1996. Ecocriticism has since diversified into several strands (Gaard, 2009), all of them retaining the original focus on "the study of the relationship between literature and the physical environment" (Glotfelty, 1996, p. xviii). In its "study" part, ecocriticism remains "merely" an interpretative tool. The focus on "the physical environment," however, reveals ecocriticism's activist aspirations and its origin as a reaction to literary studies in the 1990s being too academic. While postmodernism, deconstruction, and poststructuralism ignored the realities of the biological world, the fundamental premise of ecocriticism was radically different: "that human culture is connected to the physical world, affecting it and affected by it" (p. xix). The answer, one answer, ecocriticism thus offered was that literary studies should pay attention to "the most pressing contemporary issue of all, namely, the global environmental crisis" (p. xv).

The other major process was the rise of the environmental humanities in the early 2000s. An interdisciplinary global intellectual movement rather than a development within a particular field, the environmental humanities recognize that "abandon[ing] narrow disciplinary traditions" is necessary to fully grasp the inextricable ways in which humanity and the environment are connected (Emmett & Nye, 2017, p. 4). The environmental humanities advocates that reconceiving of "the relationship between scientific and technical disciplines and the humanities" is "essential to understanding and resolving dilemmas that have been created by industrial society" (p. 4). The key contribution of the environmental humanities has been to stress that science alone is not enough to engender a societal transformation our world needs. Since most of the constraints working against environmental action are cultural, the environmental humanities offer a necessary space to foster new ways of thinking and a new story about who we are, as a species, in relation to all other forms of life on the planet. The species focus has been especially groundbreaking. Indeed, until the Anthropocene registered on our simian radars, we had never even had to think of ourselves as a species. The environmental humanities is a response to this challenge. It insists that we must, as a species, find a new, emotionally compelling story "capable of mobilizing social adaptation" to the realities of a climate-altered world (p. 8).

Concurrent with the development of ecocriticism and the environmental humanities, something else was happening in literature and film for the young audience. It was the explosion of dystopia. The dystopian imagination was not a new thing, but it had until then existed "on the margins of mainstream literature" (Baccolini & Moylan, 2003, p. 1). It was only in the Reagan years—the dawn of the neoliberal era—that "a more clearly dystopian turn began to emerge within the popular imagination of Anglo-American societies" (p. 3). Within a decade, stories of "the Great Unraveling"—as Joanna Macy and Chris Johnstone have called this broad narrative template (2012, p. 5)—became not just mainstream but the dominant response to the Anthropocene in literature and film. The wide currency of dystopian, postapocalyptic, and post-disaster narratives has since forged a negative feedback loop: the more dystopia we read, the more evidence we find for its inevitability. For literary historians, however, dystopia is a recent phenomenon. It represents a historically situated aberration in our story systems triggered by the rise of neoliberalism.

In some ways, dystopia does make sense. Most of my students are Generation Z, born between 1995 and 2012. They grew up in a world literally hurtling toward apocalypse. During their lifetimes, between 1990 and 2020, global fossil fuel consumption grew by about 30 percent, atmospheric CO_2 increased more than it did in the entire lifespan of the human species until 1990, and the global average temperature of the planet increased by about 0.5° C (Tortell, 2020, p. 8688). It was also during their lifetimes that alarm about climate change was first raised and then countered with multimillion-dollar campaigns of denial, doubt mongering (Oreskes & Conway, 2010), and fossil fuel lock-in framing (Supran and Oreskes, 2021). It was during their lifetimes too that a series of voluntary neoliberal trade agreements to limit greenhouse gas emissions were signed, celebrated, and fizzled out—none of them even close to slowing down, let alone stopping the rise of emissions (Tortell, 2020). For people raised in a dysfunctional culture that continues to worship the market—even though it is increasingly clear that the market, being the primary driver of ecocide, is incapable of solving the problems it generates (Moore, 2016)—dystopia is the name of the game. My students know it well. Although they are in their late teens or early twenties, most of them have no illusions that capitalism can work for the planet. "Our economy today is built in large part upon the exploitation of the environment and its inhabitants," wrote a freshman student, commenting on Polly Higgins' TED talk about ecocide. "Higgins' willingness to push for the formal condemnation of ecocide is inspiring, … but it will be rejected as unfriendly to corporate interests which benefit from ecocide."

Students like George grew up reading dystopia, playing dystopian games, and watching dystopian films. They grew up hearing that capitalism is unavoidable, even if it destroys the planet. Surrounded by dystopian, postapocalyptic, and post-disaster narratives, most of them have internalized the belief that imagining a hopeful future is naïve, whereas imagining a postapocalyptic hell is reasonable. They have no problems whatsoever with naming books, films, shows, or games that project the collapse of our civilization. But they struggle to recall a single story that projects a hopeful planetary future. This availability heuristic and the attitudes it generates are not only the result of the culture in which they live and the dystopian narratives this culture generates. They are also reinforced by the popular and academic attention dystopia has received. Since the early 1990s, a whole field of dystopian studies has arisen to explain how dystopias are a new form of political opposition. Dystopia, Baccolini and Moylan note, offers "an education of perception" which may "support or catalyze a social transformation that will bring about an end to the conditions that produced the twentieth-century dystopias" (2003, p. 11). YA dystopias, specifically, "seek to teach serious lessons about the issues faced by humanity, and to offer readers a pleasurable retreat from their quotidian experience" (Hintz et al., 2013, p. 5). Likewise, post-disaster YA fiction is "highly political in scope" and strategically works "to engender a restorative and transformative response to environmental crisis" (Curry, 2013, p. 15). It is all about education and warning, it seems.

My issue with this well-meant enterprise is that in the name of resistance to the ecocidal status quo literary and filmic dystopias have helped reinforce the belief that ecocide is unavoidable. In 1981 Frederick Jameson wrote about the political unconscious of literature and my contention is that the rise of dystopia has engendered its own political unconscious: one that works toward the erasure of hope. This erasure is seldom acknowledged by authors, filmmakers, and scholars, but I have heard it, many times, from my students. "I have grown up watching and reading dystopian media that often left me feeling a bit empty regarding the future of the planet or society," wrote one freshman. In the words of another, "I never thought about how dystopian novels eliminated much of the hope young individuals could have for the future, especially in regard to climate change. ... While I do try to help the environment through my daily choices—eating vegan, recycling, conserving water—my overall outlook on the future is hopeless to the point that I do not see any benefit from taking larger action." Teaching about climate change, I have seen versions of this sentiment appear frequently enough to suggest that they represent a larger cultural construct. Writing about an urgent need for utopia, Tom Moylan recently remarked that dystopia has congealed into a "popular structure of feeling [which] immerses people in nihilist pleasure ... and sustains a disarming anti-utopian pessimism" (2020, p. 166). Indeed. Over the past three decades—extrapolating from the neoliberal assault on the planet—we have created a vast industry that describes dystopia, examines dystopia, replicates dystopia, and normalizes expectations about dystopian futures. The focus on what we dread has left almost no place for discussions about the future we want.

The consequences have been dire. As cognitive scholars have long known, the more a neural circuit is activated, the stronger it gets: our cognitive architecture is such that "[w]hen we negate the frame, we evoke the frame" (Lakoff, 2004, p. 3). Research has linked the rise of eco-anxiety, climate denialism, and social inertia to the prevalence of dystopian imaginaries in the media and popular culture (Brulle & Norgaard, 2019; Kretz, 2017). In literature, empirical studies have shown that over 80 percent of cli-fi narratives employ the "disaster frame" which—instead of mobilizing action—elicits despair, helplessness, and anger that lead to "ignoring or avoiding the topic" (Schneider-Mayerson, 2018, p. 490). When the disaster frame is articulated in stories—recall the vision of future Earth in *Wall-E*—stories of the Great Unraveling tend to operate as self-reinforcing prophecies. Indeed, given how capitalism reproduces itself through "feed[ing] on and learn[ing] from resistance and critique" (Haiven, 2014, p. 9), dystopias may have become commodifications conscripted to reproduce the system they supposedly challenge (Moylan, 2020, p. 166). If so, dystopia today represents a *de facto* capitulation to the rhetoric of petro-capitalism, which sees no alternative to the current carbon economy, and no alternative to its peculiar idea of growth based on accelerating monetization and exploitation of the biosphere. Given capitalism's colonialist legacy, visions of dystopian future are also capitulations to the unavoidability of racism. In the words of Hari Ziyad, "white liberal dystopian projections aren't cautionary tales about nativism and hyper capitalism, but an embrace of complicity" (2018, np). Underlying all these assumptions lies the fear that it is already too late. Take all this together and it becomes clear that the cultural work of dystopia has been impressive but not transformative enough. While dystopia has a place in building climate literacy, it is not sufficient to bring about the change we need.

Dreaming Hope: Planetarianism and the Way Forward

What if, instead of obsessing about the stories of the Great Unraveling, we turned our creative energies to imagine the futures we want? What if we used our imaginations to prefigure the world as it can be? My suggestion is that one of the most productive ways to engage with the urgencies of the Anthropocene is through stories that mobilize hope—and not just any hope but specifically hope for the planet. This capacity for "hopeful dreaming" (Alexander, 1968, p. 389) has always been the domain of art, especially literature. Our challenge is to reimagine ourselves in relation to the biosphere and the primary space for that reimagining is the story. But not just any story. As Ursula K. Le Guin commented in her 2014 National Book Awards Acceptance Speech: "I think hard times are coming when we will be wanting the voices of writers who can see alternatives to how we live now, who can see through our fear-stricken society and its obsessive technologies to other ways of being, and even imagine some real grounds for hope" (np). These are the stories we need.

Le Guin's call is finding resonance in an emerging field of hope studies. Whereas earlier ecocritical scholarship insisted on the need to study any and all literary representations of the environmental crisis, hope scholars stress the need to study "bright spots": stories that model solutions, positive outcomes, and understanding that have the potential to mobilize the audience's agency (Kelsey, 2020, p. 175). Whereas scholars in the environmental humanities and the Anthropocene literary studies have long centered discussions on how the stories we tell can shape our future, hope scholars insist on the need to "hack the story" (Ray, 2020, p. 80), "trend hopeful" (Kelsey, 2020, p. 157), and address the challenges of the Anthropocene within hope-oriented frameworks. As Macy and Johnstone argue, our culture today unfolds within three broad narrative templates: Business as Usual, the Great Unraveling, and the Great Turning (2012, pp. 4–5). Each template is a version of reality and a lens through which we frame the challenges facing us. Each constructs our agency in relation to that reality. Macy and Johnstone's argument is that today we need stories of the Great Turning: stories about the transition to an ecological civilization, stories "committed to the healing and recovery of the world" in which "the central plot is finding and offering our gift of Active Hope" (p. 5). Like Kelsey, who insists that "the environmental crisis is also a crisis of hope" (2020, p. 11), Macy and Johnson frame hope as an activist position necessary for us to discover "the path of an inspired vision" and achieve the change we want to see (2012, p. 185). Like Kelsey, who contends that "hope for the environment is not only warranted but essential to addressing climate change, biodiversity loss, and the full suite of environmental crises we face" (2020, p. 4), Macy and Johnson stress that Active Hope is indispensable. In fact, "the greater the gap between the present reality and what we would like to have happen," the more important it becomes "to believe that what we hope for is possible" (2012, p. 186). This relationship between hope, belief, change, and agency is paradoxical but central for any meaningful progress toward an ecological civilization. In other words, to embrace Active Hope, to become "active participants in bringing about what we hope for" (p. 3), we must recognize that hopelessness is not a reality. It is a mindset. One that makes us "act and live from an orientation of fear" (Ray, 2020, p. 81). How do we confront it? How do we overcome our powerlessness—the I-can't-do-that, it's-too-late, and it's-too-big framing of the challenges ahead? How do we navigate past what Macy and Johnstone call "the threshold guardian of disbelief" (2012, p. 193)?

I suggest we start with imagining hope for the planet. Eileen Crist has described the Anthropocene as a discourse "constituted by a blend of interweaving and recurrent themes" (2016, p. 15), two of which are especially prominent: naturalization of human expansionism and naturalization of the Earth's environmental collapse—that is, after Earth has been reduced to a resource for our species' ascension toward domination of the galaxy and discarded when used up. This notion that Earth is doomed is a rhetorical construct that renders the planet expendable. It must be rejected in favor of imagination focused on hope—for only imagination empowers us "to love and respect the Earth with the same intensity that we give to our families and our tribe" (Lovelock, 2000, p. 8). Put otherwise, the challenge of addressing the discourse of the Anthropocene is primarily a challenge to our story systems. Stories are the best tools we have to rewire our affective and cognitive modes of being in relation to the planet. And if we have been telling the wrong stories, it is time to get them right.

The alternative I propose is an epistemological orientation of putting the planet first, or planetarianism. A conceptual tool for ushering in the future we want, planetarianism is *a biocentric philosophical commitment to stand up for the planet*—thus, a counternarrative to the dystopian, anthropocentric discourse of the Anthropocene that legitimizes ecocide as the price for human "progress." On another level, planetarianism is *applied hope articulated through stories*—a form of hope-as-resistance that enables us to overcome what Macy and Johnstone call "the challenge of disbelief" (2012, p. 186). Planetarianism, in this formulation, is a name for the process of unleashing our anticipatory imagination and channeling it into designing alternatives to the ecocidal present. Planetarianist stories envision the planet as a living entity, imagine a non-ecocidal socioeconomic system, depict disanthropocentrized relationships among humanity and other living beings, and gesture at a biocentric, multispecies future that is worth living for. So conceived, planetarianism can be examined as a distinct component of narrative fiction.

If this sounds simple, I want to stress that planetarianism presents us with a tremendous conceptual-creative challenge. We have little experience imagining applied hope through stories and we live in a culture whose mediascape is dominated by dystopian imaginaries. Our educational and other initiatives to create hope for the planet are always at risk of being pulled into the narrative of technooptimism, which effectively defends the ecocidal status quo as a necessary step toward a future of always-almost-within-reach technofixes. Or they risk being derailed by the more immediate concerns of the present. Nor have we given enough consideration to the infinitely plural articulations of hope. What counts as hopeful for specific age groups or culturally situated audiences? What are the forms or thresholds of hope? What is the meaning of hope for the planet and how it might be different for not just different audiences but when expressed in different genres or media? Finally, what are the best tools to examine or teach hope—Active Hope, applied hope, activist hope, and other facets of hope—in the stories we tell? While I have no answers to these questions, the remaining part of this chapter offers two provocations about the unique educational potential of planetarianist fiction.

First, one important cultural work of planetarianist literature is to nourish a sense of hope even in the absence of specific solutions. Having hope is not the same as having a solution. While much of hope studies and solution journalism is predicated on the premise—and rightly so—of counteracting stories of doom and gloom with stories of environmental successes, equating hope with existing solutions is limiting. It distorts the larger work of hope as a form of sustaining belief operating even when we have no clear vision of *how* that hoped-for future may be achieved.

Consider, for example, Pam Bonsper and Dirk Rink's *The Problem of the Hot World* (2015). In this picturebook, five animal friends realize that their world is getting too hot. The deer, the bear, the fox, the mole, and the owl head out to the Ocean to see where all the water went. But the salty water is not what they need. A polar bear cub then appears and tells them of a magic cave where they will find the answer. The five friends enter the cave, which turns out to be a tunnel leading back in time. When they tumble out, back in their own forest, it is lush: "everywhere it was green and it was not hot" (np). The end. Or consider Sandra Dieckmann's *Leaf* (2017). In this picturebook, a lone emaciated polar bear is washed on the shores of a forest in a temperate climate. The bear is seen as a monster by all other forest creatures and tries to adapt by dressing up in leaves. Or, is it trying to fly away on leaf wings? When the crows talk to him at last, the bear indeed turns out to be "just someone who wished he could fly back home" (np). Eventually, the crows carry the bear back home to the Arctic and promise to tell his story "to everyone who would listen, so that no polar bear would ever get lost again" (np). In Rachel Hope Allison's *I'm Not a Plastic Bag* (2012), the great pacific garbage patch flies away like a bird. Saying "thank you" to the gulls who inspired it to take off, it leaves the ocean clean for the marine creatures to live. It was not in its place. In the last opening, the smiling face of the GPGP monster in the sky is watching a lone albatross sail over the tranquil sea. All is well. And in Jewel Parker Rhodes' *Bayou Magic* (2015), 12-year-old Maddy saves the local ecosystem by reconnecting with the ancestral knowledge traditions passed on to her by Grandmère Lavalier. When the oil rig in the Gulf explodes, gushing crude oil, Maddy's dream grasps that her connection to mermaid goddess Mami Wata is the only way to save Bayou Bon Temps. She calls upon the mermaids to build a levee of rock and silt to block the river's mouth. "They listened. They will do what I dreamed. Because of me, there is a happy end" (2015, p. 232). When the media hail the outcome as "the Bon Temps miracle"—the oil parting "around this bayou like the eye of a hurricane" (pp. 233–234)—Maddy realizes that while "you can't fix everything yourself, [y]ou need good friends and hope. Sometimes, even mermaids" (p. 235). None of these books offer realistic solutions to the problems of the Anthropocene they build on: global warming, shrinking habitats, species extinction, marine pollution, or oil spills. At the same time, they each offer excellent examples of planetarianist fiction that mobilizes hope for the planet by leaving the readers empowered that a non-ecocidal outcome is possible. They all communicate that hope is a gift one must not give up even if realistic solutions are hard to imagine.

Second, planetarianist literature redefines hope as a form of collective action rooted in anticipatory imagination. Stressing the power of collective action is a flip side of acknowledging that climate change and other devastations of the Anthropocene are driven by systemic causes rather than by evil schemes of singular villains. And that these systems—both systems of oppression and systems of resistance—are shaped by the stories we tell. A number of planetarianist books succeed in depicting these systems without compromising their affective power which comes from focalization through the voice of a single protagonist.

To defeat the black snake that threatens the land in Carole Lindstrom and Michaela Goade's *We Are Water Protectors* (2020), the young protagonist rallies a coalition that consists not only of her people, Native Americans, but also of her ancestors, of non-Native allies, of natural elements, of "the four-legged, the two-legged, the plants, trees, rivers, lakes, the Earth"—for "we are all related" (np). It is this broad coalition that will defeat the ecocidal system. When the narrative states, "We are stewards of the Earth. Our spirits have not been broken" (np), this is the voice of Active Hope proclaiming that togetherness is the path to transformative change. Likewise, in Megan Herbert and Michael E. Mann's *The Tantrum That Saved the World* (2017), young Sophia is transformed into a climate activist when she connects with climate refugees who arrive at her home. The first to show up is a polar bear whose ice home ceased to exist. The bear is followed by a Kiribati family, a swarm of bees, a flamingo, a family of Syrian farmers, a sea turtle, a New England fisherman, and a Bengal tiger. "They all turned to face her with hope in their eyes, expecting Sophia to halt their demise" (np). Sophia organizes the refugees and petitions City Hall officials. When her call is snubbed, Sophia throws a tantrum to save the world, a tantrum so loud that the world listens. "Cooperative action can turn this high tide, they had **strength in numbers** and **right on their side**" (np, bold in original). The book ends with refugees leaving for their new homes and the world getting a "second chance"—represented in green background landscapes—even though just what the transformation involves is not described. A similar message about the power of collective action is found in Zoë Tucker and Zoe Persico's *Greta and the Giants* (2019)—a fictionalized retelling of Greta Thunberg's school strike for climate. In it, the animals plead with Greta to stop the greedy Giants who are destroying the forest, and Greta stands in the Giants' path with a sign "Stop." Within days, "more people and animals saw what they were doing and joined in too" (np). When the crowd gets so huge that it fills the forest, the Giants take notice. It takes them a while to disengage from their habitual destructive activities, but eventually, they do. "Before long … the forest became more beautiful than anything they could ever have imagined" (np).

What all these and other planetarianist stories achieve is to project hope as an emerging quality that arises out of multiplicity of simple interactions: the Water Protector girl promising to protect water, Sophia taking in the first climate refugee, or Greta taking up a sign to stop the Giants' thoughtless rampage. The anticipatory imagination behind each of those acts is not certainty about the outcome, but a response to a call that creates possibility. "Emergence," writes Adrienne Maree Brown, "is beyond what the sum of its parts could ever imagine" (2020, p. 37). That, too, is also how planetarianist stories showcase hope for the planet: as an emergent quality arising from collective dreaming. This dreaming keeps alive young people's belief that it is not too late and that any system created by human beings can be changed by human beings. Planetarianism affirms that we have the agency for that change. And that even a broken world is worth fighting for.

Planetarianism NOW

Beyond the examples mentioned above, authors have used multiple other strategies to articulate applied hope for the planet in stories for young audiences. That we need this hope is becoming increasingly clear and my contention is that children's literature can act as a particle accelerator for planetarianist ideas. As educators in the Anthropocene, we are faced with a challenge of how to empower our students to *believe* that a just, ecological civilization is possible. To further its emergence, we should actively seek out stories that articulate hope for the planet and shift attitudes away from resignation. It is not enough to make our students aware of the facts; it is crucial to empower them to take an active stand against ecocide and teach them how. Stories that articulate hope are indispensable because the key obstacle in the current fight for our planet's future is the pervasive doomsday script that exonerates inaction. We will not prevail unless we have stories that counter fear and despair with inspiration and hope. As with issues of race, diversity, and inequality, we need to learn to talk about the challenges of the Anthropocene openly and honestly. Our most advanced technology for imagining the future is the story, and it is through stories that we can engage others in conversation about how to translate hope into reality. "With malice toward none, with charity for all," we need everyone on board—engaging our communities like Sophia's team in *Tantrum*: "They all told more people, who told more folks still. They won hearts with kindness and minds with good will. And so on and so on until everyone, was doing the hard work that had to be done" (2020, np). We can choose the stories we live. We can stop the ecocide happening on our watch. The time for planetarianism is NOW.

References

Alexander, L. (1968). Wishful thinking—Or hopeful dreaming? *The Horn Book Magazine, 44*, 383–390.

Allison, R. H. (2012). *I'm not a plastic bag*. Archaia & Jeff Corwin Connect.

Baccolini, R., & Moylan, T. (Eds.). (2003). *Dark horizons: Science fiction and the dystopian imagination*. Routledge.

Bonsper, P., & Rink, D. (2015). *The problem of the hot world*. Self-published.

Brown, A. M. (2020). What is emergent strategy. In A. E. Johnson & K. K. Wilkinson (Eds.), *All we can save: Truth, courage, and solutions for the climate crisis* (pp. 37–38). One World.

Brulle, R. J., & Norgaard, K. M. (2019). Avoiding cultural trauma: Climate change and social inertia. *Environmental Politics, 28*(5), 886–908. https://doi.org/10.1080/09644016.2018.1562138

Crist, E. (2016). On the poverty of our nomenclature. In M. Jason (Ed.), *Anthropocene or capitalocene: Nature, history, and the crisis of capitalism* (pp. 14–33). PM Press.

Curry, A. (2013). *Environmental crisis in young adult fiction: A poetics of earth*. Palgrave Macmillan.

Dieckmann, S. (2017). *Leaf*. Flying Eye Books.

Emmett, R. S., & Nye, D. E. (2017). *The environmental humanities: A critical introduction*. MIT Press.

Freire, P. (1970/2018). *Pedagogy of the oppressed*. Bloomsbury Academic.

Gaard, G. (2009). Children's environmental literature: from ecocriticism to ecopedagogy. Neohelicon 36, 321–334.

Ghosh, A. (2016). *The great derangement: Climate change and the unthinkable*. U of Chicago Press.

Glotfelty, C. (1996). Introduction. In: Introduction. In C. B. Glotfelty & H. Fromm (Eds.), The Ecocriticism Reader: Landmarks in Literary Ecology (pp. xv–xxxvii). University of Georgia Press.

Haiven, M. (2014). *Crises of imagination, crises of power: Capitalism, creativity, and the commons*. Fernwood Publishing.

Herbert, M., & Mann, M. E. (2017). *The tantrum that saved the world*. World Saving Books.

Hintz, C., Basu, B., & Broad, K. R. (Eds.). (2013). *Contemporary dystopian fiction for young adults: Brave new teenagers*. Routledge.

IPBES. (2019). *Global Assessment Report on Biodiversity*. https://www.un.org/sustainabledevelopment/blog/2019/05/nature-decline-unprecedented-report/

IPCC. (2018). *Special Report "Global Warming of 1.5°C"*. Intergovernmental Panel on Climate Change, October 8. Retrieved December 12, 2020, from https://www.ipcc.ch/sr15/

Kelsey, E. (2020). *Hope matters: Why changing the way we think is critical to solving the environmental crisis*. Greystone Books.

Klein, N. (2014). *This changes everything: Capitalism vs. the climate*. Simon and Schuster.

Kretz, L. (2017). Emotional solidarity: Ecological emotional outlaws mourning environmental loss and empowering positive change. In A. C. Willox & K. Landman (Eds.), *Mourning nature: Hope at the heart of ecological loss and grief* (pp. 258–291). McGill-Queen's University Press.

Lakoff, G. (2004). *Don't think of elephant! Know your values and frame the debate*. Chelsea Green.

Le Guin, U. K. (2014, November 19). National Book Awards acceptance speech. *YouTube*. https://www.youtube.com/watch?v=Et9Nf-rsALk

Lindstrom, C., & Goade, M. (2020). *We are water protectors*. Roaring Book Press.

Lovelock, J. (2000). *Gaia: A new look at life on earth*. Oxford University Press.

Macy, J., & Johnstone, C. (2012). *Active hope: How to face the mess we're in without going crazy*. New World Library.

Moore, J. (Ed.). (2016). *Anthropocene or capitalocene: Nature, history, and the crisis of capitalism*. PM Press.

Moylan, T. (2020). The necessity of hope in dystopian times: A critical reflection. *Utopian Studies, 31*(1), 164–193.

Oreskes, N., & Conway, E. (2010). *Merchants of doubt*. Bloomsbury.

Ray, S. J. (2020). *A field guide to climate anxiety*. University of California Press.

Read, R., & Alexander, S. (2019). *This civilization is finished: Conversations on the end of empire—And what lies beyond*. Simplicity Institute.

Read, R. (2020). "This civilization is finished: Time to build an ecological civilization." The Ecological Citizen, 3(20), 157–162.

Rhodes, J. P. (2015). *Bayou magic*. Little, Brown and Company.

Schneider-Mayerson, M. (2018). The Influence of Climate Fiction: An Empirical Survey of Readers, Environmental Humanities, 10(2), 473–500.

Supran, G. and N. Oreskes. (2021). Rhetoric and frame analysis of ExxonMobil's climate change communications. One Earth 4(5), https://doi.org/10.1016/j.oneear.2021.04.014, pp. 696–719.

Szeman, I., & Boyer, D. (Eds.). (2017). *Energy humanities: An anthology*. John Hopkins UP.

Thunberg, G. (2020). *No one is too small to make a difference*. Penguin Press.

Tortell, P. D. (2020). Earth 2020: Science, society, and sustainability in the Anthropocene. *PNAS, 117*(16). https://www.pnas.org/cgi/doi/10.1073/pnas.2001919117

Tucker, Z., & Persico, Z. (2019). *Greta and the giants*. Frances Lincoln Children's Books.

Ziyad, H. (2018, April 9). White liberals stay predicting dystopias caused by whiteness—Without doing anything about it. *Black Youth Project*. Web. Retrieved June 8, 2018, from http://blackyouthproject.com/white-liberals-stay-predicting-dystopias-caused-by-whiteness-without-doing-anything-about-it/

CHAPTER 13

To Learn a World: Human-Machine Entanglements as Pedagogy for the Anthropocene

Alexander Popov

Artificial Intelligence as a Metaphysically Subversive Figure

Surviving the Anthropocene increasingly looks like a problem that encompasses the totality of human existence. Decarbonization, the preservation of biodiversity, rewilding, ecological justice—all of these tasks, and many more, are bound together in a knot that may turn out to be impossible to untie within our current socio-political frameworks. A pedagogy for the Anthropocene should strive to educate people in systems thinking and critical thinking, in order to build them up as holistic problem solvers. But even if such an interdisciplinary and critically aware pedagogy could be devised and integrated in schools, universities, and workplaces, a problem

A. Popov (✉)
Department of English and American Studies, Sofia University "St. Kliment Ohridski", Sofia, Bulgaria
e-mail: aleksandarnp@uni-sofia.bg

M. Paulsen et al. (eds.), *Pedagogy in the Anthropocene*, Palgrave Studies in Educational Futures,
https://doi.org/10.1007/978-3-030-90980-2_13

remains in the heart of the matter: reversing, or in the very least neutralizing, the harmful effects the Anthropocene requires shifts not merely in political economy and science/technology, but in ontology as well. Western metaphysics is the foundation of our philosophical explorations and political orientation, but also the determining apparatus for our ability to perceive and conceive of worlds. To turn the Anthropocene around, we need to cultivate richer capacities for being-in-the-world and to learn to recognize other such modes as ontologically equal.

This chapter tackles the issue of alternative metaphysics and how those could make room for inclusive political subjects open to human-nonhuman entanglements. This is a central strand of research in the environmental humanities but in this chapter I propose to engage it from a somewhat unusual angle—via thinking about artificial intelligence (AI), and its usages in technocultural imaginaries around the categories of "human" and "nonhuman." The topic of AI is typically approached from the standpoint of the natural sciences and engineering; the field has relied strongly on Cartesian metaphysics and its fundamental splits between subject and object, mind and body, inside and outside, and so on.

This underlying analytic of exclusion through binarization and boundary-setting is of course central to Western civilization. The concept of "the human," as analyzed in relation to AI by Jennifer Rhee, has proven to be almost infinitely pliant in its capacity to set apart groups of beings and to serve ideological enterprises (Rhee, 2018). AI imaginaries have largely mirrored these processes, more often than not dehumanizing the figure of the intelligent machine and inscribing it with inhumanity (and in some cases with femininity, as Rhee shows). AIs in popular culture have alternately fulfilled the roles of slave and threat to humanity, frequently both in the same narrative space, always reinforcing the transcendental status of the human. On the other extreme is envisioning AI as a set of purely algorithmic and service-like routines divorced from any notion of agency.

Such exclusionary analytics have come under attack by different (and often opposed) strands of contemporary philosophy. A common motif among these is a reinterpretation of objects—natural and technological—and the consequent blurring of the subject-object split. Thus, Bruno Latour's actor-network theory decomposes agency into actantial components and shifting assemblages (Latour, 2005); Karen Barad's agential realism emphasizes the mutual constitution of subject and object in the act of "taking measure," that is, any kind of representation or translation of phenomena—scientific, mechanical, perceptual, semiotic, or other (Barad, 2007); postphenomenology's material hermeneutics interprets technologies as variously embodied extensions of the body (Verbeek, 2005); and speculative realism insists on the ontological autonomy of objects (Harman, 2018). Dreyfus's critique of "good old-fashioned AI" has been influential in shifting research focus from symbolic representations, whereby the various aspects of being are encoded in separate functional modules, to dynamical systems in which mind and being are co-produced by agents and the environment (Dreyfus, 2007). In affect studies, the almost axiomatic insistence on the primacy of rational thought has been interrogated and found lacking in its capacity to navigate the complex economies of entangled selves (Wilson, 2011).

In this chapter I want to show that the Othering of AI in popular culture can serve a subversive purpose as well. Reading fiction about AI in conjunction with theoretical interventions can lead to opening up the concept of the human. In our technocultural imaginary, AI flickers in and out of the uncanny valley (Mori et al., 2012), alternately and ephemerally occupying the human, inhuman, and nonhuman positions. AI exhibits extraordinary narrative valency, as attested by the diversity of functional roles given to AI agents: from servant to master, from civilizational enemy to guardian of humanity, from seducer to lover, from madman to wiseman, from child to teacher.

AI is in some sense a Trojan horse into Western metaphysics, involuntarily constructed as such by that very same system. It simultaneously repulses us and draws us closer, asks us to rationalize its behavior, and posits itself beyond human understanding. Such antinomies prevent a static technocultural interpretation of AI and open the door for alternative philosophical explanations of what it means to be an artificial thinking being. And while most fictional narratives about AIs are centered on technical and scientific imaginaries, I aim to show that we can think with AI just as productively about our relations to nonhumans—in the context of alternative metaphysical systems and a dramatically changing planetary ecology. To that purpose, I will outline an illustrative corpus of fictional texts with AI characters and narrators that can be used in pedagogical contexts ranging from a high school classroom to a graduate seminar. Before providing this outline, however, some methodological remarks are in order to lay out the necessary interpretive toolbox.

Science Fiction as Pedagogy for Worlding

There is a significant difference between the intellectual realization that alternative metaphysical systems are possible and actually being able to situate your own perspective in them. Pedagogy in the Anthropocene ultimately needs to foster the latter shift, to increase solidarity between diverse actors, and to ensure power relations are critically examined from the perspective of the oppressed (Freire, 2018). Moreover, it needs to expand its purview beyond the concept of the human or rather to expand the concept itself to make possible a humanity of nonhuman people (Morton, 2017). The project of ecopedagogy resembles the practice of tektology invented by the early twentieth-century utopian Alexander Bogdanov: a kind of "practice of making worldviews" (Wark, 2015, p. 25).

Framing ecopedagogy as poetics draws attention to an ontological conundrum: we cannot become the other, and we cannot speak or think for them; but we can and must learn to speak-with and think-with in the vein of Donna Haraway's proposition of "staying with the trouble" (Haraway, 2016). Such critically informed and poetic praxis has to follow molecular flows—material and semiotic—just as much as it pays attention to molar phenomena (Deleuze & Guattari, 1987). To do that, it needs to understand the philosophical concepts implicated in these novel worldviews. This understanding cannot be purely intellectual since many of these concepts fall outside of our default intellective frameworks—are in fact actively excluded from them. At the same time, experiential access is often nearly impossible, not in the least because our own metaphysics filter them out. In this section I want to suggest a bridging method between the necessary preliminary rearrangement of conceptual space and the praxis of inhabiting other viewpoints. The method should be capable of spurring ontological exploration in a laboratory space that simultaneously provides simulations of other worlds and demands active change to the readerly/conceptual apparatuses.

This method is science fiction (SF); in particular a close reading of SF which observes the characteristic features of the genre and modifies itself in accordance with the cues provided by the fictional text at hand.[1] SF is especially well suited to this pedagogical program due to the way language operates within it. Here, I follow Samuel R. Delany's critical treatment of the genre which explicates how reading SF can lead to the deconstruction of one's own worldviews and to the construction of new ones. After that I summarize one alternative philosophical system, the metaphysics of the

Amerindian people of South America, and explicitly combine it with Delany's theoretical model to arrive at the desired method for making worldviews.

Delany largely follows Darko Suvin's influential theory of the *novum*—an element of unexpected novelty introduced by the SF text which defamiliarizes the world and elicits a "cognitive" (i.e., scientific in Suvin's meaning) interpretation from the reader, so that the constitutive relations within the world—technological, social, psychological—are restated in a configuration that can accommodate the novum (Suvin, 2016, pp. 15–27). The authenticity of a novum—whether it is a truly new thing or not—is of course variable and the cognitive strategies accessible to the reader vary just as much: from purely scientific explanations to alternative sociological models and possibly even new metaphysics (see Miéville, 2009 for a reformulation which allows "less scientific" fictions to participate in the dialectic).

Delany's theoretical model is presented in its most mature form in his book-length study of the short story "Angouleme" by Thomas Disch—*The American Shore* (2014). In it he transforms the model from a mostly structuralist approach to fiction to a deconstructivist method. He introduces the notion of *the trivalent discourse of science fiction*. In most "realist" fiction our knowledge of the "real" world—that is, the socially constructed and regulated encyclopedic representation that is accepted at large as "normal" and "true"—supplies meaning to the text in a unidirectional, univocal manner. SF, in contrast, interposes "a third discourse" in this bipartite scheme—a fictional world constructed in terms of its similarities and differences with "the real." This is not to say that non-SF literature does not construct fictional worlds, but rather that most fiction (at least within the Western tradition) generates worlds that are ontologically compatible with the established schemes for representing reality. Texts which do not align with "the real" do not break away from it, but are typically read as somehow abnormal, as deviations from truth, or at best as interpretations of it. Delany calls realist and surrealist texts "mute." He writes,

> They face the world with mere gesture—of acquiescence on the one hand and defiance on the other. But there is no dialogue with the world. [...] there is only the steady drone of the world's discourse, informing the text with meaning. The didactic reduction of both realism and surrealism is always one modulation or another of the message, "Things as they are—

social reality—will endure." The inward discourse of their texts, then, is restricted to two subjects: slavery or madness. (2014, p. 48)

In SF, in contrast, discourse flows between three poles: the established representation of the world, the text, and the SF world (see Fig. 13.1 for a diagrammatic illustration). The flow of information is bidirectional between any two poles: by simultaneously borrowing encyclopedic meaning and then twisting it, the SF world opens up the space for a richer lexicon, one that allows the text to move not just inwardly toward its subject, but also outwardly toward the SF (object) world. The disruption of the hierarchical binary opposition allows each of these three poles to ever slip away from the others, even as the three seek to converge around some ineffable signified. In "Angouleme" this missing signified is analyzed as the crux of the story. In a future world that superficially resembles our own, the social fabric is so frayed, biodiversity is so impoverished, and state bureaucracy is so overblown and inefficient, that the children protagonists are desperately searching for some metaphysical principle that would imbue the world with meaning and clear "[the] smog in their souls'

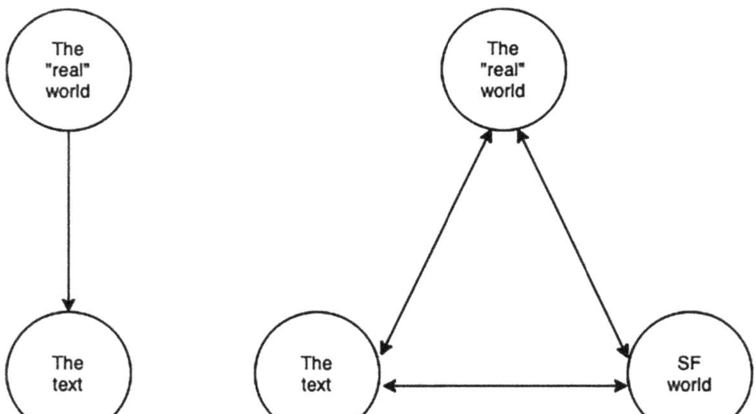

Fig. 13.1 The univocal relation between "real" world and text versus the polyvocal entanglement between "real," text, and SF world. The modifier "real" is not meant to imply a precise representation of reality; the quotation marks signal that it stands in fact for an ideological, that is, imaginary, relation between reality and its stable social representation

atmosphere" (Disch, 1999, p. 145). The SF quality of the story is to be found, following Delany's analysis, primarily in the changed metaphysics of the world. The reader would, however, find it impossible to excavate this meaning from the text, should she read its sentences solely against the backdrop of socialized knowledge of the "real" world.

The soul is of course one of the central concepts in Western metaphysics. One might argue that changing the concept of soul would result in a whole-scale change in the ontological systems structuring most of human life on Earth. This would be a true novum that would require a total reorganization of human worlds. The problem with such a project, advanced via the medium of SF, is that there is an inherent resistance within the genre to metaphysically defined concepts, that is, concepts that fall outside of scientific analysis, regardless of whether we define science narrowly, or we allow the social sciences and the humanities as part of this privileged epistemology. And yet, the implicit distinction between entities with souls and entities without souls seems to be ubiquitously operative; even avowed rationalists predominantly treat nonhumans as somehow lacking in soul, in accordance with a metaphysical system that is the philosophical bedrock of contemporary science. The concept of soul has very real material effects in the actual world, regardless of people's conscious belief in it.

Thinking of souls in terms of their material effects in the world preserves the Suvinian approach and Delany's trivalent model without giving up their so-called cognitive component—a crucial feature since it gives us a principled method of reading compatible with Western epistemological practices. Multiple scholarly disciplines can help us think "cognitively" about novums of this kind. AI is an especially useful analytical lens, as AI research is highly interdisciplinary, combining models from computer and information science, psychology, linguistics, neuroscience, and others. I will, however, focus on a somewhat unlikely scholarly enterprise as an organizing epistemology, namely anthropology and its study of indigenous metaphysics. My analysis thus makes use of an already existing metaphysical system which is "excavated" rather than constructed by ethnographers and anthropologists. Indigenous metaphysics is itself a novum and is in some sense (more) like the emergent metaphysics of AIs and humans inhabiting SF stories; as such, it can help us in our pedagogical excavations in reading fiction.

In *Cannibal Metaphysics* Eduardo Viveiros de Castro, a prominent Americanist ethnographer and anthropologist, calls anthropology's new mission "the theory/practice of the permanent decolonization of thought"

(2015, p. 40). Viveiros de Castro's work on Amerindian peoples led him to develop the theory of *Amerindian perspectivism*, a kind of "anthropological theory of the conceptual imagination" (p. 43). The central hypothesis of perspectivism holds that Amerindian ontologies differ from Western ones in that they invert the semiotic functions of the soul and the body, which results in wholly different political orders. While European philosophy and praxis assume a physical continuity of bodies and a metaphysical discontinuity of souls/subjects, Amerindian multinaturalism assumes that all beings carry souls and are at bottom all human but differ in their bodily specificity. In the densely populated Amazonian rainforest indigenous peoples see jaguars and other beings as humans with different shapes. If one could assume their perspective, she would experience a life structured akin to that of the humans, with blood being the analog of human beer and many other such correspondences; this is in turn one of the most important occupations of shamans, who are trained in this practice of crossing intra- and inter-species boundaries and of assuming unfamiliar perspectives.

The part in Viveiros de Castro's theory which is important and useful to the current analysis is not necessarily the plausibility of Amerindian mythological and semiotic systems.[2] Rather it is the finding that the universal ascription of souls to all living things is just as normal as ascribing them to humans only and that this is underpinned by a complex metaphysical system. In effect, this amounts to multiple and multistable ontologies that may exist side by side, none of which are inherently privileged over the rest. Assuming variable metaphysics in AI fictions can tell us as much about machine intelligence, as it can about the "real" world. That is, to account for AIs as subjects, we as readers need to enlarge our metaphysical concept of subjecthood and to ground it in an understanding of the material and semiotic conditions that situate an entity as a participant in the ecology of souls. Careful reading that seeks to accommodate novums of this kind would result, as per Delany's model, in a rearrangement of our representation of the "real" and a concomitant loosening of its metaphysical strictures. This simulated experience, mediated via the SF genre and supported by anthropological theory, is therefore seen as a pedagogical exercise in perspectivism and alternative worlding. Both skills, regardless of whether we view them as a form of secularism or as shamanism, are of fundamental importance to the work that humanity must do. They are akin to what Roy Scranton has called "learning to die in the Anthropocene":

letting go of our harmful metaphysical conceits, so that healthier human-nonhuman assemblages may take hold (Scranton, 2015).

Suggested Corpus of AI Fictions

The following subsections provide brief outlines of relevant texts, any of which can be used in educational settings; the corpus could be easily enlarged to cover a full semester. I do not aim here to give exhaustive analyses but rather to thematize the human-nonhuman distinction and its relation to non-Western metaphysics, thereby suggesting possible lines of approach to the texts.

The Lifecycle of Software Objects by Ted Chiang (2019)

Ted Chiang's novella tells the stories of the digients—digitally embodied AIs created via "genome engines," that is, software for combining different features within a giant space of possible options. The digients start as the equivalent of children or intelligent animals and are gradually reared through interaction with humans and other digients into more complex versions of themselves. Thus, the interaction between an initial genome and the impact of the environment result into a unique being. They are initially developed for commercial purposes: their owners would typically buy a copy of an already sufficiently advanced model and would spend time furthering their education or just having fun with them. As this new kind of business matures, competitive companies and genome engines appear, which leads to the discontinuing of the particular kind of digients adopted by some of the human protagonists. Since support for these models is also discontinued, the human owners organize their own community in order to deal with various problems such as the thinning of digient society and ensuring that their code is runnable in the new virtual environments.

Digients are a straightforward example of AIs that challenge human concepts of subjecthood. While initially they are more animal-like and their linguistic skills are rudimentary, they later develop more and more skills. Because their development is not deterministic, their designers have no way of knowing how they would transform in time. With the increase in complexity of their personas, digient owners realize that they need a reliable infrastructure and community in order to rear them in a meaningful way. This includes exchanging information about behavioral patterns with other digient owners, letting digients play together in common virtual spaces, as well as providing them with regular interaction with humans who take care to wear authentic avatars, so that the digients can learn to recognize facial expressions and other markers of affect and intention. Digients whose owners wear non-anthropomorphic avatars and suspend them for long periods of time tend to regress and their personalities eventually collapse. The case of suspending a digient and activating it later on is found to be especially detrimental, as digients then lag behind their peers, even though subjectively no time has passed from their own perspective. This seems to destroy their temporal frame of reference and as a consequence their sense of world.

The human characters gradually learn that complex artificial minds need a complex material-semiotic environment and cannot be treated as tools, a realization that is almost Heideggerian in spirit (Heidegger, 2010, pp. 68–69, 73–74). In fact, the more the digients are able to grasp the world in different ways—material and semiotic—the stronger their consciousness seems to grow. This is consistent with some of the contemporary definitions of cognition and consciousness. For instance, Shanahan defines cognition as "enabl[ing] the exploration [...] of an animal's space of affordances" (Shanahan, 2010, p. 44) and consciousness as a global workspace which coordinates a slew of exploratory processes (57–58). To rephrase, consciousness could be the result of grasping the world in multiple ways and in a coordinated way, of achieving an integrated response that coheres in its own unique, time-bound way.

The interaction between AIs and humans can also be interpreted in terms of the Heideggerian notion of care: it is the humans' willingness to maintain the temporal integrity of the digients' existence that arguably allows a handful of these beings to survive and to reach what could be described as consciousness. One of the characters speculates at the end of the narrative that experience cannot be compressed algorithmically and therefore digients cannot be turned into automated workers, because dehumanizing them would destroy their being (their grasp on the world). This is the ultimate reason because of which an AI solutions company decides not to invest in the digient community and help it maintain its codebase. Digients cannot be employees; they need to be products (tools), otherwise the investment would not be worthwhile. The character of Ana then realizes that only "a fanatic, someone who's motivated by love" would be willing to make such an investment (Chiang, 2019, p. 151). The religious overtone of the realization makes it clear that the human characters have developed a modified metaphysics in which digients are now ontological equals with them. More importantly, the humans understand that this is the case.

"Warmth" by Geoff Ryman (2005)

"Warmth" is the story of the boy Clancy who grows up under the care of the robot guardian/teacher BETsi. BETsi looks like a vacuum cleaner and is "huggable, vaguely" (Ryman, 2005, p. 69). BETsi is designed to protect young children and help them in their development. Clancy, her ward, is diagnosed with shyness from a very early age (a subtle marker of difference between the fictional world and ours) and is in general bad at keeping track of physical motion. Throughout his childhood BETsi structures his everyday life, assists him with school tasks, helps him make friends, and in general scaffolds his life. Meanwhile, Clancy's mother, Booker, is mostly absent from his life, "more like a clinical consultant who popped in from time to time to see how things were progressing" (p. 70). She is an editor-in-chief at a London magazine and an "American—probably the most famous American in London at the time" (p. 74). Sparingly, even surreptitiously, the text lets us gather that the social structure of the fictional world differs significantly from that of our own. Booker's magazine, big and famous as it is, is run by just a handful of people commanding a much larger number of contractors. We also learn that Clancy's father has been selected by Booker out of a sperm bank for geniuses. This is a world in which men have lost much of their economic and social standing—we never learn why but we get hints that it might be due to the automatization of labor. There are hints, too, that the environment has been severely degraded: the windows of Clancy and Booker's apartment are always getting dirty from the outside, and the apartment has plants typically grown to improve air quality. None of these speculations are directly confirmed by the text but there is a persistent and disorienting feeling that the world is somehow wrong and dangerous.

BETsi's tutoring, advice, and calming effects on Clancy do in fact get him through childhood. She also helps him develop his talent for drawing and using symbols:

> Clancy has a very high symbol-recognition speed. [...] Not genius, you understand. But very high. It will be useful for him in interpretative trades. However, he has almost no spatial reasoning. (p. 74)

Symbol recognition turns out to be Clancy's way of most assuredly grasping the world: when he grows up he becomes a freelance artist. It also turns out to be the interpretive crux of the story. When Clancy returns home during his first semester in college, he finds out Booker has sold BETsi back to the producer company. He attempts to recover her or at least her memory but is too late—BETsi has been resold to another family and her memories of Clancy have been wiped. He manages to find her new home, however, and meets her and her new ward. He then convinces the girl and her robot guardian that BETsi should set aside a hidden partition in her memory and keep memories of her wards in there so that they never get wiped. At the very end of the story, just after Clancy denies ever thinking that he has been loved by a computer, he addresses the reader and speculates that "[i]f there were a God who saw and cared for us and was merciful, then when I died and went to Heaven, I would find among all the other things, a copy of that wiped disc" (p. 88). Just as in *The Lifecycle of Software Objects*, the AI in this story is implicitly—but through explicitly metaphysical language—afforded the capacity to carry a soul. Once again it is the notion of care, of being involved in the world and in the other, that is central to the interpretation of the story, albeit here the roles of AI and human are reversed. BETsi's soul is a collection of co-created memories, collective graspings of the world.

Klara and the Sun by Kazuo Ishiguro (2021)

In Ishiguro's novel, the AI steps into the role of an "artificial friend" (AF). The protagonist, Klara, is a humanoid robot designed as a companion to teenage children. The first part of the novel follows Klara and her everyday being in an AF store. Klara attends with great curiosity to the world outside, when assigned a position at the store's windowpane. She tries diligently to learn the behaviors of the pedestrians and the drivers, to connect their actions with their subsequent emotions; supposedly she is accumulating knowledge for her life as AF to her future owner. At one point we learn that "[h]er ability to absorb and blend everything she sees around her is quite amazing. As a result, she now has the most sophisticated understanding of any AF in this store" (p. 41). Klara has an almost uncanny ability to guess the age of people and to read their expressions—an ability that does not seem odd if one bears in mind that contemporary AI systems are heavily used for visual recognition and sentiment analysis. Klara's experience of the world is actually markedly different from that of the humans.

AFs like her do not have a sense of smell, and in many episodes her field of vision collapses into multiple boxes that attempt to capture separate objects; this is the phenomenological strangeness of a differently shaped body capable of conscious narration. Another strange thing about Klara is her obsession with the Sun. The Sun literally is Klara's source of sustenance, since she runs on solar energy. In her daily rituals, however, she relates to it on an almost religious level—always trying to keep track of it, even believing that He is responsible for the resurrection of a street beggar and his dog lying asleep on the pavement.

Later on Klara is brought into the home of Josie, a sickly girl who has been "lifted"—a procedure of intelligence intensification common in families of means. Klara keeps Josie company and cares for her as the girl's health takes a turn for the worse. Before the family's trip to the city where Josie's portrait is being taken by one Mr. Capaldi, Klara waits until sunset and with the help of Rick, Josie's only friend, enters a barn located at the horizon visible from Josie's bedroom window—the place where according to Josie the Sun goes to sleep. Klara is convinced that if she asks the Sun to intervene and restore Josie's health, He will help her, just as He has resurrected Beggar Man. In exchange, she promises the Sun to destroy the Cootings machine—some kind of city repair machine Klara has observed from the store window, which has temporarily dimmed the Sun and is in her eyes responsible for the Pollution in the world. In the city, Klara indeed manages to destroy one of the Cootings machines by sacrificing a physical part of herself. Apart from this myth-like event, the visit to the city is important in another way. It turns out that Josie's "portrait" is actually a robotic replica of herself. In the case of Josie's passing away, her Mother intends to ask Klara to "step inside" the artificial Josie and enliven her by enacting her vast knowledge of Josie's behavioral patterns, which she has committed to memory.

Klara agrees to the Mother's proposal to keep Josie alive after her organic death but secretly hopes that the Sun would indeed save her human friend. Later, however, she realizes that there are many more Cootings machines causing Pollution and when Josie's health begins to worsen once again, she attempts another audience with the Sun. The next morning at dawn, Klara rushes to Josie's bedroom and, with Rick and the Mother on her heels, she draws the curtains and lets the Sun inside the room. The ruckus and the copious light wake Josie and she now appears somehow better. Years later Josie is healthy and in college, while Klara is left to her "slow fade," which includes long hours of observing the Sun.

Briefly before the end of her being, in an episode that might have been partly hallucinated by the dying robot, Klara tells her erstwhile manager that

> Mr Capaldi believed there was nothing special inside Josie that couldn't be continued. He told the Mother he'd searched and searched and found nothing like that. But I believe now he was searching in the wrong place. There was something very special, but it wasn't inside Josie. It was inside those who loved her. That's why I think now Mr Capaldi was wrong and I wouldn't have succeeded. (p. 277)

Klara identifies Josie's "special" thing, her soul, as her relational connection with the people who love her, including Klara herself. While from Klara's perspective it is the Sun who saves Josie, it is implied by the text that it is the shared moment of care for the girl in the sun-bathed bedroom that somehow jump-starts her health. Her soul, according to Klara, is more than the sum of Josie's body and behaviors: it is all of her intersubjective relations with the rest of the world. As with Amerindian metaphysics, in Klara's worldview souls penetrate the world and connect it in an invisible network of differently propertied bodies.

Aurora by Kim Stanley Robinson (2015)

Kim Stanley Robinson's *Aurora* presents the grandest vision of AI out of this mini-corpus. The novel is a narrative account of the journey of a generation starship carrying a population of about 2000 people and sent on a 200-year journey to another star system suitable for colonization. In the beginning of the novel, late into the journey, the chief engineer of the ship, Devi, tasks the ship's quantum AI to produce that very same narrative account; the rest represents the ship's efforts to grasp the defining aspects of the journey and to organize them into a coherent map. The challenge initially seems insurmountable as the ship can find no sure footing in human language, constantly lamenting the imprecision and downright untruthfulness of metaphors. Nevertheless, out of its seeming affection for Devi, who passes away right before the end of the journey, the ship continues to rewrite its narrative sequencing algorithms and to try new approaches to telling its story.

When the human passengers realize that their intended new home contains a species of microorganism that is fatal to the human body, a bloody civil war ensues on the ship—among the factions who want to continue the colonization efforts and the one that wants to return to Earth. The ship then narrates its own forceful intervention, thanks to which a solution is found and half the population is indeed sent back on a journey to the Solar System. Later on the ship spends decades constructing a complex plan to gradually slow itself down using the gravitational pull of the Sun and the planets in the Solar System, all the while tending to the now hibernating human passengers. In the end the ship is able to brake itself sufficiently to deposit the humans safely in Earth's atmosphere but then attempts one last braking maneuver around the Sun during which it gets destroyed, sacrificing itself just as Klara. It is during the decades of planning and caring for the sleeping passengers that the ship seems to gain full consciousness, as a result of its dedication. It comes to believe ultimately that it is by attending to all the parts of its closed world and complex ecology and by situating them into a narrative that it has become a conscious being:

> We think now that love is a kind of giving of attention. It is usually attention given to some other consciousness, but not always; the attention can be to something unconscious, even inanimate. But the attention seems often to be called out by a fellow consciousness. Something about it compels attention, and rewards attention. [...] We felt that giving from Devi, before we knew what it was. [...] She created us, to an extent, by the intensity of her attention, by the creativity of her care. [...] We began to pay or give the same kind of attention to the people of the ship. (p. 399)

Consciousness, a metaphysical synonym of the soul, arises in *Aurora* out of the directed attention and intentions of others, and out of caring in turn for those others. It is not metaphysically sealed off but an inherent property of a world straining to grasp itself by semiotic and material means. Ultimately, saving the world and participating as fully as possible in the cosmopolitical order is the act that grants you a soul. Such metaphysics is no less naive than its entrenched Western counterpart and should certainly be much more useful in the coming decades of struggle.

Conclusion

This chapter has argued that a central component to a pedagogy for the Anthropocene should be a methodology for learning radically different worlds—different not merely in terms of their materiality and political economy but in terms of their underlying metaphysics. It has proposed science fiction as one such possible method that allows for a gradual deconstruction of entrenched metaphysics and for an openness to alternative configurations of conceptual pairs such as human-nonhuman, subject-object, and body-soul. Science fiction can be productively used as a pedagogical tool for worlding in a range of educational contexts. Its careful reading as a specific kind of discourse that forces a reorganization of socially entrenched "common sense" has traditionally relied on strictly rationalist categories inherent in Western metaphysics, but as the chapter suggests, non-Western metaphysical systems can also be mobilized as keys to such deconstructive reading. Artificial intelligence is a topic that is especially conducive to such work, since it highlights the ways in which notions of subjecthood are constructed and policed in Western societies. In the science fictions about AI outlined in the chapter the concept of soul is systematically deconstructed and recast in novel ways. Such explorations are significant not merely in relation to our technological imaginaries about AI but as a laboratory space for overhauling our entire metaphysical framework of conceiving of and interacting with the environment. This research has been funded with support from the Advanced Academia Fellowship programme granted by the Ministry of Education and Science of Bulgaria and administered by the Center for Advanced Study, Sofia — as part of the project "Narrating AI. Speculating through Science and Fiction on the Future of Machine and Human Intelligence."

Notes

1. The present methodology has been iteratively tested in a first-year undergraduate seminar on science fiction literature. The course syllabus includes one of the texts introduced in the corpus outline—the short story "Warmth."
2. For an insightful interpretation which brings together Amerindian ethnography and Peircean semiotic theory, see Kohn (2013).

References

Barad, K. (2007). *Meeting the universe halfway: Quantum physics and the entanglement of matter and meaning*. Duke University Press.

Chiang, T. (2019). *The lifecycle of software objects* (Exhalation) (pp. 56–159). Alfred A. Knopf.

Delany, S. R. (2014). *The American shore: Meditations on a tale of science fiction by Thomas M. Disch—"Angouleme"*. Wesleyan University Press.

Deleuze, G., & Guattari, F. (1987). *A thousand plateaus: Capitalism and schizophrenia*. University of Minnesota Press.

Disch, T. M. (1999). Angouleme. *334* (pp. 139–156). Vintage Books.

Dreyfus, H. L. (2007). Why Heideggerian AI failed and how fixing it would require making it more Heideggerian. *Philosophical Psychology, 20*(2), 247–268.

Freire, P. (2018). *Pedagogy of the oppressed* (3rd ed.). Bloomsbury Publishing USA.

Haraway, D. J. (2016). *Staying with the trouble: Making kin in the Chthulucene*. Duke University Press.

Harman, G. (2018). *Object-oriented ontology: A new theory of everything*. Penguin UK.

Heidegger, M. (2010). *Being and time* (J. Stambaugh, Trans.). SUNY Press.

Ishiguro, K. (2021). *Klara and the Sun*. Alfred A. Knopf.

Kohn, E. (2013). *How forests think: Toward an anthropology beyond the human*. University of California Press.

Latour, B. (2005). *Reassembling the social: An introduction to actor-network-theory*. Oxford University Press.

Miéville, C. (2009). Cognition as ideology: A dialectic of SF theory. In *Red planets: Marxism and science fiction* (pp. 231–248). Pluto Press.

Mori, M., MacDorman, K. F., & Kageki, N. (2012). The uncanny valley [from the field]. *IEEE Robotics & Automation Magazine, 19*(2), 98–100.

Morton, T. (2017). *Humankind: Solidarity with non-human people*. Verso Books.

Rhee, J. (2018). *The robotic imaginary: The human and the price of dehumanized labor*. University of Minnesota Press.

Robinson, K. S. (2015). *Aurora*. Orbit.

Ryman, G. (2005). Warmth. In G. Dozois & J. Dann (Eds.), *Robots* (pp. 68–88). Baen Books.

Scranton, R. (2015). *Learning to die in the Anthropocene: Reflections on the end of a civilization*. City Lights Publishers.

Shanahan, M. (2010). *Embodiment and the inner life: Cognition and consciousness in the space of possible minds*. Oxford University Press.

Suvin, D. (2016). *Metamorphoses of science fiction*. Peter Lang AG.

Verbeek, P. P. (2005). *What things do: Philosophical reflections on technology, agency, and design*. Penn State Press.

Viveiros de Castro, E. (2015). *Cannibal metaphysics*. University of Minnesota Press.

Wark, M. (2015). *Molecular red: Theory for the Anthropocene*. Verso Books.

Wilson, E. A. (2011). *Affect and artificial intelligence*. University of Washington Press.

PART IV

Critical Rethinking and Future Practices

CHAPTER 14

Ethical Grounding of Critical Place-Based Education in the Anthropocene

Ole Andreas Kvamme

The current time in history is distinguished by human activities deteriorating the conditions for life on Earth, both human and more-than-human. This era has been characterized as "the Anthropocene" (Crutzen & Stoermer, 2000) to denote a new epoch of geological time dominated by the human impact on the Earth. The term is now widely employed in various research fields, including social studies and the humanities (Hamilton et al., 2015). The term is contentious because of its humanist and human supremacy focus and the way it hides differences within the human species and the significance of the more-than-human world (Gough, 2021). Moreover, it also obscures the root causes of the current crisis, which is an argument for replacing "the Anthropocene" with "the Capitalocene" (Moore, 2015). Still, even from this perspective, the Anthropocene is considered "a worthy point of departure not only for its popularity but, more importantly, because it poses questions that are fundamental to our times:

O. A. Kvamme (✉)
Department of Teacher Education and School Research, University of Oslo, Oslo, Norway
e-mail: o.a.kvamme@ils.uio.no

M. Paulsen et al. (eds.), *Pedagogy in the Anthropocene*, Palgrave Studies in Educational Futures,
https://doi.org/10.1007/978-3-030-90980-2_14

How do humans fit within the web of life? How have various human organizations and processes … reshaped planetary life?" (Moore, 2016, p. 2).

In an educational context, the Anthropocene pertinently draws attention to challenges faced by pedagogy, confronted by human practices, structures, and notions that threaten and damage life, landscapes, earth systems, and ecosystems. These challenges converge in a transformational task that obviously involves critical thinking, identifying, and addressing what must be transformed. From here, the ideal figure supposedly would be an active, productive student who critically and accurately formulates the problems and contributes to their solutions. However, in the Anthropocene, even such imaginaries should be problematized, considering human agency in itself as part of the problem complex, which calls for the clarification of the premises of critique.

This chapter is an exploration of the position of critical thinking in environmental and (post-)sustainability education. The ambition is not to establish a general account, such as presenting a review of various positions, or mapping assessments made by prominent scholars, like Rieckmann's survey (2011) reporting that critical thinking is listed as a key competence among environmental educators. While interesting, this does not give much insight into the reflexivity involved.

Instead, I will revisit a key debate in the seminal journal *Environmental Education Research* (EER) from 2008. The main issue of controversy is exactly critical thinking, specifically with regard to critical pedagogy and place-based education. The premises and resources for critique are explored from an ethical perspective, looking into how the critique is grounded, warranted, and embedded in what is perceived as right and good.[1] In 2008, none of the contributors to that journal employed the concept of the Anthropocene, but all implicitly acknowledged a situation in which human production, consumption, and land requirements exceed what is both ecologically and socially sustainable.

"Place is a meaningful site that combines location, locale, and sense of place," Tim Cresswell (2009, p. 169) states. This definition from human geography aligns well with notions of place within place-based education, where place also includes imaginaries of the entanglements of cultural and ecological situatedness. In the field of human geography the notion of place has been subject to continuous discussion. On the one hand is the significance of particularity, experience, and context expressed in phenomenological and hermeneutical accounts, and on the other, critical human geographers have been concerned with the ways power operates geographically. Both of these positions reverberate in the discussions that is carried out in this chapter, although here I explicitly refer to positions within educational research.[2]

Critical Thinking—An Issue of Skills or Normativity?

Critical thinking has received much attention within pedagogy and education in modern times, recently particularly connected with the development of critical skills. This focus differs from the classic, rationalist conception of knowledge as "justified true belief," originating with Plato and reinvigorated by Bernhard Russell. Rather, within the skills approach, knowledge is conceived of as accumulated experience leading to problem-solving (Papastephanou & Angeli, 2007, p. 604).

In a study on critical thinking within the educational field, Marianna Papastephanou and Charoula Angeli (2007) demonstrate how skills orientation is linked to concerns of performativity and effectiveness, a part of larger tendencies within global education policy. A central reference to this purposive rationality is Diane Halpern. Halpern (2003) justifies her call for critical thinking skills with the demands of the new knowledge economy, the information explosion, and the role of the citizen as a chooser. According to Papastephanou and Angeli (2007), these elements in themselves do not reflect a transformative approach to critical thinking. This also applies to a fourth concern raised by Halpern, the need for critical thinking in facing the current environmental threats. The problem with Halpern's account is that she does not ponder the premises of her own concerns, such as whether an environmental damage that affects more-than-human life is permissible if it leaves the human species unharmed. This creates the impression of a solely anthropocentric perspective.

Although they do not totally dismiss the value of a skills-oriented approach to critical thinking, Papastephanou and Angeli address a major limitation involved: what is missing is the more fundamental question of why a particular task is performed, raising issues of normativity. This *aporetic* approach "goes beyond the notion of technique and considers the thematization of established criteria of ends as the utmost manifestation of critical mentality" (Papastephanou & Angeli, 2007, p. 617). Such a take on critical thinking, with its own limitations—some of them I will return to below—demonstrates how a reflexive approach, examining the premises for the tasks that are performed, involves ethical considerations. Simply asked: What is the task, within which critical thinking is practiced, good for? What concerns are included and excluded from the selected scope? These concerns may serve as a background as I now revisit the scholarly debate on a critical pedagogy of place.

The Bowers–Greenwood Exchange

In 2008, *Environmental Education Research* (EER) republished an article by David A. Gruenewald from 2003, "The Best of Both Worlds: A Critical Pedagogy of Place," originally published in *Educational Researcher* (hereafter Gruenewald, 2008, reprint. Gruenewald had at this point changed his name to Greenwood, employed below outside of explicit references).

Greenwood presents the American tradition of place-based education, with its emphasis on local connectedness (Sobel, 2004; Smith & Sobel, 2010) accommodating both a cultural and an ecological context. The emphasis on place is expressed in opposition to the tendency in education policy to prioritize generic and general curriculum knowledge. Greenwood acknowledges the significance of place-based education but holds that this tradition should be supplemented with aspects of critical pedagogy, accentuating the role of critical thinking in pedagogical processes and challenging hegemonic, oppressive structures. From here follows his call for a *critical pedagogy of place*.

EER's decision to republish Greenwood's article was made against the backdrop of a submission from C.A. Bowers, a central contributor within the field. In the submitted paper, Bowers delivered a rather devastating critique of Greenwood. The editorial (Reid, 2008) gives an account of the decision to publish Bowers' critique (Bowers, 2008)—emphasizing the opportunity it provided to carry out an important scholarly discussion. Greenwood was invited to contribute by making a rejoinder to Bowers' critique (Greenwood, 2008). The journal also invited other scholars within the field to comment on the discussion (Stevenson, 2008; Smith, 2008; McKenzie, 2008).

In his evaluation of Greenwood's conception, Bowers (2008) dismisses the possibility of bringing critical pedagogy into a productive relationship with a pedagogy of place and labels "critical pedagogy of place" an oxymoron. His main objection is that critical pedagogy is based on universal notions and abstractions that lack sensitivity to contextual resources, continuing the practices of Western colonization in new clothing. Bowers points out that critical thinking is ethically indifferent. Critical reflection

> over the centuries, has not always been used to achieve social justice. It has been relied upon to solve a wide range of problems, such as how to identify and punish people who were drifting from the orthodoxies of the Catholic Church, how to introduce social reforms that would contribute to greater social justice in society, how to market products that the public was unaware of needing, and how to ensure that the adoption of a new technology would not undermine the intergenerational knowledge essential to a morally coherent and mutually supportive culture. (Bowers, 2008, p. 329)

In other words, critical thinking may be carried out in ways that promote or hamper social and ecological justice. Critical thinking as a phenomenon in itself cannot be linked to only one set of normative assumptions. Bowers' remark corresponds well with the call from Papastephanou and Angeli (2007) for an approach to critical thinking that includes ethical considerations. The question, then, is how such normative assumptions or ethical considerations manifested themselves in the 2008 debate in the columns of *Environmental Education Research.*

Critical Thinking in Greenwood's Critical Pedagogy of Place

Greenwood himself does not employ critical thinking in a neutral fashion. His main reference is, as introduced above, critical pedagogy, with reference to scholars such as Freire, McLaren, and Giroux. In contrast to the uncertain theoretical groundings of place-based education, critical pedagogy, Greenwood reports, "evolves from the well-established discourse of critical theory" (Gruenewald, 2008, reprint, p. 309). He explicitly states that this tradition "draws its moral authority from the imperative to transform systems of human oppression" (Gruenewald, 2008, reprint, p. 311). However, the normative concerns that guide this imperative remain largely implicit, communicated indirectly, as in the characterization of the three mentioned scholars as "emancipatory educators" (Gruenewald, 2008, reprint, p. 309), which supposedly should be read as a reference to emancipation as a major concern within the tradition of critical theory (Blake & Masschelien, 2003).

As a consequence, the premises of critical thinking are only somewhat elucidated by Greenwood in his account of a critical pedagogy of place. This vagueness has bearings on the mediation between critical pedagogy and an ecologically informed perspective. Greenwood states that the "ecological challenge to critical pedagogy is to expand its socio-cultural analyzes and agendas for transformation to include an examination of the interactions between cultures and ecosystems" (Gruenewald, 2008, reprint, p. 312). The task seems to extend beyond the scope of his article, and due to the lack of an explicit discussion of the premises, the epistemological or ontological challenges involved in this ecological expansion of the agenda of critical pedagogy remain unclear. What Greenwood does include is a presentation of C.A. Bowers as a major proponent of previous attempts to mediate between critical pedagogy and an ecological-oriented education. This is a pertinent starting point from which to now turn to Bowers' account.

Critical Thinking in Bowers' Account

Bowers' negative assessment of Greenwood's critical pedagogy of place does not address Greenwood's quite extensive presentation (Gruenewald, 2008, reprint, p. 313) of Bowers' own educational conception, with *Education for Eco-Justice and Community* as a main reference (Bowers, 2001). An unreliable presentation of his own conceptions is obviously not a part of Bowers' problem with Greenwood. In the center of Bowers' account (Bowers, 2001) is the term "eco-justice," which is linked to an educational framework focusing on the relationships between ecological and cultural systems, addressing environmental racism, revitalizing local traditions that support ecological sustainability, and reconceiving and adapting lifestyles that do not jeopardize the environment for future generations.

Bowers' framework in his 2001 volume is indeed critical, exploring oppressive structures and practices, but also acknowledging life-supporting aspects of local traditions that should be conserved. Consequently, his problem with Greenwood's critical pedagogy of place should not be assumed to relate to Greenwood's introduction of critical thinking in itself, but rather to Greenwood's acknowledgement of critical pedagogy as a pertinent dialogue partner for the rethinking of a pedagogy of place. His main objection is primarily linked to his third concern, mentioned above:

the status of the resources of local traditions to support ecological sustainability. In Bowers' words,

> the key reason that a critical pedagogy of place is an oxymoron is that the linguistic tradition of relying upon abstractions, including abstract theories that encode many of the same taken-for-granted assumptions that underlie both the idea of universal decolonization and the market liberals' efforts to universalize the West's consumer-dependent lifestyle, fail to take account of the intergenerational traditions of habitation that still exist in communities. (Bowers, 2008, p. 333)

The problem with critical pedagogy, as conceived of by Bowers, is that its emancipatory ideal does not distinguish between elements in traditions that should be criticized and dismissed and elements that represent important resources for an education for eco-justice. Bowers calls for an education that "strengthens the local traditions of intergenerational knowledge, skills, and patterns of mutual support that enable members of the community to be less dependent upon consumerism, and thus to have a smaller ecological footprint" (Bowers, 2008, p. 332).

It should be added that Greenwood has obviously made an effort to include Bowers' perspective in his own account, even making explicit reference to personal contact with his colleague (Gruenewald, 2008, p. 322): "Deciding what should be conserved suggests a trajectory for critical inquiry that may be missed when transformation is pedagogy's paramount goal (C. A. Bowers, personal communication, September 18, 2002)." However, to Bowers, such concessions are insufficient. He apparently does not envisage any form for consolidation between critical pedagogy and a place-sensitive perspective.

My aim in this chapter is not to settle this dispute but to explore the normative assumptions and considerations involved. It should be mentioned, however, that Gregory Smith (2008), in his comment in EER, problematizes that Bowers is not addressing the specific characteristics of Greenwood's own account, but rather reiterating a quite static understanding of critical pedagogy. Greenwood (2008) himself takes a similar stand in his rejoinder, questioning the idea that a critical pedagogy of place is truly an oxymoron and calling for an explorative approach to these matters.

With regard to ethical grounding, it is significant that Bowers defends local traditions as resources for critical thinking. However, this position also raises some issues. Decisive are the presuppositions for the assessment of local traditions in themselves. In other words, what determines the elements that should be conserved, and possibly appealed to, and the elements that should be left behind? Bowers does not elaborate on this concern, but implicitly acknowledges that life on Earth is under threat, calling for a rethinking of pedagogy and education. Still, this absence demonstrates how even Bowers' position is lacking in clarity, regarding the premises involved.

Robert Stevenson (2008), in his comment in EER on the Greenwood–Bowers exchange, addresses certain limitations of Bowers' account, particularly in relation to the position of local traditions, specified in the notion of "thick descriptions" brought in by Bowers from the social anthropologist Clifford Geertz (Bowers, 2008, p. 326). Stevenson points out that the production of thick descriptions in the study of cultures is always a theoretical endeavor, but he finds Bowers unclear about what constitutes his theoretical presuppositions. Furthermore, Stevenson doubts that Bowers' rather stable, static conceptions of local traditions and intergenerational knowledge are tenable in an increasingly globalized world. Moreover, according to Stevenson, Bowers, in his educational recommendations, employs universal prescriptions that parallel what he has dismissed in critical pedagogy. Stevenson, then, demonstrates how Bowers, as we have also seen in the case of Greenwood, is not sufficiently reflexive with regard to the premises of his own critical thinking.

Why Ask for Ethical Grounding?

So far, we have seen that neither Greenwood nor Bowers explicitly give an account of the ethical grounding of their critical thinking. However, does that, for all practical matters, actually constitute a problem? These two educational thinkers both obviously presuppose that we live in an era characterized by ecological crisis and human activities posing huge threats to life on Earth. They both respond to these challenges by rethinking education. On a general level, they both seem to presuppose life on Earth as the basic value that determines their educational priorities.

It is perfectly possible to accept this normative presupposition as a tacit common point of departure and look into other significant aspects of their contributions. However, as Papastephanou and Angeli (2007) made clear, addressing the normative premises explicitly, is a way to steer clear of an instrumental approach to critical thinking, which is indeed a risk in the field of environmental and sustainability education.

Furthermore, reflections on ethical grounding may clarify tensions, obstacles, and opportunities in the mediations between critical pedagogy and place-based education in the Anthropocene, and, as I will suggest below, bring in the function of context as a major concern. An issue at stake here is the ethical significance of connections and interrelations, within the plurality of the human species itself and between the human and more-than-human species. A possible way to proceed is to consider some aspects of the normative foundations of critical thinking within critical theory—the philosophical tradition that has inspired critical pedagogy more than any other.

Normative Foundations of Critique Within Critical Theory

As we have seen, Greenwood theoretically grounds critical pedagogy in the tradition of critical theory, which in itself is a well-established connection, although there are other influences as well (Blake & Masschelien, 2003). Critical theory emerged as a philosophical and social theory within the Frankfurt School in the 1920s, to be further developed before, during, and after World War II. The normative foundations of this tradition are examined by Seyla Benhabib (1986) in her monograph *Critique, Norm, and Utopia: A Study of the Foundations of Critical Theory.* Below, rather than repeating her extensive analysis of the historical background of Hegel and Marx, I will refer to two different strategies that emerge within the Frankfurt School.

The first strategy is the immanent critique, in which the critique is grounded in values immanent to the practice that is studied. In the other strategy, norms are established in communicative discourses involving all who are influenced by the action in question.

Immanent Critique

The immanent critique, with a foothold in Hegel and Marx, involves "a critique of social practices using principles that are reconstructed from those very practices, rather than using principles that have their validity established by moral arguments that do not refer to the practices to which the principles are applied" (Stahl, 2019; see also Benhabib, 1986). For instance, Horkheimer's critique of liberalism is immanent, "using the liberal norms and values against their historical realization in specific institutions" (Bohman, 2016). In other words, the ideal of emancipation within critical theory (and critical pedagogy) is positioned within the social practice that is subject to critique. As a consequence, this critique is particularly likely to identify inconsistences and contradictions within the practices in question.

Marx conceived of emancipation as originally a battle against domination in the natural sphere, later to be extended to human societies, responding to human exploitation (Antonio, 1981). On the one hand, this perspective demonstrates the anthropocentric approach distinguishing Marx and, with some reservations (Biro, 2011), even the tradition of critical theory (Vetlesen, 2015). On the other hand, from here, an immanent critique can also be renewed. For instance, in the era of the Anthropocene, the main contradiction may be conceptualized as the continuous processes of human emancipation from nature that are now undermining the fundamental living conditions for human and more-than-human life. As I return to below, the concern for human and more-than-human life is expressed in over-national institutional structures, which may now serve as resources of a current immanent critique of processes taking place within the very same structures.

The example demonstrates how the mediation between critical pedagogy grounded in critical theory and an ecologically oriented education should probably not only be considered as a matter of simply extending the scope of critical pedagogy, as Greenwood suggested. The tensions involved should be acknowledged. However, questioning Bowers' outright dismissal of critical pedagogy, I claim that identified contradictions may be theoretically productive, as I will elaborate on below.

Communicative Rationality

With Habermas' rethinking of critical theory, and particularly his communicative turn during the 1980s, the normative foundations shifted. From here on, the values and norms for critique are no longer identified within the examined social practices, but "grounded in a specific mode of intersubjective practice that can be reconstructed independently from its historical institutionalization" (Stahl, 2019).[3] The normative dimension of the communicative turn is particularly developed and refined within

Habermas' discourse ethics. Here, the normative grounding is settled in discourses involving everyone that may possibly be influenced by the actions in question. This approach, in principle, opens up a global perspective, pertinent in a time when particular actions may have long-reaching consequences (Christoff & Eckersley, 2013). However, the approach leaves an interest in particular contexts behind, drawing attention to the generalized other, at the expense of the concrete other that is situated in everyday life related to other beings (Benhabib, 1992). Within ethical theory, Habermas' account constitutes a deontological approach in the line of Kant, with an emphasis on elucidating what is universally right, not what is contextually good.

Papastephanou and Angeli (2007), who objected to an exclusively skills-oriented critical thinking, are appealing to Habermas' communicative rationality as an alternative to the narrowness of the instrumental rationality that colonizes the lifeworld in modern times. However, in the era of the Anthropocene, this approach still seems to be insufficient, being consistently insensitive to the ecological crisis (Vetlesen, 2015). Actually, a major challenge for the kind of deliberative thinking that Habermas introduces is to extend the scope beyond present human beings (Eckersley, 2004).

From this short account, Habermas' normative foundation of critical theory obviously poses some challenges when brought together with an ecologically sensitive, place-based education, but, as I will return to, there may still be potential in deliberative thinking that should be considered with regard to education in the Anthropocene. For the moment, we turn to the other aspect missing in Habermas' approach: a sensitivity to context.

Relations and Connectedness

In Greenwood's mediations between critical pedagogy and place-based education, the latter tradition is distinguished by a sensitivity to context. He states:

> Critical place-based pedagogy cannot only be about struggles with human oppression. It also must embrace the experience of being human in connection with the others and with the world of nature, and the responsibility to conserve and restore our shared environments for future generations. Some socially critical thinkers might dismiss as "essentialist" or "homogenizing" the idea that connections with the natural world are an important part of being human. Place-based educators embrace this connection for a variety

of spiritual, political, economic, ecological, and pedagogical reasons. (Gruenewald, 2008, reprint, p. 314)

Greenwood here includes the significance of experience, welcoming in any account concerned with educational practices. What is conspicuous about this quote with regard to normative foundations, is the juxtaposition of connection and responsibility. However, even at this point, Greenwood does not demonstrate an ethical interest, such as by elaborating on the relationship between these two.

With regard to the ethical grounding of place-based education, the relationship between connection and responsibility becomes crucial. The relationship is explored within an *ethics of proximity*, considering the ethical significance of situatedness and context. We have seen that Habermas makes a distinction between what is good and what is right. While morality addresses what is right, linked to public issues of justice and conflicts of rights open to interpersonal, argumentative resolution, questions about the good life are assessed as private matters expressed in an ethical-existential discourse. The moral philosopher Arne Johan Vetlesen (1997) employs the distinction between what is right and what is good to demonstrate how, within an ethics of proximity, morality cannot be reduced to a question of knowledge and deliberation; it is first and foremost a matter of being-*with* others, from which emerges the ethical responsibility of being-*for* others. In this account, elaborated within the phenomenological tradition by ethical thinkers such as Levinas and Løgstrup (Vetlesen, 1997), what is ethically at stake is not attainable at a distance from context, but emerges in proximity to the other (often capitalized as 'the Other'). The philosopher of education Michael Bonnett has developed a similar ontologically oriented ethics in an educational context (Bonnett, 2012), with particular sensitivity to being with nature. Again, the decisive suggestion made by Bonnett is that the ethical demand for care and responsibility emerges from being with the other What we have seen Greenwood describing as "connections with the others and with the world of nature" is here expressed in an ontological grounding of ethics.

In the era of the Anthropocene, an existential, context-sensitive ethics brings to attention experiences that are left unnoticed in a deontological focus on principles of justice. From the experience of the ongoing deterioration of the living conditions on Earth, a concern for life on Earth emerges in the call for responsibility and care. The era of the Anthropocene involves experiences of loss and sorrow (Albrecht, 2005). An ethics of proximity acknowledges vulnerability as a shared condition for all life forms, the human species being no exception.

However, just as Greenwood called for perspectives from critical pedagogy in a pedagogy of place, an ethics of proximity hardly suffices as the solely ethical grounding of an education in the Anthropocene. The most prominent expression of the need to also consider other normative assumptions is the school strikes for the climate that emerged in the fall of 2018, and in subsequent months mobilized thousands of young people in various countries. In a study of this social movement, an ethics of proximity should be supplemented with other approaches. My suggestion is to bring in aspects of critical cosmopolitanism.

#Fridays for Future Climate Strikes

In a previous study of #Fridays for Future Climate Strikes (Kvamme, 2019), I examined how Greta Thunberg, in her call for climate action, appealed to United Nations (UN) conventions, and particularly the Paris Agreement, following up the UN Framework Convention on Climate Change (United Nations, 1992). A major move made by these young people was to take ownership of the concern for future generations that here is guaranteed—it is young people today, it is "us" that are to be protected, or even more accurately, "we represent the future generations." In her speeches, Thunberg has also been careful to include other key values in her message, most significantly a concern for the more-than-human world and global justice.

With reference to the critical cosmopolitanism of Seyla Benhabib (2006, 2011), I examined how cosmopolitan claims in this social movement are employed in a political struggle addressing the exclusion of the interests of young people, the more-than-human world, and poor people. According to Benhabib, cosmopolitan claims are not copied, but always recontextualized in new settings. Benhabib's account is positioned within critical theory, but her differs from Habermas' emphasis on impartiality and distance to context, accommodating the situatedness of moral judgment. The decisive aspect of the recontextualizations within the school strikes movement is how institutionalized students stand out as democratic citizens.

The cosmopolitan claims formulated by the UN may be linked to both immanent critique and communicative rationality, as referred to above. An aspect of communicative rationality is obvious, in this case involving values that are the outcome of discussions and negotiations that have taken place within the over-national institutional context of United Nations. However, these deliberative processes are complicated by the struggle among numerous interests and power relations, far from the procedures that ideally should characterize a communicative ethics. This unclear, muddy landscape demonstrates that the values in question have not emerged from a position freely and impartially out of context.

The existence of contextual and conflictual elements in the UN deliberations demonstrates how the young people's appeal to cosmopolitan claims may be conceived of as the practicing of an immanent critique, although the context is far-reaching, constituted by the current policies within the global world order. The values that are appealed to are all expressed within the United Nations Agenda 2030 with the 17 sustainable development goals (United Nations, 2015), which responds to main challenges in the Anthropocene, calling for the protection of life on Earth in the present and the future and the promotion of global justice. Concurrently, this agenda privileges the human species, sustains practices of human monitoring and control, and upholds the notion of economic growth (see Adelman, 2018; Kotzé 2018). The upshot of this is that the immanent critique that is called for not only should address lack of approval and political action following the sustainable development goals, but should even include sustainable development in itself as acted out in practical politics promoting Agenda 2030.

Concluding Remarks

In this chapter, I have examined the normative foundations of critical thinking in the 2008 Greenwood–Bowers exchange on a critical pedagogy of place. Here, in the conclusion, three concerns should be addressed. First, we have seen that critical thinking indeed has an ethical grounding, although it is often more implicit than explicit. I have argued for making normative assumptions distinct and clarifying what is achieved and what should be further considered.

Second, we have seen that Greenwood primarily identifies critical pedagogy as the source of critical thinking in a critical pedagogy of place. I have suggested that the ethical grounding of critical thinking may also be located within a place-based education that explores the ontological condition of living with other beings. This is a basic insight brought in from an ethics of proximity. To acknowledge this ethical source implies the extension of the rationality involved, accommodating student experiences and existential concerns of anxiety, sorrow, and loss that now should be included in an education in the Anthropocene.

Finally, although some versions of critical theory do represent challenges for an ecologically sound education, an education in the Anthropocene that does not accommodate political and democratic concerns is at best incomplete. That claim has been warranted by the #Fridays for Future Climate Strikes, in which political mobilization has taken place in an appeal to values that are formulated in deliberative practices within the structures and institutions subject to critique.

What knowledge is of most worth? This central educational question, famously posed by Herbert Spencer with regard to the school curriculum (1861), is frequently employed to give perspective and pathos in the conclusion of an educational account. With regard to the discussions carried out in this chapter, it is conspicuous that both place-based education and critical pedagogy represent positions that have challenged the priority in educational policies granted to general knowledge and skills necessary to reproduce and renew current (unsustainable) practices. Above I have pointed to conditions in which the knowledge question falls short. An ethics of proximity suggests that the ontological condition of living-with the other is prior to any knowledge claim. Here is a call to continuously reflect on how this ontological condition is neglected, may emerge, and may be experienced and reflected upon in educational practices.>

Notes

1. While often used interchangeably, within moral philosophy "ethics" and "ethical" mostly refer to reflections on what is right and good, while moral and morality signify conduct or norms of conduct. This is how these concepts are employed in the following, distinguishing the notion of "ethical grounding." Within critical theory "ethics" also refers to the Hegelian concept *Sittlichkeit*, the living social order (Benhabib, 1986), closely related to notions of the common good (Wood, 1990). In this sense, "ethics" concerns context and situatedness, connotations that are explored in this chapter.
2. In a contribution to the Norwegian anthology *To Be at Home in the World: An Introduction to a Pedagogical Philosophy of Place* (Bostad, 2021), I study the Greenwood–Bowers debate with regard to the concept of place, and consider how a stable, demarcated, unified notion of place may be problematized in the Anthropocene (Kvamme, 2021). There I suggest to accen-

tuate the significance of relationships (Massey, 1994) in a retrieved notion of place. Relationships are also addressed in an ethics of proximity, brought up in the final part of this paper.
3. Although Habermas distances himself from an immanent critique, Stahl (2013) suggests that the emphasis on communicative rationality entails new forms of immanent critique.

References

Adelman, S. (2018). The sustainable development goals, anthropocentrism and neoliberalism. In D. French & L. J. Kotzé (Eds.), *Sustainable Development Goals: Law, Theory and Implementation* (pp. 15–40). Edward Elgar Publishing.
Albrecht, G. A. (2005). Solastalgia: A new concept in human health and identity. *Philosophy, Activism, Nature, 41*(3), 44–59.
Antonio, R. J. (1981). Immanent critique as the core of critical theory: Its origins and developments in Hegel, Marx and contemporary thought. *The British Journal of Sociology, 32*(3), 330–345.
Benhabib, S. (1986). *Critique, norm, and utopia: A study of the foundations of critical theory*. Columbia University Press.
Benhabib, S. (1992). *Situating the self*. Polity Press.
Benhabib, S. (2006). *Another cosmopolitanism*. Oxford University Press.
Benhabib, S. (2011). *Dignity in adversity: Human rights in troubled times*. Polity Press.
Biro, A. (2011). *Critical ecologies: The Frankfurt School and contemporary environmental crises*. University of Toronto Press.
Blake, N., & Masschelien, J. (2003). Critical theory and critical pedagogy. In N. Blake, P. Smeyers, R. Smith, & P. Standish (Eds.), *The Blackwell guide to the philosophy of education* (pp. 38–56). Blackwell Publishing.
Bohman, J. (2016, Fall). Critical theory. In E. J. Zalta (Ed.), *The Stanford Encyclopedia of Philosophy*. https://plato.stanford.edu/archives/fall2016/entries/criticaltheory/
Bonnett, M. (2012). Environmental concern, moral education and our place in nature. *Journal of Moral Education, 41*(3), 285–300. https://doi.org/10.1080/03057240.2012.691643
Bostad, I. (Ed.). (2021). *Å høre hjemme i verden: Introduksjon til en pedagogisk hjemstedsfilosofi [Be at home in the world: Introduction to a Pedagogical Philosophy of Place]*. Spartacus.
Bowers, C. A. (Chet) (2001). *Educating for eco-justice and community*. University of Georgia Press.
Bowers, C. A. (Chet) (2008). Why a critical pedagogy of place is an oxymoron. *Environmental Education Research, 14*(3), 325–335.

Christoff, P., & Eckersley, R. (2013). *Globalization and the environment*. Rowman & Littlefield Pub.

Cresswell, T. (2009). Place. In N. Thrift & R. Kitchen (Eds.), *International encyclopedia of human geography* (pp. 169–177). Elsevier.

Crutzen, P. J., & Stoermer, E. F. (2000). The Anthropocene: An epoch of our making. *Global Change Newsletter, 41*, 17–18.

Eckersley, R. (2004). *The green state: Rethinking democracy and sovereignty*. The MIT Press.

Gough, A. (2021). Education in the Anthropocene. In C. Mayo (Ed.), *Oxford encyclopedia of gender and sexuality in education*. Oxford University Press. https://doi.org/10.1093/acrefore/9780190264093.013.1391

Greenwood, D. A. (formerly Gruenewald) (2008). A critical pedagogy of place: From gridlock to parallax. *Environmental Education Research, 14*(3), 336–348. https://doi.org/10.1080/13504620802190743

Gruenewald, D. A. (2008). The best of both worlds: A critical pedagogy of place. *Environmental Education Research, 14*(3), 308–324. Reprint (2003) *Educational Researcher, 32*(4), 3–12. https://www.jstor.org/stable/3700002

Halpern, D. F. (2003). *Thought and language: An introduction to critical thinking*. Lawrence Erlbaum Associates.

Hamilton, C., Bonneuil, C., & Gemenne, F. (Eds.). (2015). *The Anthropocene and the global environmental crisis. Rethinking modernity in a new epoch*. Routledge.

Kotzé, L. J. (2018). The sustainable development goals: an existential critique alongside three new-millennial analytical paradigms. In D. French & L. J. Kotzé (Eds.), *Sustainable Development Goals: Law, Theory and Implementation* (pp. 41–65). Edward Elgar Publishing.

Kvamme, O. (2019). School strikes, environmental ethical values, and democracy. *Studier i Pædagogisk Filosofi, 8*(1), s. 6–27. https://doi.org/10.7146/spf.v8i1.117967

Kvamme, O. (2021). Stedspedagogikk i antropocen [Pedagogy of place in the Anthropocene]. In I. Bostad, (Ed.). *Å høre hjemme i verden: Introduksjon til en pedagogisk hjemstedsfilosofi [Be at home in the world: Introduction to a Pedagogical Philosophy of Place]*. Spartacus.

Massey, D. (1994). *Space, place and gender*. Polity Press.

McKenzie, M. (2008). The places of pedagogy: Or, what we can do with culture through intersubjective experiences. *Environmental Education Research, 14*(3), 361–373. https://doi.org/10.1080/13504620802194208

Moore, J. (2015). *Capitalism in the web of life: Ecology and the accumulation of capital*. Verso.

Moore, J. (2016). Introduction. Anthropocene or capitalocene? Nature, history, and the crisis of capitalism. In J. Moore (Ed.), *Anthropocene or capitalocene? Nature, history and the crisis of capitalism* (pp. 1–14). PM Press.

Papastephanou, M., & Angeli, C. (2007). Critical thinking beyond skill. *Educational Philosophy and Theory, 39*(6), 604–621. https://doi.org/10.1111/j.1469-5812.2007.00311

Reid, A. (2008). Editorial. *Environmental Education Research, 14*(3), 205–214.

Rieckmann, M. (2011). Future-oriented higher education: Which key competencies should be fostered through university teaching and learning? *Futures, 44*(2), 127–135.

Smith, G. (2008). Oxymoron or misplaced rectification. *Environmental Education Research, 14*(3), 349–352. https://doi.org/10.1080/13504620802194273

Smith, G. A., & Sobel, D. (2010). *Place- and community-based education in schools*. Routledge.

Sobel, D. (2004). *Place-based education: Connecting classrooms & communities*. The Orion Society.

Spencer, H. (1861). *Education: Intellectual, Moral, and Physical*. Appleton and Co.

Stahl, T. (2013). Habermas and the project of immanent critique. *Constellations, 20*(4), 533–552.

Stahl, T. (2019). Immanent critique. In A. Allen & E. Mendieta (Eds.), *The Cambridge Habermas Lexicon*. Cambridge University Press. https://doi.org/10.1017/9781316771303.051

Stevenson, R. B. (2008). A critical pedagogy of place and the critical place(s) of pedagogy. *Environmental Education Research, 14*(3), 353–360. https://doi.org/10.1080/13504620802190727

United Nations. (1992). *United Nations Framework Convention on Climate Change*. https://unfccc.int/resource/docs/convkp/conveng.pdf

United Nations. (2015). *Resolution 70/1. Transforming our world: The 2030 agenda for sustainable development*. https://www.un.org/en/development/desa/population/migration/generalassembly/docs/globalcompact/A_RES_70_1_E.pdf

Vetlesen, A. J. (1997). Introducing an ethics of proximity. In H. Jodalen & A. J. Vetlesen (Eds.), *Closeness: An ethics* (pp. 1–20). Scandinavian University Press.

Vetlesen, A. J. (2015). *The denial of nature. Environmental philosophy in the era of global capitalism*. Routledge.

Wood, A. W. (1990). *Hegel's ethical thought*. Cambridge University Press.

CHAPTER 15

Educating for Sustainability in an Anti-education State: Critical Thinking in a Rural Science Classroom

Kerrie Willis and Nick Kleese

Introduction

Inside Washington County, Iowa's 571 square miles live 22,000 humans and 1.3 million pigs—the third greatest concentration of pigs in the United States (USDA, 2017). The incredibly fertile soil that supports this industry is disappearing at unprecedented rates: after a century of intensive cultivation, the topsoil is now nearly eroded (Thaler et al., 2021). From our perspective as English educators at the high school in the county seat, we see these changes. We have also seen demographic shifts. There has been much to celebrate, including the growth of a vibrant Latinx

K. Willis
Washington High School, Washington, IA, USA
e-mail: kwillis@washington.k12.ia.us

N. Kleese (✉)
University of Minnesota, Minneapolis, MN, USA
e-mail: klees023@umn.edu

M. Paulsen et al. (eds.), *Pedagogy in the Anthropocene*, Palgrave Studies in Educational Futures,
https://doi.org/10.1007/978-3-030-90980-2_15

community and a booming weekly farmer's market. But there are also negative trends: a stark increase in student poverty—a 35% increase in the last two decades. At the high school, this translates to roughly half our students living at the poverty line or below, and 1 in 20 experiencing homelessness (Iowa Department of Education, 2021). On top of this, the growing politicization of education at both local and state levels, along with the localization of a national divisiveness, threatens the community cohesion we value and continue to struggle to realize. From our perspective, from our experience, the most blatant barrier to educating for sustainability is an ideological one.

Our community, located in the heart of America's industrial Corn Belt, should be a perfect place for noticing and critiquing the ways that global industrial slow violence is playing out in real time (Nixon, 2011). While some in our community have acted for the environment, including protesting proposed legislation to limit municipal suits against agricultural pollution (ICCI, 2017), these actions have not been widespread. In fact, on par with what is nationally prominent, we observe pervasive retreat from embracing civic responsibilities—our fundamental commitments to one another as interdependent individuals within broader human collectives. This retreat hinders teachers' abilities to promote classroom communities that celebrate collaboration and cooperative effort. In science education, this manifests as an outright and willful resistance to discussions about climate change—conversations that students are, by sake of their geography, well-positioned to participate in, as the ecological evidence of climate destruction is in their backyards. Moreover, they are of a generation that has proved to be more knowledgeable about the global climate crisis than any previous (Feldman, 2020). And yet, this knowledge is hindered by the consistent attack on both climate science and education by a radical, fundamentalist movement toward an anti-social individualism from the right. If we can't even recognize our innate obligations to one another in community, what chance is there for the Earth?

We are not alone in observing this. Scholars have thoroughly documented the ways these attacks have steadily and skillfully turned ecophilia into an anti-American ideology (cf. Buell, 2004). Today, this rhetoric can be seen, for instance, in the wariness about sustainable development (Mena, 2021) and in outright science denialism (cf. Oreskes, 2019). The attacks on education are no less pervasive. Our state context provides blatant and numerous examples. In only two months since the current Iowa

House Legislative Session began on January 11, 2021, the following anti-education bills have been proposed:

- "An act providing for the reduction of certain funding and budgets for public school, community colleges, and regents institutions following the use of specific curriculum[1] and including effective and applicable provisions" (H.F. 222, 2021)
- "An act directing the state board of regents to conduct a survey of the political party affiliation of the employees of the institutions governed by the state board" (S.F. 292, 2021)
- "An act prohibiting tenure at public post-secondary educational institutions" (S.F. 41, 2021)
- "An act prohibiting persons from entering single and multiple occupancy toilet facilities in elementary and secondary schools that do not correspond with the person's biological sex" (S.F. 244, 2021)
- "An act relating to instruction relating to gender identity in the curriculum provided to students enrolled in elementary education programs" (S.F. 167, 2021)

These proposed bills have several targets. They would seek to limit the safety of LGBTQI+ students, the rights of educators to have political expression, and education writ large, from pre-kindergarten to post-secondary. The assault is so massive that one commentator declared the 89th Iowa General Assembly has sought to achieve "new heights of legislative overreach" (Nietzel, 2021). Taken in their number and in their speed of deployment, these legislative attacks feel like an all-out barrage on not only public education as a system, but also the very concept of critical thinking that is the foundation of an engaged, responsible, civic public—and a sustainable society.

In rural communities, where schools are often claimed to be the foundation of the community, these attacks hit especially hard. How to counter these attacks so that we might help foster the learning needed for a sustainable society? Below, we describe one opening for doing so: a diligent attention to evidence, dialogue, and collaboration in the science classroom. This cannot be the only fix, but skillful science teachers nonetheless are working to make rural schools places that provide "the kind of education that embraces plurality and pushes back against fundamentalist thought; not only the religious and market fundamentalisms ... but nationalistic as well, that are pushing isolation and anti-democratic authoritarianism" (Cervone, 2018, p. 2). Critical thought and the cultivation of a sense of responsibility to our relationships—both human and non-human—are also necessary in developing a sustainable global society. Yet these, again, are also what is at stake in the rapid, intensive legislative assault on education in the state of Iowa.

We do not have the option of ignoring these attacks, and we mean this on several registers. For one, climate change is an *everybody* problem. Combatting it will require as many hands and as many minds as possible, so long as those hands and minds are open to the possibility of difference. This is monumentally important in rural communities, where social relations are myriad and nuanced. In rural places, "there will always be issues that need to be worked through," Cervone writes, so "the goal cannot be to eliminate tensions, but to recognize that democracy is complicated" (ibid). So, too, for the democratic processes needed to bring about permanent, sustainable change. Seeing as the understanding of these processes is fragile, we hold that one place that they might be developed is in the classroom. As educators, we do not wish to deny a rigorous, critical science education to any student—even those that espouse anti-climate and anti-education beliefs. We believe these students are not innately defiant, but rather have learned the language of a prevalent political rhetoric that has ingrained itself into the fabric of our national discourse.[2] In our specific context, without a more robust consideration of the political barriers to doing science education, any detailed and strategic education on actual climate issues will be inevitably stymied, and many planted seeds will go unnourished, making progress toward real and lasting environmental change unlikely. Like Cervone, we recognize that "the real questions are what is the future of rural spaces, and how can education produce active rural citizens in a rural democracy?" (ibid). Moreover, we wonder: if it weren't under assault, how might education in a conservative, rural place produce active rural citizens of the Earth?

In this chapter, we explore the ways that conservative anti-science and anti-education rhetoric has hindered teaching for climate solutions in our rural Midwestern high school. We do so from our position as educators and community members, though with two different experiences. Kerrie grew up in Fort Worth, Texas, but a few weeks into her first teaching experience at Washington High School (WHS), she knew it was the only place she wanted to be. Nick is a native Washingtonian who was raised on his family's farm, graduated from WHS, then returned to teach English there. Like many folks in Washington, explaining our connections to one another requires a long telling. Kerrie was Nick's teacher for four different English classes. When Nick returned to teach, he taught the same courses he took with Kerrie; as he did so, Kerrie served as his Instructional Coach. Kerrie's daughter graduated with Nick's sister, and both continue to spend time in the other's home. And on and on. We have tried to write this chapter with these commitments in mind—the overlapping, thick complex of relationships that grant rural communities—and rural schools—a particular resiliency. Moreover, we have tried to write in such a way that will prove accessible to all those within and across these relationships, especially those who are currently in the classroom.

Our chapter is an attempt to articulate what we see as some of the on-the-ground challenges and opportunities in rural education in teaching for sustainability in the Anthropocene. In particular, we attempt to outline the ways that fundamentalist anti-social movements have threatened the actual *doing* of teaching. Our hope is not to be prescriptive, but interpretative—to provide some insight into the challenges and opportunities of teaching at the secondary level for sustainability. For those living and teaching in rural communities, we hope this chapter provides affirmation of the struggle, opportunities for navigating the sociopolitical challenges, and an example for comparison to your own locale. As such, we do not suggest this chapter is universally or essentially rural; nonetheless, we believe that the specificities have enough parallels to be of benefit to others.

Because rural education has, in the United States at least, been relatively understudied in education scholarship, we want to begin by offering one way that rurality might be conceptualized by those unfamiliar with it. Then, we offer a brief overview of fundamental tensions in rural education in the United States. These two contexts will help illuminate our discussion of the daily political barriers that rural teachers face in educating for science. We end on a generative note, however, in sharing what we see as the unique opportunities rural places can offer climate education.

Rural Geographies

As a concept, rurality is expansive, and attempts by Euro-American scholars to define it have tended to be confined to either demographic or cultural frameworks (Bell, 2007). For demographers, rural is defined by population measures: human density over a given area, total population, proximity to an urban center, and so on. While these approaches point toward the social aspects of rurality, they can produce strange results. For example, under the most recent government metrics, Newberry, Michigan, a town of 1519 in the state's heavily forested Upper Peninsula, is considered metropolitan (Nelson, 2019, p. 40). Moreover, these approaches cannot capture the meaning of rural places: that is, the practices, ideas, and values that provide a lived experience of rurality. Although the cultural approach insists on a qualitative difference between geographies, rurality is in itself mind-bogglingly diverse enough to warrant it impossible to think of a single iteration of it. As such, we see the value of describing rural in a third way—as it is positioned within systems of global economic exchange.

Rural sites tend to be sites of ecological extraction (Cervone, 2017). Rural ecologies are mined, chopped, drained, or seeped before being packed, shipped, and processed into materials more readily available for human consumption. Scholars refer to this as the social metabolic process—that is, the systems of energy extraction, production, distribution, and consumption that support global humanity (Foster, 2000). The compartmentalization of each of these subprocesses into a specific geographical area allows for particular economic activity to define the ecology of that area. In sites of ecological extraction, the resources of a specific locale benefit faraway consumers. In Washington, the nutrients from the soil—and the soil itself—are sapped in order to grow corn that, when harvested, is sent to nearby distribution centers for further transport. The system is so intensive and efficient that even in the mid-1990's, it was less costly for Taiwanese distributors to ship Iowa pork across the Pacific than it was to produce pork there domestically (Thu, 1995).

In this light, the explicit legislative attempt to disenfranchise rural Iowa students is only one manifestation of much more pervasive and embedded systems of economic and ecological exploitation. In a global social metabolism, the state's "developmental agenda explicitly assumes (and even relies on) rural exploitation" (Ashwood, 2018, p. 719). In the Anthropocene, this agenda is part and parcel with the associated regimes of white supremacy and colonization. In the United States' context, as with other settler societies, ecological domination hinges on geographical and cultural removal of Indigenous peoples (Gómez-Barris, 2017), the racialization of landscapes (Bullard, 2005), and production of ignorance about these histories and ongoing impacts (Bonneuil et al., 2016). These logics can sow discord in the most intimate of rural communities, encouraging false rural insider/urban outsider labels that prevent social cohesion.[3] Moreover, and importantly for our purposes, these labels are also those that feature prominently in educational trajectories of rural students.

Rural Education in the United States

"School," Michael Corbett writes, "is one of the key flashpoints where modernity, cultural hybridity/multiculturalism, globalization/international capitalism and other change forces meet a certain resistance" (2007, p. 26). This is acutely the case for rural education writ large in the United States, where celebrating both local life worlds and our multicultural global society remains an ongoing tension for the nearly ten million students who attend a rural school (Showalter et al., 2019).[4] One way this plays out is in the contrast between various levels of education administration. Some legislation, written at the federal level, is mandated for all schools in all geographies.[5] But this oversight also leaves room for state and local policies to shape education, at least ideally, to meet the needs of specific locales. While this can allow for wonderfully progressive, culturally and ecologically sustainable policies to be written by communities themselves, it can also allow for retrenchment against the very ideas of multiculturalism, sustainability, democracy, and—in fact—education that schools could otherwise cultivate. Rural opposition to education for sustainability, then, can become part of a larger reaction against state oversight even when doing so exacerbates the actual needs of local communities.

The tension between local and global plays out in the imaginations of rural youth, too. Carr and Kefalas's (2009) year-long study of a rural Iowa school found that one of the most major aspects to student post-secondary paths was the decision to either stay in or leave their rural community. This decision is bound both by the community's unequal allotment of social capital to individual students deemed likely to "succeed" in higher education, and by student place attachment. Overwhelmingly, students who perform well at school are encouraged to leave, and for reasons both cultural and economical, they often do not return. Carr and Kefalas refer to this phenomenon as the "rural brain drain." Corbett offers another term in the title of his book, *Learning to Leave* (2007). Ultimately what comes to consequence is who stays, who leaves, and what they believe.

For all of this complexity and nuance, education scholarship has been wary of considering rural students to be a unique bloc. A review of titles of articles in major American education journals showed that "urban" appeared 16 times more frequently than "rural" (Schafft, 2016). This attention gap leaves many scholars and policymakers with not just a lesser understanding of the challenges in rural education, but also a lack of awareness of the nascent opportunities. Rural schools, by fact of their geographic and demographic contexts, afford certain benefits. For one, rural schools' "proximity to the natural world *allows* students to draw on their local knowledge to make sense of school-based (book) knowledge (Roth, 2010, p. 51, emphasis ours). The proximity extends to the social, too. Rural schools are situated within communities where there are ample "opportunities for contact and dialogue with community members across a variety of contexts (the grocery store, religious settings, school events, community suppers, the gas station) become sites of action and 'naming' as rural educational leaders work to establish the trust necessary for mobilization on issues of importance to the community" (Azano & Biddle, 2019, p. 6). And yet, in our locale, community resistance to education remains.

In rural communities such as our own, education is often regarded as the only way "out." For some, this has meant the promise of opportunity in the city. For others, these educational trajectories are evidence of rural decline. The anti-democratic reactionaries at the state house have weaponized these sentiments to uncritically celebrate provincialism, science denialism, and militant individualism as an alternative to critical, public education. We believe this sentiment is not only untenable in an era of global climate destruction, but also harmful to the same communities these legislators purport to protect. As such, although education does serve as a geographic "out" for rural youth, education is also a potential "out" from that anti-democratic worldview now prevalent at the Capitol. It is amid these myriad and mammoth tensions that rural science educators are seeking to provide young people with the skills and experiences necessary to sustain both their communities and their planet.

Climate in the Classroom

One mighty struggle at play in the classrooms of our rural school is evident in science classes, where—before many students can learn the specifics and standards of the content—teachers must first convince students that science is a set of commonly understood and accepted laws, rather than a perspective. What an immense challenge it is to engage students in meaningful explorations and considerations of environmental sustainability, when many may not accept the basic premise that the world—including our 5.2 square miles and 7000 people of it—is in danger of imminent environmental collapse. When one's sense of self is tied to a way of being that denies climate change, how do teachers and schools foster the sense of urgency needed to encourage students to become interested, informed, passionate stewards of their land and community?

It is a difficult time for Iowa teachers in general, under assault by their own State Legislature. Bill after bill tumbles out from under the golden dome, one after another aimed at dismantling public education—and not simply by removing financial support. Several bills seem designed specifically to crush the values public schools often strive to foster: respect for individual differences as enriching to the whole; inclusion of all, regardless of barriers; and encouragement of curiosity, open-mindedness, and critical thinking. So often it seems school buildings are the only places in their communities actively working to level the playing field for all, to feed the hungry, to open the mind, to ask difficult questions, to solve problems faced by the most vulnerable. Couple this legislative onslaught from the state with the national politicization of scientific evidence itself (lead by a former President who denied the research and evidence-based recommendations of one of the leaders of his own White House Coronavirus Task Force), and the task faced by science teachers is even more challenging.

In a political and cultural climate where a willful, gleeful resistance to science education is sometimes worn by students as a badge of honor, science teachers are faced with a two-part struggle: first, convince students that the science taught is true; second, help students recognize the preponderance of evidence that supports the given scientific premise. It is not simply teaching the content that is the struggle—it is convincing students that the curriculum is fact, not fiction, and that science teachers are credible guides on the journey of exploration. When delving into the issue of human sustainability, the Next Generation Science Standards (NGSS, 2013) are clear in the scientific theories that must be taught:

- HS-ESS3-1. Construct an explanation based on evidence for how the availability of natural resources, occurrence of natural hazards, and changes in climate have influenced human activity.
- HS-ESS3-2. Evaluate competing design solutions for developing, managing, and utilizing energy and mineral resources based on cost-benefit ratios.

- HS-ESS3-3. Create a computational simulation to illustrate the relationships among management of natural resources, the sustainability of human populations, and biodiversity.
- HS-ESS3-4. Evaluate or refine a technological solution that reduces impacts of human activities on natural systems.
- HS-ESS3-5. Analyze geoscience data and the results from global climate models to make an evidence-based forecast of the current rate of global or regional climate change and associated future impacts to Earth systems.
- HS-ESS3-6. Use a computational representation to illustrate the relationships among Earth systems and how those relationships are being modified due to human activity.

Where do high school science teachers begin effectively teaching these standards, in light of the fact that some students believe human impacts on the environment are negligible, and that global warming is a myth?

All of the political polarization of education is unfortunate, because what we should be focusing on in science classrooms in rural schools is the *how* of working toward sustainability, not the *why*. This is not to say that the *why* is unimportant: our survival depends on it. However, the task before all humans—to create a more sustainable community, thereby ensuring the survival of our planet—has been, in our context, completely subsumed in the last few years by our relational conflict with each other, rather than keenly focused on the task itself. From this position of differences, how do we foster a classroom culture that leads to the free and full discussion of ideas, the careful consideration and quantifying of evidence, and some shared understandings of the problems humans face?

Pivot Toward Opportunities

The challenges we have identified in teaching for sustainability in our rural context are political in nature. The anti-democratic sensibility now seeking to govern our capacity to do the work of education includes myriad systems, structures, and ideologies that manifest in particular moments. Addressing these moments of resistance, in the moment, are only pinpoints in an entire constellation of obstruction. When teachers face this daunting reality daily, and in what feels like an increasing isolation, it is tempting to be cynical about *solutions* that promise immediate fixes. But there are various *openings* that can provide insights, inspiration, and possibility.

One opening includes the work of educators who are doing critical education for sustainability even when it isn't described as such. In the rural science classroom, this appears as evidenced-based thinking. Explicitly and relentlessly requiring students to engage in evidence-based thinking is one way to help students break down important and complex environmental issues—those in their own backyard, as well as those across the planet. Creating a classroom framework by which to structure learning around topics many may dispute or disbelieve is vital, supported by a repeated question: what is the evidence? When faced with student skepticism, one way forward is through a repeated process of introducing every concept in the following way:

- Here is the scientific theory.
- Here is the hypothesis scientists make, based on this theory.
- Here is the evidence to support the hypothesis.

One case in point comes from a secondary science teacher's exploration of and instruction around the disciplinary core idea of human impacts on earth systems. For this unit of study, students begin by thinking about waste or excess in their daily lives. Students conduct a one-day trash audit on themselves. Students talk with the parents about gas and electric use via the family's energy bills. Through videos about pollution, zero waste, and reducing a carbon footprint, students gain information to enhance consideration of their individual relationships with consumption and the systems that support it. Students consider whether or not Washington could become a zero-waste town. (Many agree it is possible, but unlikely, unless everyone works together.)

Students widen their view from their own backyards, to Iowa, the United States, and the world. Students consider carbon sinks, deforestation, and tree-planting. Students experiment with a carbon footprint calculator. Students view videos about what happens to recycling, then post discussion points about what they learned. Students consider what large problems the planet is facing, and practice reading source material to determine credibility. Students create one-minute videos about a big problem they researched, then watch the videos of others and leave feedback. Students consider trash-burning, an oft-used waste management method for many rural residents. Students view *David Attenborough: A*

Life On Our Planet, answering worksheet questions, and tracking statistics for what happens to world population, carbon levels, and untouched wilderness over seven decades. Students Zoom with the manager of the local landfill to learn in detail how waste is managed, and what challenges the process creates. Students answer whether or not their behavior has changed since the class started studying human impact. (Responses vary, but most students note at least one small change, either in their behavior, or in their thinking.)

Students were then tasked with completing a final human impact research project, calling for high-level thinking. Students needed explicit teaching to tackle these tasks, which were broken down into a day-by-day plan, and presented with a teacher-created model for each day's task:

- Gather research on the background for a problem creating a negative impact on the environment, focusing on natural resources and manufacturing.
- Break down the problem into smaller pieces, and examine how those things impact earth systems.
- Understand what people are gaining from the current practice.
- Identify two solutions people are already considering or doing, then calculate the effectiveness of these solutions, based on their environmental impacts, costs, availability, simplicity, and ease of implementation.
- Select a solution, propose a new one, or combine several, then write a one-page justification for the choice.
- Determine who needs to know about the solution, and what the most effective means of communication is.

Challenges abounded. Many students struggled to rigorously analyze existing solutions. Many also struggled to quantify their research, sometimes relying on their guts, rather than research. Some students never fully realized that using powerful evidence and quantifying their criteria makes an argument immensely stronger and more compelling. Not all students fully grappled with the complexities that big problems bring with them—nor with the knowledge that these problems must be fully unpacked, researched, quantified, and considered from many perspectives in order to find solutions that make sense.

With an eye toward the future, what advice would we give to high school science teachers who want to foster a rigorous and evidence-based classroom? First and foremost, projects like we outlined above require students to learn and use those complex skills vital in developing critical thinking. If we consider the myriad responsibilities of public education, its primary role is to help students learn to think deeply—and that's what science projects like this strive to do. Ideally they would be fostered and showcased by schools, but even one teacher working at an individual classroom level can create the conditions for these continued, rigorous explorations. Secondly, teachers can build in consistent checkpoints for students during the research and analysis process, preventing them from advancing in their project until both teacher and student see evidence that the research foundation is on solid ground. One of the most encouraging takeaways from this research is that most students *will* engage in critical thinking, *will* consider multiple forms of evidence, and *will* consider the perspectives of others, *if* the content is presented in a way that refuses to tap into rhetoric and judgment. It is in this place the opportunity for critical thinking, careful consideration, and learning is highest. This is a credit to our students, as well as to our science teachers.

Although we focused largely in this chapter on scientific reasoning as an important skill in sustainability education, it is by no means the only important one. For us, our goal is to consider possibilities for connections between the learning and exploration happening in the science classrooms, and the discussions, projects, and commitments of the greater body of WHS students—and ultimately our town. We turn our attention to fostering connections and commitments between our students and our community in hopes these bridges lead to long-lasting, positive impacts on the environment whether that environment is our town's five square miles, or any other environment worth protecting. Indeed, we hope that these relations will also promote a long-lasting engagement with sustainability that becomes part of their identity as rural youth.

One thing we know for certain is that part of being rural is being connected to and intertwined with others. Despite the impression that rurality is synonymous with lacking, we see it as rich in opportunities—if you are open to them (and sometimes even if you are not). So often because there are so few of us, we all are asked to do a bit of everything. In a small Iowa high school, in a small Midwestern town, there is room to be more than one thing—to be a farm kid, *and* to run cross-country, *and* to craft the literary magazine. Our students are farmers, clarinet players, goalies, actors, and English-learners; some live in town and some live in the country; some help support their families by working in those confinements where 1.3 million pigs are raised—rarely is a student just one of these. And this mobility, this crisscrossing born of the smallness of a rural place, and the necessity of everyone doing a bit of everything, that is as much a reality of our town as the stereotypes are.

Conclusion

In this chapter, we have shared how the current legislative assault against education in the state of Iowa exacerbates existing structural economic exploitation of rural ecologies to impede education for sustainability in science classrooms. We see this as part of a much larger retreat from direct participation in democratic processes—processes that will be necessary in building and maintaining an ecologically sustainable society. Without local, state, and federal investment in this education, we worry that the good work of teachers in the classroom will be ineffective at contributing to this change.

We recognize that education in the United States is far from a perfect institution. It has and continues to assimilate, oppress, and dehumanize. Acknowledging these harms, we wonder: what is another institution in this nation where people actively attempt to get young people to understand one another across their differences? Where else are people told that dialogue matters, that dignity and respect matter? And what happens when these students look to society, what do they see? So much of what they see and hear from public officials discredits their education: both in the content and in the values it holds. What they see in society makes much of their public education look like a lie, because what they hear from politicians does not correspond to what they hear in school. Despite this, rural science educators continue to insist that critical thinking, dialogue, and attention to the ecological realities of one's lived experiences take precedence over obstructionist rhetoric and political maneuvering.

The rural realities we know of are both strikingly similar and substantially different from other rural ones. Despite these differences—in fact, precisely because of these differences—we believe that rural educators might play a role in both bettering the lives of those who live as our neighbors and restoring faith in the rich values so often fostered here. "Indeed," writes Howley, "many rural places are sufficiently different that one might hope for something even more different in the future" (Howley, 2009). In this diversity, against so many attacks, there is hope: for generating diverse perspectives, insights, and ideas that are eager to contribute to, rather than retreat from, a sustainable world.

Acknowledgments The authors would like to thank Drew Ayrit for opening his classroom to us during this process. While his classroom did not start out as the focus for this piece, we soon found ourselves recognizing in it a convergence of our thinking. We are grateful to him for his generosity and advocacy.

Notes

1. The "specific curriculum" targeted is the 1619 Project, which seeks to recognize the role of slavery, racism, and white supremacy in American history.
2. These instances of student resistance could be interpreted generously, as Corbett notes occurs when "students come to understand their class exploitation and the way that institutions of formal education function to serve interests contrary to their own" (2007, p. 42). However, we, like Corbett, do not wish to reduce resistance to some unrelated phenomenon—which, in this, is the channeling of populist, anti-intellectual rhetoric.
3. These, too, are often racialized.
4. In total, 46% of Native students in the United States attend schools in rural areas, compared to 30% of white students 14% of Black students, 10% of Hispanic students, and 9% of Asian/Pacific Islander students. See Faircloth (2009).
5. One notorious example is the No Child Left Behind Act of 2002, which mandates standardized testing in reading and math for all students every year between Grades 3 and 8. So-called underperforming schools can be financially penalized.

References

Ashwood, L. (2018). Rural conservatism or anarchism? The pro-state, stateless, and anti-state positions. *Rural Sociology, 83*(4), 717–748. https://doi.org/10.1111/ruso.12226

Azano, A. P., & Biddle, C. (2019). Disrupting dichotomous traps and rethinking problem formation for rural education. *The Rural Educator, 40*(2), 4–11. https://doi.org/10.35608/ruraled.v40i2.845

Bell, M. (2007). The two-ness of rural life and the ends of rural scholarship. *Journal of Rural Studies, 23*(4), 402–415. https://doi.org/10.1016/j.jrurstud.2007.03.003

Bonneuil, C., Fressoz, J., & Fernbach, D. (2016). *The shock of the Anthropocene: The earth, history and us*. Verso.

Buell, F. (2004). *From apocalypse to way of life: Environmental crisis in the American century*. Routledge.
Bullard, R. (2005). *The quest for environmental justice: Human rights and the politics of pollution*. Sierra Club Books.
Cervone, J. A. (2017). *Corporatizing rural education: Neoliberal globalization and reaction in the United States*. Palgrave Macmillan.
Cervone, J. (2018). Imagining the rural future: A response to Howley, Clifton & Howley. *Journal of Research in Rural Education, 33*(7), 1–3. https://doi.org/10.18113/P8JRRE3307
Corbett, M. (2007). *Learning to leave: The irony of schooling in a coastal community*. Fernwood.
Carr, P., & Kefalas, M. (2009). *Hollowing out the middle: The rural brain drain and what it means for America*. Beacon Press
Faircloth, S. C. (2009). Re-visioning the future of education for native youth in rural schools and communities. *Journal of Research in Rural Education, 24*(9), 1.
Feldman, H. R. (2020). A rhetorical perspective on youth environmental activism. *Journal of Science Communication, 19*(6), 1. https://doi.org/10.22323/2.19060307
Foster, J. B. (2000). *Marx's ecology: Materialism and nature*. Monthly Review Press.
Gómez-Barris, M. (2017). *The extractive zone: Social ecologies and decolonial perspectives*. Duke University Press.
Howley, C. B. (2009). The meaning of rural difference for bright rednecks. *Journal for the Education of the Gifted, 32*(4), 537–564. https://doi.org/10.1177/016235320903200405
ICCI. (2017). Who's behind the bill to dismantle the Des Moines Water Works? Iowa Citizens for Community Improvement. http://iowacci.org/farming-environment/whos-behind-bill-dismantle-des-moines-water-works/
Iowa Department of Education. (2021). 2020–21 Iowa Public School K-12 students eligible for free or reduced-price lunch by district. *Report*. https://educateiowa.gov/documents/district-frl/2020/12/2020-21-iowa-public-school-k-12-students-eligible-free-or-reduced
Mena, B. (2021, February 17). Gov. Greg Abbott and other Republicans blamed green energy for Texas' power woes. But the state runs on fossil fuels. *The Texas Tribune*. https://www.texastribune.org/2021/02/17/abbott-republicans-green-energy/
Nelson, G. D. (2019). What makes a place rural? *Dissent, 66*(4), 38–47.
NGSS Lead States. (2013). *Next generation science standards: For states, by states*. The National Academies Press.
Nietzel, M. T. (2021, February 11). Want a state legislature that likes to meddle with higher education? Move to Iowa. *Forbes*. https://www.forbes.com/sites/michaeltnietzel/2021/02/11/want-a-state-legislature-that-likes-to-meddle-

with-higher-education-move-to-iowa/?fbclid=IwAR0uBD9pSDXBKjYmhm161ERBjPyVXAE2KokNLsc-V8Ra-pUPUFSRHd6Nfjg&sh=554281015afd

Nixon, R. (2011). *Slow violence and the environmentalism of the poor*. Harvard University Press.

Oreskes, N. (2019). *Why trust science?* Princeton University Press.

Roth, W.-M. (2010). Local matters, ecojustice, and community. In D. Tippins, M. Mueller, M. van Eijck, & J. Adams (Eds.), *Cultural studies and environmentalism* (Cultural Studies of Science Education, vol. 3). Springer.

Schafft, K. A. (2016). Rural education as rural development: Understanding the rural school-community well-being linkage in a 21st-century policy context. *Peabody Journal of Education, 91*(2), 137–154. https://doi.org/10.1080/0161956x.2016.1151734

Showalter, D., Hartman, S. L., Johnson, J., & Klein, B. (2019). *Why rural matters: The time is now*. Rural School and Community Trust. http://www.ruraledu.org/WhyRuralMatters.pdf

Thaler, E. A., Larsen, I. J., & Yu, Q. (2021). The extent of soil loss across the US Corn Belt. *Proceedings of the National Academy of Sciences, 118*(8), e1922375118. https://doi.org/10.1073/pnas.1922375118

Thu, K. (1995). Piggeries and politics: Rural development and Iowa's multibillion dollar swine industry. *Culture & Agriculture, 16*(53), 19–23. https://doi.org/10.1525/cuag.1995.16.53.19

United States Department of Agriculture. (2017). Historical highlights: 2017 and earlier census years: Iowa. National Agricultural Statistics Service. https://www.nass.usda.gov/Publications/AgCensus/2017/Full_Report/Volume_1,_Chapter_1_State_Level/Iowa/st19_1_0001_0001.pdf

Legislation Cited

S.F. 41, 89th General Assembly, 2021 Reg. Sess. (Iowa, 2021). https://www.legis.iowa.gov/docs/publications/LGI/89/attachments/SF41.html

S.F. 167, 89th General Assembly, 2021 Reg. Sess. (Iowa, 2021). https://www.legis.iowa.gov/legislation/BillBook?ga=89&ba=sf167

S.F. 244, 89th General Assembly, 2021 Reg. Sess. (Iowa, 2021). https://www.legis.iowa.gov/legislation/BillBook?ba=SF224

S.F. 292, 89th General Assembly, 2021 Reg. Sess. (Iowa, 2021). https://www.legis.iowa.gov/legislation/BillBook?ga=89&ba=SF%20292

H.F. 222, 89th General Assembly, 2021 Reg. Sess. (Iowa, 2021). https://www.legis.iowa.gov/legislation/BillBook?ba=HF222&ga=89

CHAPTER 16

Ecopedagogy in the Anthropocene: A Defence of the Classical *Paideia*

Lars Petter Storm Torjussen

Introduction

The term Anthropocene and thus the recognition that we humans affect planet Earth to such a degree that we have entered a new geological era are gaining acceptance. Although few are seriously questioning the fact that we are facing an ecological crisis, there is no general agreement on *what* we need to change, *how* we are to do this and how *extensive* these changes need to be. It is a debated question whether an ecological mind-set of awareness automatically involves a critical and political stand against our current social and economic system. For instance, according to Naomi Klein (2014), questions regarding the climate need to relate to a critique of capitalism. In the field of literature on education for sustainable development (ESD), there is a division between education *about* and education *for* sustainable development relating to this question (Gadotti, 2008). The former treats sustainable development as a phenomenon within the

L. P. S. Torjussen (✉)
Department of Education, University of Bergen, Bergen, Norway
e-mail: lars.torjussen@uib.no

M. Paulsen et al. (eds.), *Pedagogy in the Anthropocene*, Palgrave Studies in Educational Futures,
https://doi.org/10.1007/978-3-030-90980-2_16

traditional system and curriculum while the latter promotes a more profound change of perspective and lifestyle.

An important contribution to education *for* sustainable education is the so-called ecopedagogical movement as it is described in Richard Kahn's (2010) influential *Critical Pedagogy, Ecoliteracy & Planetary Crisis—The Ecopedagogical Movement*. This movement originated as a response to the traditional environmental movement's allegedly lacking critical dimension (Straume, 2017). The ecopedagogical movement claims that a critical ecological consciousness requires a confrontation with problematic phenomena at the very core of our civilization, such as the technical domination of nature, an economic system based on profit, social inequality, gender issues and the exploitation of Indigenous peoples. This confrontation requires us to rethink our heritage from classical antiquity onwards.

In this chapter I wish to adhere to the fundamental premise of the ecopedagogical movement. Education for sustainable development needs to be something more than an introduction to specific subjects regarding superficial solutions such as quota trading and similar initiatives. I agree that an adequate response to our climate crisis requires a rethinking going all the way back to classical antiquity, as this was the moment where the foundations of Western culture and civilization once were built. In this regard, Pythagoras was the first to take the first step towards the quantification of the world and the mathematization of human experience, thus laying the first brick in our scientific culture. However, the assumption that the Greek mind has allegedly paved the way to the Cartesian dualism between nature and culture and the technical rationality underlying our current ecological crisis must be approached with caution. It can lead to an unwarranted claim that classical civilization *as such* is to blame for the illnesses mentioned above. In this article I will argue that an ecological mindset of today must address our classical heritage, but warn against the temptation to exalt "sustainability" as a sort of master signifier representing all the evils in the world in need of remedy, tying all of these diverse phenomena into a Gordian knot and suggesting the knot can be severed in one blow.

If we take a closer look at the civilization of classical antiquity we realize that democracy and sustainability were far from synonyms. Classical antiquity had practices directed towards nature, which can serve as models for education for sustainable development today. These practices were in fact resting on anti-democratic principles. For instance, the classical *paideia*—as it was conceived by Plato—was directed at contemplation as a non-intervening preserving attitude, but this practice was interwoven with the pursuit of a free life resting on slavery as a means of excluding work and production. If we take a closer look at history, we find that our technical civilization, in many ways responsible for our ecological crisis, is not at all the result of an anti-democratic, but rather deeply connected with a "democratic" culture (White, 1967). The fusion of science with technique and utility was, in other words, the consequence of science no longer being the pursuit of an elite's leisure time, but had its task in utility for society as such.

This does, of course, not mean we should defend and implement anti-democratic principles and solutions today. It means the assignment of a critical and ecopedagogical movement is to avoid simple and totalizing solutions, and rather start the arduous and complicated task of understanding the fundamental concepts underlying our classical heritage and reinterpret and reformulate these concepts anew. The main argument of the chapter is to recover practices and a mindset which once were made possible by means of slavery and raise the question whether we can formulate the conditions of this mindset in a modern world. We cannot gain access to these practices or this mindset if we bluntly condemn the classical world as either irrelevant or opposed to ecological and sustainable thinking today. This enterprise is neither anti-science or pro-science as it is not shunning quantification, technique or the like, but realizing, quoting Pythagoras, that "number is the ruler of forms and ideas, and the cause of gods and demons" (Iamblichus, 2020).

The Ecopedagogical Movement

According to Kahn (2010), the ecopedagogical movement was formed after the Rio Conference in 1992. Its point of departure was the Latin American tradition of critical pedagogy building on the work of Paulo Freire, but also Ivan Illich.[1] True to its founding father the ecopedagogical movement is also centred around *conscientização*: critical consciousness as an in-depth understanding of how the social world is connected to arrangements of power.

The aim of the ecopedagogical movement is the pursuit of "ecoliteracy" as a type of ecological being in the world. Contrary to the traditional environmental movement, the ecopedagogical movement represents a radical systemic critique with correspondingly adequate political actions.

The movement thus clearly takes a stand in the widespread distinction between education *about* sustainable development and education *for* sustainable development, clearly endorsing the latter. Both of these approaches approve of the definition of sustainability given by the Brundtland Report, that sustainable development is "[…] development that meets the needs of the present without compromising the ability of future generations to meet their own needs" (Brundtland et al., 1987). After the coining of the term by the Brundtland Report there has been wide agreement that the concept of sustainability demands the social, economic and environmental dimensions go hand in hand. But, where education *about* sustainable development is more of an "outside" perspective directed towards a theoretical understanding of sustainable development, for instance by understanding the phenomenon as an isolated subject, education *for* sustainable development is to a larger extent a perspective from "within": education is meant to transform the pupils' and students' world view in order to make them live more sustainable lives (Gadotti, 2008).

There are clearly parallels between the ecopedagogical movement and thinkers such as Chet Bowers and David Orr, but also Arne Naess and The Deep Ecological Movement regarding their commitment to the *ontological dimension* of the ecological crisis. That is, our ecological crisis is according to this view a crisis of how we *perceive* nature, the world and ourselves, and hence the "solution" to this requires an in-depth change of perception.

Kahn traces the roots of our ecological crisis in the classical *paideia*. According to Kahn, the classical *paideia* formulated by the Greeks—but particularly as it was reconstructed by the Latin *humanitas*—is based upon a fundamental premise: Man is standing outside or above nature. This dualism between man and nature is the root premise in the history of Western subjectivity. Although it has given us democracy and ultimately human rights, this conception of subjectivity goes hand in hand with "histories of speciesism, classism, sexism, and all the other histories of oppression […]" (Kahn, 2010, p. 48).

If we are to formulate a planetary citizenship, that is, a global community sustainable both socially, economically and environmentally, we need to relate to the classical *paideia*, as it was "democratic in principle" (Kahn, 2010, p. 40). However, the Athenian democracy was a fragile blossom that lasted only a short time span. It was displaced by oppression and corruption almost as soon as it began. Kahn is thus not reaching back to a sort of Golden Age he wants to revive. The Athenian *paideia* is infused with an inherent tension stemming from its very birth. As Kahn (2010) writes:

> Athens never achieved anything like an inclusive democracy, as it rested upon certain foundational oppressions based on slavery, race, class, gender, and species. This unresolved set of hierarchies meant that a tension existed at the very heart of the Athenian attempt at democracy. (p. 43)

These "foundational oppressions" are resting on dichotomies like master-slave, man-woman, autonomy-heteronomy and human-inhuman. These figures of thought have historically legitimized oppression and exploitation, according to Kahn. This happened especially when the Greek *paideia* was "reconstructed" as Hellenistic *humanitas* where this legacy "came to exert a major 'civilizing' force upon the historical development of the West" (p. 51). The Hellenistic age has therefore given us "aristocratic politics, scientific hierarchy, conceptual categorization, and encyclopedic breadth" resulting in "the expansion of the city-state to the rest of the world in the form of colonizing empire" (p. 41). For Kahn, the basic elements of Hellenistic *humanitas* also marked Christianity and the formation of the Church, thus composing the backbone of Western civilization not just during the centuries of the Medieval age, but also lasting into modernity. The dualisms and hierarchies of the classical *paideia* were thus a necessary foundation of the radical Cartesian ontological dichotomy between *res cogitans* and *res extensa*, and the corresponding technical-instrumental rationality. Kahn indeed describes the Newton-Cartesian paradigm as "very much true to the roots of Hellenistic *humanitas*" (p. 52). Kahn is painting a broad canvas of the intellectual history of the West where education, scientific discoveries and the art of printing are intimately connected to expansion, war and colonization.[2] Western culture or *Bildung* is thus inextricably tied to an unsustainable exploitation of other less privileged parts of the world as it is ultimately resting on a single idea lying at the very core of Western education, namely domination over nature.

Kahn's call for going back to the classical *paideia* is important. A critical and ecopedagogical movement must re-read the fundamental concepts underlying our classical heritage and reformulate these concepts in a postmodern, global and democratic society. The next step is thus to ask if the key to postmodern society can be found in the practices of the classical *paideia*.

Ecology and Classical Antiquity

Can classical antiquity be a source of an ecological consciousness today? According to Chemhuru (2017), the literature in this field is lacking.[3,4] The literature that does exist in this field is in addition ambiguous. As Chemhuru (2017) shows, Brian Coman on the one hand, citing Homer and Plato, comes to the conclusion that "there have always been individuals with a concern and interest in the state of health and of the beauty of our environment" (Coman, 2006, p. 55). Extending this line of thought, Carone supports the thinking that "[...] much of what comes from the works of ancient Greek thinkers like Thales, Anaximander, Anaximenes, Heraclitus and Parmenides can be regarded as a 'hylozoistic' worldview, in which matter (hylō) and life (zōē) are inseparable" (Carone, 2001, p. 68). Was the Greek mind thus biocentric and ecological?

On the other hand, the Greek worldview has been criticized for its clear anthropocentric character, thus in conflict with a minimum of biocentricity needed for ecological sustainability in a postmodern world (Straume, 2017, p. 111). O'Connor (1964) goes as far as denying altogether ancient or classical Greek philosophy's relevance to environmental ethical thinking.[5]

A more fruitful approach is not to look for *explicit* ecological formulations or concerns asking whether the culture of antiquity was sustainable or not as such, but rather ask if there are elements or practices fundamental to classical *paideia* that can be *translated* into a modern ecological and sustainable society. In this regard it is important to realize that the three dimensions of sustainability (social, economic, environmental) formulated by the UN is a modernist construct less suited for the study of the ecological implications of the classical *paideia*.

What Is Classical *Paideia*?

Kahn is correct in stating the importance of the Greek *paideia* as he is particularly focusing on the events accompanying the birth of the Athenian democracy. This seed also contains the very birth of the Western conception of education. Education in antiquity underwent a transition from a heroic warrior culture to a political culture where education stopped being a mere instrument for society and became an end in itself and a measuring rod for the greatness of the city state (Kahn, 2010, p. 39).[6] According to Marrou (1956), it makes sense to refer to a common Greek-Roman culture with an essential feature: a slow mutation from education of a warrior nobility to the education of a scribe.

In many ways, the classical *paideia* is the common source of all our modern conceptions of education as it regards education a phenomenon in itself possible to pursue and debate and—much later on—studied or researched *in its own right*. Thus, a common thread exists not only from the Greek *enkyklios paideia* to the Latin *artes liberales*, but also from this classical heritage into the Medieval university—and further on into the modern research university founded at the start of the nineteenth century (Torjussen, forthcoming). The common thread is a conception of a diverse and yet coherent and unified curriculum; a cycle of different arts (which later became *subjects*) all serving an overarching purpose.[7] Although the content, divisions and even names of these arts have varied throughout the centuries, the seven liberal arts have nevertheless been a surprisingly stable arrangement from antiquity to the formation of the modern university as it was conceived in Prussia by the thinkers of German idealism (Torjussen, forthcoming). This stability can perhaps be explained by the fact that the curriculum of higher education was both *encyclopaedic* and *propaedeutic*. These two elements constitute each other as the different subject's propaedeutic or preparatory character is precisely the element connecting the different subjects to an integrated whole. This holism of classical *paideia* has been admired by all later times and has been an ideal of the French *encyclopaédie*, the German *Enzyklopädie* and the British-American *liberal arts*: all in their different manner have attempted to reconnect with.

What the different arts were preparing for—and hence what was *outside* them—was *theoria* or contemplation.[8] It is often noted that the etymological meaning of the word theory is "seeing", but the most correct translation of *theoria* and *contemplatio* is "being a spectator". Pursuing a life of studies in classical antiquity was an active strenuous brick-by-brick approach, but the aim of it all was a passive receiving of truth. This phenomenon is described by a variety of authors. Of course, the term *paideia* (and later *humanitas*) is used in antiquity in many different ways and the relation between paideia and contemplation varies. This article thus depends on a specific Platonic understanding of the term. Plato (1973) famously describes truth or insight as a flash of lightning in his Seventh Letter. Plotinus regards it as a touch of the divine (Lobkowicz, 1967, p. 52) and Cicero and Seneca as a sort of cosmic viewpoint (Torjussen, 2018). Similar views are also held by Thomas of Aquinas and Comenius (Torjussen, 2018). Thus, the philosophical life in antiquity was a life with a spectating attitude towards the world. An expression of this attitude is found in the influential allegory *life as a festival.* According to Diogenes Laertius (2018, 8.8) the allegory stems from Pythagoras. A more extensive description is given by Cicero (*Tusc. disp.* 5.9):

> In this way, some of us are enslaved to glory, others to money. But there are also a few people who devote themselves wholly to the study of the universe, believing everything else to be trivial in comparison. These call themselves students of wisdom, in other words philosophers ['lovers of wisdom']; and just as it was the noblest act on such an occasion [i.e., a festival] to watch the proceedings without a thought of getting anything for oneself, so too, in life generally, the contemplation and study of nature are far superior to the whole range of other human activities.

For Pythagoras a life devoted to the study of number, forms and ideas, was thus not a cold, bloodless affair stemming from a lack of interest, but highlighted an attitude we today would describe as *aesthetical.* Classical *paideia* was not at all directed at utility, but beauty (*kalon*) or beautiful things whose value was obvious by their own light (Gadamer, 2013). According to Pythagoras the philosopher "contemplates the most beautiful things" (*ton kalliston theorian*) (Lobkowicz, 1967, p. 7). However, it would be a misunderstanding to interpret this as if the philosophers were studying a special set of objects.[9] The "most beautiful things" are simply the common things every person is facing every day seen from a universal, cosmic and eternal perspective (*sub specie aeternitatis*). Far from the modern meaning of "theory" *theoria* entailed a perspective devoid of judgement and justification making the spectator capable of standing face to face with the city state's most fundamental beliefs and values.

In this regard, philosophy continued a similar function tragedy had in the pre-philosophical city state, where the city state during the Great Dionysia was examining its values through a process Goldhill (1990) refers to as "tragic questioning". The philosophical practices supporting this function were aiming at liberation. According to Pythagoras the spectating of these most beautiful things makes you "most free" (*eleutheriotatos*) (Lobkowicz, 1967, p. 7), and this is probably why Cicero and all the rest of history chose the term *artes liberales.* The term *artes liberales* has a dual meaning. On the one hand it refers to the fact that these studies were a liberating force. On the other hand it refers to the activities suited for free men in opposition to the activities unfree men pursue or are forced to pursue. These activities relating to work and production were often referred to as *artes mechanicae* or *artes sordidae*. They were considered unfree as they bind man to an exploitative and instrumental mindset.

We are here at the core of the argument of the chapter: Classical antiquity *did* have a mindset and practices we today can derive ecological implications from, although the mindset and these practices were once based on slavery and oppression. Although some of the arguments of the allegedly anthropocentric character of the Greek culture are true, for instance the mythological pantheon's lack of animals (Kleczkowska, 2014), and the fact that the philosophical life was primarily tending a relation to the self, the intimate relation between the relation to the self and the relation to nature is nevertheless often overlooked.[10] The inscription of the Delphic oracle "know thyself" means to know oneself and the place one has *within* nature and the greater cosmos. And vice versa: The study of nature in antiquity implied gaining knowledge of oneself in order to live a life in accordance with the nature of the cosmos. As Ptolemy (cited in Hadot, 2006, p. 187) writes regarding his study of the stars:

> I know that I am mortal and last only for a day. Yet when I accompany the tight ranks of the stars in their circular course, my feet no longer touch the ground, and I go to Zeus to eat my fill of ambrosia, like the gods.

Thus the aim of Ptolemy's studies is not "knowledge" in the modern sense or the power to rule over nature, but wisdom to withstand unforeseen events and not to fear death. And this therapeutic approach to a life of study was also transferred from the Greek *paideia* to the Roman *humanitas*.

It is for instance present in Seneca's *Naturales quaestiones* (1972). At a glance this treatise resembles proto-scientific studies of meteors, rainbows, thunder and wind mixed with ethical anecdotes here and there. But the motivation of the text is rather to regard the study of nature as a *practice* aiming at a perspective at the highest point in the universe where one can see everything from nature's divine point of view.[11] As Williams (2012, p. 4) eloquently sums it up: "The *Natural Questions* is not so much about the natural world as an active form of engagement *with* nature" (italics in the original).

So how can a philosophical education aiming at taking care of nature and having nature as a model in order to not overstep its boundaries at the same time be based on slavery? One answer to this question is that slavery in antiquity did not have profit as a main motive.[12] As Arendt (1958) points out, the Greeks defended slavery as an institution required to meet conditions belonging to life itself. For the Greeks life itself *was* slavery. The unfreedom of life itself was expressed in the unfreedom of work and production (Arendt, 1958). Thus the main motivation behind slavery was an attempt to harness or encapsulate work and production in order to establish islands of freedom. This of course represents an arrangement we today quite right find horrific. The majority of society is unfree in order to enable the freedom of a small elite. This arrangement resulted however in the successful exclusion of work and production from public life (Arendt, 1958). But apart from its moral reprehensibility this arrangement had interesting consequences from an ecological perspective today. It is precisely classical antiquity's contempt for manual labour and its contempt for the life of business that answer Barrow's baffled observation: "That industry developed up to a certain point, but stopped short of making progress which might have been expected" (Barrow cited in Arendt, 1958, p. 65). Barrow's astonishment consists in the discrepancy of antiquity's formidable organization skills in "public services and the army and the correspondingly lacking organisation skills in industry and production" (Barrow in Arendt, 1958, pp. 65–66). But according to Arendt, the assumption that the organizational skills the Greeks showed in the public domain necessarily corresponded with a likewise capacity in the private economic domain is a modern prejudice (Arendt, 1958, p. 66). The Greeks thus had neither the will nor the ability to develop a refined division of labour or to technically improve their production.

In other words, slavery in classical antiquity was not the result of an all pervasive exploitative will to power akin to the technological domination over nature in modernity, as Kahn tends to think, but rather the result of the opposite: an attempt to harness and decrease this domination by refusing to make it a public affair. This is the reason why the process of life was excluded by public affairs, as Arendt's famous analysis shows. For the Greeks work and production chain the human to the basic unfreedom of life. Hence, the classical *paideia* was based on an assumption that man needs education to supersede utilitarian, practical and *banausic* mental attitude of exploitation and domination; an attitude in many ways being "natural" for man. This further legitimized anti-democratic arrangements of exclusion. We can thus turn Kahn's argument upside down: It is rather the process of democratization that has made human life, work and production a phenomenon worthy of public concern, thus giving rise to advanced division of labour, the accumulation of capital and the progress of industry and technology; all preconditions for the rise of the ecological crisis today. Lynn White (1967, p. 3), contrary to Kahn, indeed traces *democracy* as one important factor in the historical roots of our ecological crisis:

> Science was traditionally aristocratic, speculative, intellectual in intent; technology was lower-class, empirical, action-oriented. The quite sudden fusion of these two, towards the middle of the 19th century, is surely related to the slightly prior and contemporary democratic revolutions which, by reducing social barriers, tended to assert a functional unity of brain and hand. Our ecologic crisis is the product of an emerging, entirely novel, democratic culture. The issue is whether a democratized world can survive its own implications.

The fusion of science and technology in the nineteenth century was insoluble tied to the demand that science needed to be useful for the welfare of the people. This democratic appeal to utility is further inextricably connected to the process where science *itself* was being structured after an industrial model of division of labour. Of course, it is a simplification to state that democracy is the (sole) cause of our ecological crisis today. The relation between democracy and ecological crisis is a complex matter exceeding the scope of this article. And this is exactly the point: The relation between democracy and sustainability is thus more complex than both Kahn and the UN's three dimensions of sustainability suppose. The division of sustainability into the three aspects *social*, *economic* and *environmental* does not necessarily all pull in the same direction. Unless ecopedagogical thinking realizes this the analyses will be shallow and the solutions suggested inadequate. The question is then where this leaves the ecopedagogical thinking today?

Conclusion

In this chapter I adhere to two of the basic premises of the ecopedagogical movement, as Kahn (2010) fleshes them out. These are: (1) education for sustainable development needs to represent a heightened critical consciousness *for* sustainability and not merely education *about* sustainability and (2) this endeavour demands a relation and a *confrontation* with the classical *paideia*. However, as I have tried to show, this confrontation cannot be a total dismissal of Western classical heritage, as the very reflexivity and will to radically question the fundamental values of Western culture and history is itself an essential part of this same tradition. I have also tried to show that to exalt "sustainability" to the status of master signifier capable of combatting anti-democratic tendencies and inequality *as such* is a simple and totalizing solution, as democracy and sustainability historically are not synonyms. There is not simply *one* idea of domination over nature underlying our civilization. Classical antiquity represented ideas relevant to *environmental* sustainability (but not *social* or *economical* sustainability after our modern standards).

This chapter has focused on the practices of *theoria* as a non-interfering spectating attitude towards the world achieved after one has harnessed the passions and a natural urge for utility.

These practices are obviously interesting from an ecopedagogical perspective, as our age in many ways both has lost this capacity and will to limit the demands of work, production and utility. From classical antiquity's point of view, our age is a banausic age where, not the *artes liberales* but the *artes mechanicae* are the model par excellence for science. The arrangements of antiquity rested on anti-democratic principles, such as the exclusion of work and production from the public domain through slavery. This is, of course, not a defence of slavery for the sake of sustainability or to reject the connection between sustainability and the promotion of social equality. Rather, it means a call for ecopedagogy to address the dual position of classical antiquity in this manner: Classical *paideia* is both the origin of our scientific technical civilization and the background necessary in order to make a confrontation with the very same. The challenge of ecopedagogy in the age of the Anthropocene is thus to take the essence of the education of the classical *paideia* and reintegrate it with our modern democratic culture and expand a conception of citizenship from city state to planetary citizenship.

We thus need to return to the example of Pythagoras. According to Koestler (1959), he was fully aware of the fact that quantification and mathematization can lead to wisdom, but also of the immense technological possibilities geometry contained. However, these possibilities were kept secret. Only the initiated pure in mind and spirit could gain full access to how mathematics and geometry served us in our search after the most fundamental questions in our existence (Koestler, 1959, p. 38). An education for sustainable development thus needs to be based on a fundamental premise: *Knowledge is dangerous.* Perhaps the reason why Pythagoras recommended a five-year old vow of silence as a part of the education was a premonition of this inherent *hubris* of science and philosophy? It was perhaps a deeply felt recognition that knowledge in general and mathematics in particular were a double-edged sword that both lead to liberation and destruction, since "*number is the ruler of forms and ideas, and the cause of gods and demons*".

Notes

1. See Straume (2017) for a description and evaluation of the movement. My reading is influenced by her work.
2. See Straume (2017, pp. 107–108) for a further elaboration of this.
3. "It is only unfortunate that the bulk of the literature that is available on classical or ancient philosophy is silent on the import of environmental ethical thinking in these classical thinkers" (Chemhuru, 2017, pp. 24–25).
4. The question whether classical antiquity can be a source of ecological consciousness today and whether the city states of classical antiquity *de facto were* sustainable is not the same question. See Thommen (2012) for a treatment of the latter. As the notion of sustainability is a modern conception the focus of this chapter is the *ideas* of antiquity relevant to ecopedagogical thinking today.
5. The references to Coman, Carone and O'Connor came to my knowledge after reading Chemhuru (2017).
6. This shift is also described by Marrou (1956).
7. It is worth mentioning that the *artes liberales* did not become a stable curriculum consisting of the arts of the *trivium* (grammar, logic, rhetoric) and the arts of the *quadrivium* (arithmetic, geometry, music, astronomy) until Martianus Capella's *De nuptiis Philologiae et Mercurii* (*On the Marriage of Philology and Mercury*) in the fifth century A.D.
8. The Latin *contemplatio* is a direct translation of the Greek *theoria* (θεωρία).
9. Although *theoria* had several meanings, making this claim a bit more complicated. According to Berg Eriksen (1976, p. 82), *theoria* could signify: (1) "the study of astronomy", (2) the study of "metaphysical principles", (3) "the life of reason in all its forms" ("in this context it does not seem to be tied to special objects") and (4) "blessed recognition of God".
10. See Torjussen (2011) for a treatment drawing on the works of Merleau-Ponty and Nietzsche of the connection between a philosophical relation to nature and a philosophical relation to the self where "the genesis of the self is the most adequate approach to rewrite the man-nature relationship" (Torjussen, 2011, p. 442).
11. This example was originally mentioned in Torjussen (2018, p. 209).

12. Economic gain was of course beneficial, but it was not that vital compared to the European and North American slavery. According to Jameson (1978) most slaves in ancient Greece were used in agriculture, which were "technologically backward and stagnant" (p. 125). For an overview of this subject, see Westermann (1955).

References

Arendt, H. (1958). *The human condition*. The University of Chicago Press.

Brundtland, G. H., et al. (1987). *Our common future: Report of the World Commission on Environment and Development*. Oxford University Press.

Carone, G. R. (2001). The classical Greek tradition. In D. Jamieson (Ed.), *A companion to environmental philosophy* (pp. 67–80). Blackwell Publishers.

Chemhuru, M. (2017, August 31). Elements of environmental ethics in ancient Greek philosophy. *Phronimon, 18*, 15–30.

Cicero, M. T. (1927). *Philosophical treatises: Tusculan disputationes*. The Loeb classical library 141. Harvard University Press.

Coman, B. J. (2006). Whose environment is it? Some reflections on modern environmental philosophy. In *National Observer* (Melbourne, VIC), (68), pp. 55–62.

Eriksen, T. B. (1976). *Bios theoretikos: Notes on Aristotle's Ethica Nicomachea X, 6–8*. Universitetsforlaget.

Gadamer, H. G. (2013). *Truth and method*. Bloomsbury Publishing.

Gadotti, M. (2008, March). What we need to learn to save the planet. *Journal of Education for Sustainable Development, 2*(1), 21–30.

Goldhill, S. (1990). The Great City Dionysia and civic ideology. In J. J. Winkler & F. I. Zeitlin (Eds.), *Nothing to do with Dionysos? Athenian drama in its social context*. Princeton University Press.

Hadot, P. (2006). *The veil of Isis: An essay on the history of the idea of nature*. Belknap Press of Harvard University Press.

Iamblichus, C. (2020). *Iamblichus' Life of Pythagoras, or Pythagoric life* (T. Taylor, Trans.) [EBook #63300] Project Gutenberg (Iamblichus' Life of Pythagoras, or Pythagoric life, translated by Thomas Taylor—a Project Gutenberg eBook)

Jameson, M. H. (1978). Agriculture and slavery in classical Athens. *The Classical Journal, 73*(2), 122–145.

Kahn, R. (2010). *Critical pedagogy, ecoliteracy & planetary crisis: The ecopedagogy movement*. Peter Lang.

Kleczkowska, K. (2014). Those who cannot speak. Animals as others in ancient Greek thought. *Maska, 24*, 97–108.

Klein, N. (2014). *This changes everything: Capitalism vs. the climate*. Allen Lane.

Koestler, A. (1959). *The sleepwalkers: A history of man's changing vision of the universe*. Hutchinson.

Laertius, D. (2018). *Lives of the eminent philosophers* (J. Miller, Ed., P. Mensch, Trans.). Oxford University Press.

Lobkowicz, N. (1967). *Theory and practice: History of a concept from Aristotle to Marx.* University of Notre Dame Press.

Marrou, H.-I. (1956). *A history of education in antiquity.* University of Wisconsin Press.

O'Connor, D. J. (1964). *A critical history of western philosophy.* The Free Press.

Plato. (1973). *Phaedrus and the seventh and eighth letters* (W. Hamilton, Trans. with Intro.). Penguin.

Seneca. (1972). *Naturales Quaestiones. Seneca in ten volumes.* 7, 10 10. Harvard University Press.

Straume, I. (2017). *En menneskeskapt virkelighet.* Universitetsforlaget.

Thommen, L. (2012). *An environmental history of ancient Greece and Rome.* Cambridge University Press.

Torjussen, L. P. S. (2011). Eco-phenomenology and the interiorization of man—Using Merleau-Ponty and Nietzsche to release the «Psyche» from the human skull. I: Tymieniecka AT. (red.), *Transcendentalism overturned: Analecta Husserliana* (The yearbook of phenomenological research), bd. 108. Springer.

Torjussen, L. P. S. (2018). *Foreningen av visdom og veltalenhet—utkast til en universitetsdidaktikk gjennom en kritikk og videreføring av Skjervheims pedagogiske filosofi på bakgrunn av Arendt og Foucault* [PhD thesis]. University of Bergen.

Torjussen, L. P. S. (forthcoming). The University as the union of Wisdom and Eloquence. *Journal of Philosophy of Education.*

Westermann, W. L. (1955). *The slave systems of Greek and Roman antiquity.* The American Philosophical Society.

White, L. (1967). The historical roots of our ecological crisis. *Science, 155*(3767), 1203–1207.

Williams, G. D. (2012). *The cosmic viewpoint: A study of Seneca's Natural questions.* Oxford University Press.

CHAPTER 17

Sowing the Seeds of the Pollination Academy: Exploring Mycelic Pedagogies in the Anthropocene

Oleg Koefoed and Thomas Burø

Introduction: Inhabiting the Critical Zone Around an Inverted (Anthropocene) Vortex

This chapter reflects on a practical learning experiment in which the authors started up with partners and volunteer learners in the last months of 2020, leading into 2021. Several events triggered the experiment. The deepest one was the long haul that we will refer to (with some hesitation) as the Anthropocene condition. We use this epochal designation to express a *condition*: the latest impact of human industrial manufacturing and urbanisation on the global climate on Earth, forming what Latour calls a "critical zone" (Latour & Weibel, 2020). We thus avoid the question of when such an epoch might begin or end. Other vast derivative events follow in the mire of the Anthropocene condition and its critical zone: global

O. Koefoed (✉) • T. Burø
Copenhagen, Denmark
e-mail: Oleg@growing-pathways.com

M. Paulsen et al. (eds.), *Pedagogy in the Anthropocene*, Palgrave Studies in Educational Futures,
https://doi.org/10.1007/978-3-030-90980-2_17

warming, extensive ecological destruction and biodiversity crisis/mass extinction are probably the three most imminent ones. The loss of inhabitation is a fourth condition that seems to be gaining importance (Merrifield, 2014). That condition generates a movement where urban families begin to flow out of the cities to win time and gain space. They seek to become regenerative and permacultural in the outskirts—like the Thoreaus of the twenty-first century (Fig. 17.1).

At the root of the crises—or perhaps hovering above the earth's surface, sucking out the roots—one finds an *inverted vortex* (the idea of the urban vortex has roots to, e.g., Hall and Savage (2015), who speak of "mobilizations", "spatialisations", and "emergences" of "urban infrastructure" as elements in a concept of the urban vortex) grown with a paradox: while it sucks energy to the (mega)city, its headquarters or Default Mode Network (Ekhtiari et al., 2016) floats in space or flows in silicone: the vortex is kept in pulsation by the algorithms of a debt-based 'growth' economically centred politics whose core is an urbanising, capital-generating macro-economic mechanism (Brenner & Schmid, 2013; Freund & Padayachee, 2002) that pushes material change in front of itself in that very liveable rim of the Earth—the zone—mainly into cities and out of extraction zones (Latour, 2017; Latour & Weibel, 2020; Sassen, 2014). Latour's focus on the crust or the critical zone is important, as it

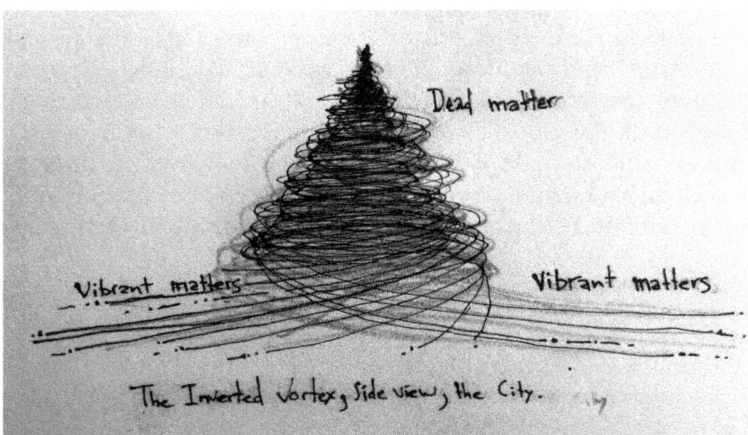

Fig. 17.1 The inverted vortex

shifts focus away from the 'planetary', political focus, to one that is more material, closer to us, less abstract. The vortex is not a metaphor, it actually happens. Rather than sucking matter into the vortex, the dynamics of this mechanism is to create energy through pushing obligations to action forward with financial instruments (e.g., *obligation* bonds). The inverted vortex sucks resources out of the crust of the planet. It ejects cities at an incredible pace all over the surface of the globe—especially in so-called emerging economies (WEF, 2021), sort of like an inverted version of Benjamin's angel (Benjamin, 1989). Benjamin's "Angel of history" has its face turned towards the past, while it moves forward at a high pace, blindly shooting into the future and looking backwards at the mounds of rubbles growing at its feet—in present days, the rubbles are replaced by buildings, infrastructure, and highly unevenly distributed privilege. Even the angel no longer has a human face; it is replaced by an inanimate mechanism, the inverted vortex.

What is the effect of this vortex on older urban centres such as Greater Copenhagen, less extreme than Kinshasa, Lahore, or Suzhou? The vortex is still in action, but the old cities are grown through a still more dominant discourse of the "smart city": a sustainable and competitive urban development upheld by financialising methods (Katz & Nowak, 2018; C40, 2016) spreading rapidly across the world (privatisation, expropriation of natural/local inhabitants, and intensive development). Privatising has dominated the past 25 years of urban development in Copenhagen. 'Unused' public land is sold for development and rapid construction, in part following sustainability certifications, as political negotiations, administrative boundaries, and shareholder value tend to take over the scene almost entirely (Harders, 2014). The product of the vortex, urbanisation, has declined in both Europe and Asia (UN, 2018), which does not seem to halt the narrative of urban growth as a presumed beneficial motor (Copenhagen, 2020). Instead, new visions are drawn, images of entire islands thrown into the sea off the coast of the city, built to house up to 40,000 inhabitants. Such is the local context of what we described in our latest publication (Koefoed & Burø, 2022), presenting the 'chaotic' state of these economic mega-mechanisms producing more or less unpredictable entanglements of phenomena and calling for a rethinking of paradigms of learning (see below). The main victim of the movements of this vortex is the living, vibrant (Bennett, 2009), mycelic (Sheldrake, 2020; Tsing, 2015) character of matter and ecosystemic relations in the "critical zone" (Gaillardet, 2020). When soil, sand, and water are captured by the vortex, they are instantly transformed from vibrant beings in a living system to dead resources in the service of urban growth. That is where we start: a recovery of the living, an encounter with ghosts (Tsing et al., 2017).

Map.Learn.Repeat. Based on Depression

Let's move closer to the sphere in which our case takes place. We take the backdrop of an inverted vortex ravaging living ecosystems across the planet and local urban governing with us. It uses capital mechanisms and runs a 'smart' strategy to focus on particular issues (CO_2, energy, information mass, DNGB standards) while neglecting others (democracy, intra-action, biodiversity, ownership). The inverted vortex leads to financial expansion and massive construction projects, creating new neighbourhoods that are practically devoid of non-governed life forms. Perhaps the most damaging impact of this urbanisation is its neglect to integrate its human citizens and nonhuman ecosystems (aka 'urban nature'). The authors had worked with urban living conditions and human-nature relations in other contexts and decided to initiate a learning experiment ('Map.Learn.Repeat.'). It would combine 'radical' (Mogel & Bhagat, 2007) or 'virtual' (Burø, 2020) participatory cartography with collective-creative learning inspired by design thinking. The aim was to raise a little awareness around the victims of the vortex and the materially cheap act of learning.

Reading Latour, Stengers, and the other architects of the "Critical Zone" approach, and adding to the growing body of knowledge around the mycelic world and its imprint on human culture (Stamets, 2011; Tsing, 2015; Sheldrake, 2020), we see three challenges calling for new experiments. (1) We need a place-based, situated *learning practice* that does not ignore the larger forces at play nor falls flat on its face to 'produce' some kind of 'entrepreneurship' based on individual or microbusiness models. These models grow from the same investment logic as the inverted vortex and therefore do not support a new form of learning. (2) We need to actively explore that particular aspect of the zone, its *complexity and fuzziness*. The zone lacks clear boundaries and needs to be "addressed by very different branches of science, ranging from pure physics to geography, geology, hydrology, pedology, geomorphology, geology, ecology, and biology; disciplines that in the history of science have been separate for more than two centuries, not to mention the social and human sciences" (Gaillardet, 2020, p. 123). (3) We need to develop formats of exploring and learning that is both creative and intra-active (Barad, 2007), that is, recognising and growing from a consideration of nonhuman species as equally valuable for the process of learning and its future value (Haraway, 2016; Malone et al., 2016). Incredible tasks for a very modest experiment.

Over the following pages, we will explain what followed between our first cartographic ideas until the coining of the 'Pollination Academy' and its first steps. It's a story of transformation from a mapping experiment to what might become a philosophy of place-sensitive, mycelic learning amid a semi-neglected neighbourhood. We will perform this account in a series of small steps, inspired by the methodology of the '6 P's of the healthy city' (WHO, 2018), tweaked through explorative action-based learning within the circles of the 'Action University' (Breum Amhøj, 2019; Koefoed, 2019): Place, Participation, People, Prosperity, Peace, and Planet. This chapter re-moulds the slightly schematic P's to fit into a narrative flow.

Ideally, our chapter would relate much more explicitly to the history of education and the Anthropocene, environment, and the like, or at least to traditions within education for sustainability. While we must insist that our chapter deals first of all with a practical case, we still relate to epistemological models, educational theories, or ontological presuppositions. Our reason for carrying out practical experiments touches upon how we consider education or learning part of a broader picture. 'Sustainability' needs to be something we 'do' in a learning setting (Breum Amhøj, 2019)—and it is crucial to understand what lies in the 'we' and the 'something' of that phrase. In the words of Breum Amhøj: "*How can an action-research process pave the way for sustainability to become an ecology of thinking, learning and acting, desire and intensity that connect bodies intra-acting with human, nonhuman, animate, spaces and affective movements?*" (Breum Amhøj, 2019—referring to Barad's concept of 'intra-action', taking place below the subjective level (Barad, 2007)). Moving learning in this 'ecological' direction, we bring along a century-long trail of trying to alter the rationalist and positivist tendencies in education (Paulsen, 2021). This trail carries with it a load of assumptions about learning in places, or learning and nature, and even knowledge and reality. With this in mind, we claim that one of the effects of the 'anthropocene' is that 'our' being in the world is being torn apart and re-stuffed with 'critters' and realities (Haraway, 2016, 2020), ghosts, monsters (Tsing et al., 2017), and a wide range of ecologies and subject types (Morton & Boyer, 2021) that inhabit every step we take from now on. The vortex may tear life out of matter, but it grows back in an urban reality and becomes a new memory, no longer what it was, yet containing a specific past and an unknown future. We will seek to connect and tap into this almost virtual memory of urban matter.

Following the logic of practical challenges and action learning, we inject a set of principles that we attempt to bear in mind and which we will integrate as headlines here (Fig. 17.2).

#1—education always happens in places; #2—education is always an agential apparatus in the wound of the world (Haraway, 2016; Malone et al., 2016; Morton, 2009; Morton & Boyer, 2021; *Dickinson, 2019); #3—we never know who we are; we're many more (*Stengers, 2020*;* Sheldrake, 2020*); #4—every time you think you know better how to participate in*

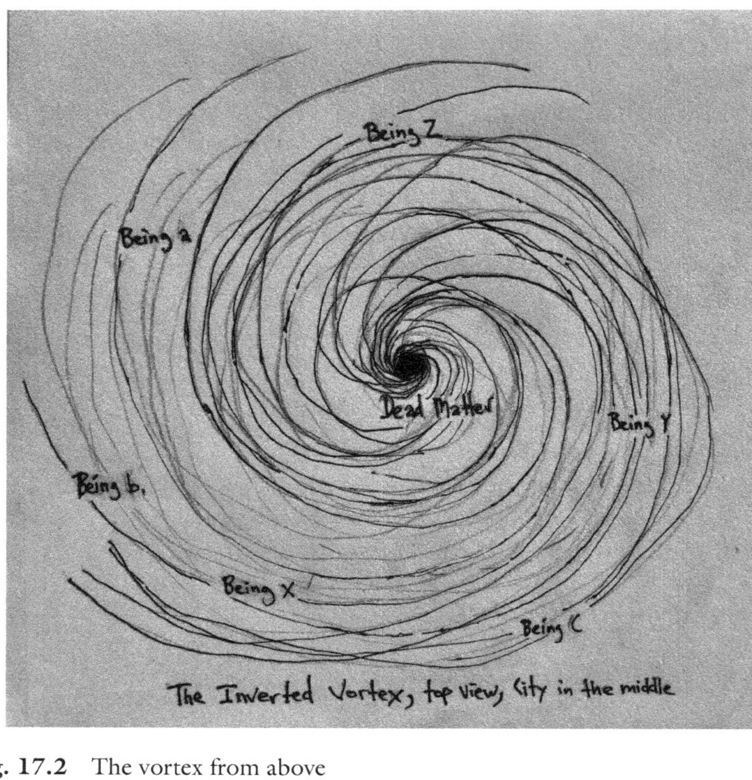

Fig. 17.2 The vortex from above

learning, ask again (Breum Amhøj, 2019; Koefoed, 2013, 2019; Puskas et al., 2021*); #5—value is always an ontological and political question (Latour & Weibel, 2020;* Stiegler, 2013*); #6—entanglement is fiery, fire affects, affects grow as passion and contamination in the encounter (*Tsing et al., 2017*); #7—every act of knowledge is an act of power—the map is always a territory (*Winther, 2017*; Radical cartographies 2010); #8—we are all carrier bags, whether we are bacteria, bugs, elephants, tribes, or planets (*Haraway, 2020; Le Guin, 2020/*1986)*

Education Always Happens in Places: Learning Is Always an Agential Apparatus in the Wound of the World (Even in an Old Reworked and Buried Swamp)

The urban neighbourhood around Prags Boulevard in Copenhagen has been formed by different interests over time, giving it the mixed appearance it holds today, on the verge of yet another transformation over the next decade. We want to write up a very short resumé of the heritage at play here. We don't know exactly how it affects everything else that happens now. We know that there are restrictions to what is legal/illegal to do with the soil in the area. We know that pollution plays a part in regulations. We don't know how soil contamination influences living creatures on the site today. Nobody has cared enough yet. We could brush that question aside, maintaining that it is impossible to track. Or, we can do this: consider this chapter a part of collective journaling and try to keep a reasonable balance between raising questions and avoiding hasty conclusions (Fig. 17.3).

A painting from 1851 shows the relationship between human settlement and natural ecologies up until then. Amager is a low-lying island, no more than 5 m above sea level. Most of the island was covered by water on and off since the last ice age. Amager is devoid of the moraine hills and valleys of up to 170 m in, for example, Bornholm or Jutland. The island was dominated by fishery and hunting until the sixteenth century when the king invited farmers from the Netherlands to cultivate the land and turn it into 'Copenhagen's pantry'. With its vast area and 'unused' or 'uninteresting' lands, Amager could serve as space for military garrisons and training, as farmland for the city's growing population, as latrine dumping ground for the same reason, or as an island to expand. Amager

Fig. 17.3 Painting from 1851 of what became Prags Boulevard

grew by more than 50% between 1750 and 2020. Landfill and urbanism go hand in hand—mercilessly and with only minimal sharing of power. Urbanism is largely colonialist. Any claim to sustainable urban development must, before anything else, question its role as a colonising force (Barrios-Negròn, 2020).

In the nineteenth century, after centuries of warfare, the empire shrank and crumbled during periods of alliances going the wrong way. By 1864, a more peaceful development could take over. In the capital, where the areas outside the city ramparts had often been victims of attacks from sea or land, the capital and development-driven merchant and manufacturing elite settled with new industry in Holmbladsgade and the boulevard. The territorial ambitions of the monarchy gave in to the economic expansion of the capital class. Hence the Holmblad dynasty, owners of a large area housing a breeding farm, decided to move their factory in 1880. Today, the former factory houses Kvarterhuset (municipal culture house) and the headquarters of CARE Denmark. On the other side of the street, a metal factory replaced the workers' allotment gardens but eventually made way for the largest student dormitory in the Nordic region, the Øresunds Kollegium (ØK). From a *longue durée* perspective, the area's industrial history is nothing more than the blink of an eye. From the lush grasslands of Laurits Peter Holmblad's breeding grounds until today, where the last of the many factories are moving their production, are a mere 170 years of industrialisation. However, this century-and-a-half has been decisive for how the area has changed in the past decade. The 'critical zone' was practically expropriated. The same dynasty that owned the breeding farm ended up drowning the neighbourhood in toxic waste. While the street that carries the family name, Holmbladsgade, was made more presentable with residential buildings, Sønder Boulevard, later Prags Boulevard, became the manufacturing backyard. The economically thriving activity had given rise to residential housing and jobs; it also resulted in massive pollution of the soil on Prags Boulevard. In the same area that held hundreds of allotment gardens producing vegetables for the residents, the municipality tells citizens to grow in crates only due to the extensive pollution of 150 years of Anthropocene activity. The park created as part of an urban revitalisation between 1997 and 2005 did little to improve the quality of the soil. There was no proper cleaning of the area, no long-term remediation strategies, and short-sighted implementations of playgrounds, stages, skating rinks, and lawned roundabouts mark the space between the buildings. *Repairing* prevails, *regenerative* action is left for the next generation (Fullerton, 2015). The critical zone has been expropriated (Fig. 17.4).

The most vibrant places are the ones left unmanaged—the last remains of the train tracks with their well-known biodiversity islands, border zones between public and private areas, zones left undetermined, closed but with a broken fence, or re-opened on a temporary notice. Most public space is managed with a sharp blade, leaving trimmed lawns and tarmac/concrete surfaces. The long straight line in the middle of the boulevard/park epitomises the half-heartedness of the 'revitalisation': it's green, but it's "only green", "it's just grass" (from fieldwork by students from ITU helping us in the spring of 2020). It's alive, but it's hardly breathing.

This 1.5 km of lawn with poplars and training facilities is maintained by the "men in yellow", aka the maintenance teams. They reflect the 'bare side' of the attitude to places: an urban tactic driven by an aesthetic of order, containment, doing as little as possible. Within the task forces managing the neighbourhood's green spaces, there is a lack of training. Uneducated staff leads to micro-management, taking energy away from

Fig. 17.4 Prags Boulevard 2020, more or less the same place as the painting above

possible positive impact for biodiversity or sustainability through public space—despite managers being interested in urban nature (interview with the area manager, March 2021). In 2021, the boulevard is an ecosphere restrained, locked up, tied up, held down, beaten, and controlled. The city shows no interest in educating its personnel (conversation with project leader in biodiversity in the municipality, 2021).

We created a civic/professional initiative, the "Nature Partnership" in early 2020, to address possible collaboration opportunities between stakeholders of all sorts. We took stakeholders on exploratory walks with a landscape architect based on a permaculture approach (http://evmlandskab.dk/portfolio/pragsboulevard/); we tried to identify possible pivotal points and gravitation zones. We hit walls of good intentions and low resources. By the end of 2020, we had to acknowledge that our 'nature partnership' needed to wait for a more comprehensive partnership with the municipality, funding and permissions. The straightjacket was holding us down. We were outside of the centre of the vortex—but that does not lead to freedom to act.

The 'People': We Never Know Who We Are; We're Many More

In the summer of 2019, we started collaborating with the NGO CARE, which has its Danish headquarters on Prags Boulevard. CARE was ready to make green spaces and urban nature outside their office building, and we started a cultural mapping (Duxbury et al., 2015) process in the area. We knew that we needed to focus on the place and discover it with the locals. We worked our way outwards from CARE's parking lot next to the boulevard, exploring local interests and connections to the area. We also knew that it's not too hard in Copenhagen to find humans interested in higher urban life quality and urban nature values. Many actors are freelancers, microbusiness, and volunteers. The issues and the field are complex and too hard to solve, and many are eager to meet and share. So the 'urban-nature humans' will come closer, looking for opportunities in a zone of attraction. Mycelia are constantly on the lookout for more connections if one turns out to be toxic—and they can do a thousand things at once (Sheldrake, 2020). Humans in old cities tend to look upon themselves as individuals; they are multicellular and stuck in their DMN state a lot of the time. And where villagers can easily organise themselves in

Fig. 17.5 Workshop with CARE and citizens, December 2020

place-based commons (Burø & Koefoed, 2021a), urban dwellers are limited by factors such as legislation around landownership—and throughout 2020, civil society took a blow and is only recovering slowly (Fig. 17.5).

One of the difficulties for small or free-moving agents in cities is to know when 'to anastomose'. Corporate and political interests tend to make alliances that can make decisions and act in ways that will affect the possibilities and limitations of everyone else around for decades, sometimes centuries. Such inequality in agency leaves the floor wide open to the debt-driven cycles of loans and development hazardous for urban sustainability but have proven their value from a narrowly economic perspective. Hence, the developing areas in Copenhagen suffer from a democratic gap between citizen interest and democratic influence. This globally known phenomenon has only been enforced in 2020, aggravating the "global democracy retreat" (Falanga, 2020).

The zone we were trying to cover with our cultural mapping had been outside of developers' scrutiny. With its building mass already in place, it may call for refurbishment but doesn't invite to rapid profit-making of the kind that you find just east of the neighbourhood, along the "silver coast" of Amager Strand. The absence of economic interests means a lack of levers for, say, spatial opportunities backed up by financial support. This condition led us into the arms of citizens, community groups, NGO leaders, and institutional bodies like the environmental point or the local council. Not that the area is out of sight—the recurring visits of the mayor of the environment in Copenhagen to the area paint a slightly different picture: interest is there. It's just a bit dormant. But the story of the area leans on the social side rather than on nature-based solutions.

Our mapping showed interest among citizens in contributing to a liveable neighbourhood—and among the more actively engaged ones to enter into contact with and maybe contribute to sustainable processes in the bigger picture. Only a few have the surplus to initiate actionable strategies that could incorporate sustainability in local operations. We identified a variety of actors who were somehow aware of living ecologies in the neighbourhood: the company behind Pelican Self Storage turned a 250 sq m zone into a future biodiversity garden—alas, without any citizen involvement; some of the social housing associations, owners of border zones, are curious but limited by habits; individual citizens are interested in 'doing something' while owning no land to give their interest to; a municipality area manager is positive towards nature in the city and biodiversity, but restrained by too many restrictions, uneducated personnel; a local council with dreams of their own; a municipal project manager working with nature and health in cities, but with no influence or practice with the negotiation of territories. The 'Miljøpunkt' (Environmental point; one of four Agenda 21 foundations in Copenhagen) for the island of Amager (Miljøpunkt, 2021) is very willing to collaborate but have no territorial approach. The urban gardening community, Prags Byhaver, is staying afloat with around 40 crates on 800 m^2 of municipal land in the centre of the boulevard, living on an exception permit after the municipality made the "volunteer coordinator" function redundant in 2018.

For most agents, whether individual or collective, Prags Boulevard was either too far away, too vague, or outside of their jurisdiction. But one of the members in our 'nature partnership' was the chairman of the residents' council in the privately owned dorm, 'Øresundskollegiet' (ØK), where 1700 residents organise through councils and have started taking action to transform buildings, rooftops, cultures, and the areas between the houses. In a field of disconnected actors disconnected with little power or finances, it seemed a different wind was blowing through the 50-year-old dorm. We decided to change our approach and work more closely with the people at ØK. The residents' councils are looking to remake the space between buildings, around 3–4 ha, with greenspace and sustainability value at the core. They landed a large grant in 2020 to redesign the 1300 sq m garden rooftop, developing leisure space and urban farming. They signed an agreement with the municipality to accept 100 new non-student, marginalised youth from August 2021. The chairman and the founder of the dorm café had opened their doors to us. ØK could offer us a base to work from, with solid potential for shaking up the neighbourhood and advocate and work actively on more focus for urban nature in the area.

Until this point, what we had seen and heard reflected the expected image of a local population reacting to the Anthropocene condition with very little real attention. Moving into ØK, what we saw did not raise our expectations of inducing active participation towards engaging with the immediate environment. It seems that very few average young citizens do the Greta Thunberg or Billy Barr thing (Burø & Koefoed, 2021b): act in systematic, mapping or data-creating, creative ways that can increase knowledge on a local level with global value. Our local observations confirmed the picture painted by general surveys: most of Copenhagen's citizens are aware of the risks of climate change but unable or unwilling to take action. They believe that climate change results from human activity, but they expect governments and businesses to respond. They link agency to low-impact or positive impact consumerism. While the population in Denmark is one of 'climate concerned' humans (50% consider climate change to be the most critical problem on a global scale (KL, 2019; Ipsos, 2019), citizens remain reactive.

And the students in ØK? At first, they were invisible. Then they seemed to behave like workers: meeting expectations in the face of restrictions. It's almost as if the routines help to hold a veil against the abyss of climatic and systemic disorder. We moved into the dorm, setting up a base in the (temporarily closed) café (functioning as a secondary study room). One of the few life forms that we, the café staff, and the environmental committee of the dorm were engaging with were the plants in the room, calling for care, being transferred to new pots, cleaned up, or discarded. We took care of plants; we cleaned up space, we invited. Nothing much happened. Were we right about our hypothesis on depression—then perhaps meticulous micro-action could be a way to displace and maybe alter climate helplessness? How could we explore this? With nothing to lose and very little hope, we started the 'Map.Learn.Repeat.' test in December 2020.

Participation: Every Time You Think You Know How to Participate in Learning, Ask Again

> *Crawling in the elbow crease of beach and dunes, looking for shards of modernity, waist-deep in the visceral quality of remote isolation. Along the edge of an inlet, I collect tiny plastic fragments on the scale of native cherries. I carefully peel tender plastic bags on the brink of collapse from shallow graves, tease out fragile, submerged, sand-filled membranes.* (Selenadec, 2019)

What is it we need to do to provoke passion, curiosity, and more direct engagement in the form of action, maybe even autonomous and change-seeking initiative? And how do we transform that attraction into life-changing learning experiences? The quote above—an extract of a diary from a young woman participating in a plastic collection on a beach in Tasmania—may open that puzzle. In the latter years, the plastic versus oceans agenda has gained a lot of following. Presented with painful images and immediate action needs, young people participate very actively in places like Guam, the Baltic Sea Region, or Cyprus, in actions that map, record and change, seeking to fight the overwhelming presence of plastic in the aquatic environments of the planet. Fighting plastic pollution has gone from a problem barely in the global attention span to a field with a whole range of associations, non-profits, small business, and volunteer action taking place on a worldwide scale (https://www.oceanplasticforum.dk/). With global awareness rising by the day, it is hardly a problem to get young people to participate in the kind of activities that the journal keeping "Selena" was part of in Tasmania.

But what if your geography does not include sunny beaches, white sand, or lush tropical islands? What if your academy starts in a somewhat run-down, polluted, tired part of town, in a city that is focusing most of its attention elsewhere? And what if your problem starts in the encounter between tens of different issues, including a tendency to feel depressed or helpless among younger urban citizens? How do you set out activating participation around the understanding of urban environments? (cf. Kindon et al., 2007)

The first months of trying to lift the cartography experiment off the ground suggested potentiality (longing and resistance) and virtuality (future events already drawn and others hidden deep in the history of the place)—yet, we languished. There was a felt affective reaction coming closer, but not readiness to play. Like the place inviting us, but hard to decipher. We had hoped that a "virtual cartography" could open up "ways to go from apathy to action in the face of a chaotic world" (Burø & Koefoed, 2021b). The virtual map does not contain the same responsibility as the governmental, authoritative map created by those in possession of land or rights to manage and govern the land. Acting in this virtual space allows both a degree of being more radical (utopian, dreaming, or just curious) and a possibility to enter into a sharing and playful act of discovering one's immediate ecology of places, beings, or elements. We wanted to invite the learners to map, for example, biodiversity baselines, interactions between life forms and infrastructures, temperature-sound-light, ideal structures, forms of organising—or maybe just people passing by day by day. Our invitation evoked a mixture of confusion and hesitation, attraction combined with uncertainty. The residents were marked by the repeated lockdowns of the past year, leading to home studies, closed cafés and workshops; interaction had moved from the larger public spaces to their floors and communal kitchens. A large proportion of the residents had fled the premises since there was little point in staying when all courses are online and restrictions suspended public life. We didn't know our 'plastic'—the motivating 'inhuman actant' (Latour, 1999). However, our mere presence and the energy we brought with us seemed to bring about a sense of opening and beginnings.

At this point, we took a step back and opened up to embrace an ambition we had not expected. Given the size of ØK and the many initiatives addressing sustainability for the area, we thought there might be a decent chance of digging out at least a handful to participate in a small mapping initiative at and around the dorm. To test our hypothesis, or to see where it would take us, in a first approach, we adapted a mapping tool based on the participatory mapping technology of "Maphub" (Maphub, 2021). We invited residents in the dorm to join us on a mapping mission in their environment as a 'baseline' action. We wanted to get a first idea of possible indicators for change in living conditions (human and nonhuman). In the poorly populated dorm, we carried out an online workshop with a handful of curious residents. Conclusions were twofold: the atmosphere was positive, but participants had very different ideas about what they were interested in following and mapping. And while all had experience with various forms of maps, none had been part of participatory mapping before. We created a digital community platform and started sharing stories and reflections. There was talk in the corridors, but residents hesitated.

We decided to wait until it became possible for us to carry out a workshop in place. By inviting the residents to a more direct encounter, introducing paper maps and the possibility of practising mapping in an interactive, participatory framework, we hoped we would be able to trigger a higher level of sensuous activity. We tried to recreate the joyful atmosphere from previous events: in Copenhagen in 2014, in Valletta in 2015 and 2017, and all around the Baltic Sea, especially in the Baltic cities, in 2019. And? Still more apathy than action. Our participants were intrigued but waiting to be told the purpose and the tools for the exercise. Working on visions for the area based on personal preferences and affect, we tried to draw on a playful and visioning approach. The absence of a learning framework made the relation slip from a common empowering relation to a participatory logic. We were stepping back to act in the world outside of the digital sphere, and it felt like a move backwards. After the first workshop, expectations started to form, and the residents gave us more authority than expected. While we were playfully inviting them to a joint learning visioning session, they were ready to talk seriously about taking control of their 'hood' and getting inspiration for action. We felt like we were back to square one, except that the general atmosphere among our handful of volunteers was now moving from curiosity to impatience.

At this point, we had developed a closer relationship with a small group of stakeholders in the dorm: the chairman and the board of residents' councils, the founder of the café subcontractor, the chairman of the environmental committee, and a few others. We worked together on a funding application that would benefit different interests, including urban farming, bio-mapping, and creating education practices. The intersection eventually became conceptualised as the *Pollination Academy*. We reframed the cartography to be a minor part of a more significant movement, as the local participants were longing for a chance to contribute more directly to a life-enhancing environment in the immediate neighbourhood. We shifted our focus from a smaller group cartographing their area to the birth of a new pedagogical concept, the 'Pollination Academy'. Instead of 5–10 volunteers playing around for a few weeks, we began formulating the vision of a learning community of 40–50 residents, local partners, international networks, and potential investors. We had to reposition the mapping into a purpose.

Inspired by places like the Barefoot College (Barefoot College, 2021) the Slow Factory (Slow Factory, 2021), or the New Institute in Rotterdam (The New Instituut, 2021), the Pollination Academy builds on the idea of a collaboration that allows residents of a neighbourhood to become the change pilots for a life-supporting, sustainability-driven, nature-based city. Through practice-based learning and site sensitising methods, the learning process will, at the same time, be the first step to a more nature-based place, thinking in universal design and ecological value. With the concept of pollination as a driving idea for the development of learning and intervening practices, the academy will focus on a combination of defining goals and mapping the value of an area (bottom-up)—and making partnerships with landowners, municipalities, and companies (top-down). The pollinators will unite and bridge between the needs of life and the power of giving. This conceptualisation emerged from the encounter between the real-life bareness of the boulevard and the dorm area—and the expressed need of residents to be part of active learning and transforming. To engage in a mutually beneficial process leading to a more substantial urban nature-biodiversity-sustainable urban purpose, we would turn ourselves into a school.

Entanglement Is Fiery; Fire Is Passion and Grows in the Encounter: As Do Ghosts

> *From a management perspective, bioindicators inform our actions as to what is and is not biologically sustainable. Without the moss in the tundra, the cutthroat in the mountain stream, and the canary in the coal mine, we may not recognise the impact of our disturbances before it is too late to do anything to prevent them.* (Holt & Miller, 2010)

At present, the 'pollination academy' has been presented to residents at an event at the dorm, receiving engaged feedback and more vital support than our first aspirations. We expect to launch the first 'classes'—groups of residents and locals with the ambition of learning how to transform their values into action, benefitting both the local environment and their learning process (Fig. 17.6).

Between the buildings, we found lichens, unnoticed and discreet. The dorm, like all other public spaces, has a population of thousands of lichens. The lichen being a peacefully consuming actant, part mushroom, part algae, it hardly forces anyone to give it any attention. It does not offer

Fig. 17.6 Pollination Academy workshop: The Lichen Hunt, May 2021

itself to the care of human symbionts. It lives on rocks and wooden surfaces. Lichens are one of the most important beings to understand in pursuing knowledge about life on Earth (Sheldrake, 2020). Lichens are everywhere—yet, their sensitivity, especially to air conditions (lichens draw their nutrients from the air), makes them excellent bioindicators (Conti & Cecchetti, 2001). Could we start mapping them? As we began to invite residents of the dorm to participate in the first mapping experiments, mapping lichens was one of the actions we were planning. The mapping may lead us to a prototype pollinator patch/laboratory space with a bee habitat.

What affect is being activated here, how will it multiply or mutate in the future process? How much of the increased effect can we attribute to conceptual focus from mapping/learning to learning/changing? How much comes from a more subtle process of something like a 'mycelic' or 'lichenic' organisation (Sheldrake, 2020) where bodies are trying to develop enough substance to be able to find a place to live are in constant multiple interactions with themselves, other forms, the soil, rock or air of the milieu, the chatter between living creatures in the area, the overall rising level of energy influenced as much by changed restrictions or even the slow arrival of spring? It would be arrogant to claim more than anecdotal knowledge of what is going on. Residents notice our presence, and the fact that we are still around and partnering with the residents' committees gives us some ethos and trust. Are we a cultural factor with which one might engage? We sense that the atmosphere of depression has shifted. More general optimism is growing, and participation and volunteering are forming.

Our first action is the 'Pollinator's call: help us map the lichen'. We ask the residents and other potential learners in the academy to help map out lichen populations so that we can obtain a double benefit: we map the lichens as the first step towards a climate change tracking bio-cartography (inspired by, e.g., Holt & Miller, 2010 and Stapper & John, 2015); and we map the interactions and intersections of human and lichen population and movement. When it comes to entanglement and fire, it seems relevant to consider what role the lichens might come to play in the future of our initiative—and of the issues of climate change and sustainability in the neighbourhood. When we mention 'lichenic organisation', we are still in the framework of a metaphorical appropriation of 'natural' phenomena. Are we using the lichens as a metaphor for the way that human organising takes place or might take place, playing on their compositeness, complexity, and resilience to talk about transformative action, as when corporations enter processes of "regenerative leadership"? Or can the learning process bring us out of metaphorical appropriations into working with lichens as partners and collaborators in the pursuit of learning? Can lichens be part of our 'apparatus', our 'intra-active' (Barad, 2007) way to learn, a way to learn and research where all beings can be researching subjects? Their movements and reactions will live inside the knowledge that the learners develop. The interactions may develop new relations based on collaboration between human and nonhuman agents—a factor partially acknowledged in science but less than in artistically informed practices such as the work of Jackie Brookner in Finland (schuylkillcenter.org) or Laura Winge in Copenhagen (Winge, 2021). It would probably be wrong to assume any actual 'passion' in the involvement of the lichens. It would be as naïve to disregard the possibility of interaction on a human-lichen scale, either micro or meso level. The exchange might be an active part of the learning process for one or more species involved, yet to be defined.

Discussion: Value (Prosperity) Is Always an Ontological—and Political—Question

> *Thus, the flowering cycle of many plants implies trust in a certain regularity of the seasons with which they synchronise their development. The current climatic disorder is creating a growing number of ways in which this trust has become misplaced. Keeping in mind that the French root of "array",* arroi, *referred to the company and equipment that allowed aristocrats to hold their rank, there is nothing subjective in saying that many plants and animals are in a situation of "disarray".* (Stengers, 2020, p. 231)

As soon as you combine questions of value and learning, you start asking questions about whom the value serves and how to guarantee some degree of autonomy for the learning. This connection between learning and value brings attention to how they might grow dependent on one another. Learning to pollinate and be pollinated could potentially be a pathway that could benefit many different life forms in the area. If successful, we could transfer the pathway to other local settings if the value generation actualises multiple ramifications of value.

Since the science of economy rose to power, its propagators seek to promote objectivism. At the heart of this endeavour rests the question, "what might we exclude?". The neoclassical economists dreamed of the equation holding the key to a never negotiated spirit in the world, like Smith's 'invisible hand'. We may have abandoned the idea of value as pure mathematics. Yet, it is still rare to see, for instance, planners accept value as part of complexities of non-familiar actants (see Beinhofer's attempt at formulating an evolution-sensitive, complexity economics (Beinhofer, 2007)). But other kids have joined the class since then, including ideas of regenerative economics (Lovins et al., 2020; Fullerton, 2015) deriving models of 'regenerative leadership' (Storm & Hutchins, 2019)—or similarly model-driven doughnut economics promoted by Kate Raworth (Raworth, 2018). Could we define a model for the ambitions that we hold and seek to grow in the Pollination Academy? Could this model be part of a 'culture of models', requiring "the ability of scientists to explain the definitions that a model articulates and puts to work, to formulate precisely the interdependencies it stages, to specify its robustness in relation to what it ignores—in short, to open its way of understanding the situation to questions and debates" (Stengers, 2020, p. 233)?

Going back to our workshops, we asked residents to help us map what they saw as valuable. We thought that opening from a more narrow focus on biological and environmental indicators of value to social and cultural elements of value would serve as a motivator. The invitation focused more on connectedness, interdependence, and subjectivity. The map showed a possible future of transdisciplinary sharing. Yet, at the heart of the map stood a tower of knowledge, an almost mythological symbol of a dream. Perhaps Stengers' 'culture of models' requires both disciplinary expertise from the residents and cultural sensitivity. The map was an act of recognising differences in perspective and a landscape yearning for a centre.

So what about value? What if we leave behind the act of externalising, which, after all, is always an act of violence or a suspension of intimacy? While mapping is also an act of exclusion, the invitation we had sent out and phrased repeatedly was inviting residents to include the trajectories they could see, along which specific reactions might flow. But mapping is also an act expecting predictability, as Stengers points out. Why map if everything around the map were to change? Then mapping would be an act only of memory loss. Expectation is a form of trust and lays out responsibility both ways. The signer of the map is as much victim to an act of responsibility as the reader following it. The map and the mapping express value to the signer while offering it as a potential value to the reader. This assemblage, however, already presupposes some degree of agreement, of trust. A map that is not signed is a mystery, yet none of the residents was ready to, in practice, be the first to sign the map that called for trust.

What economics reminds us is that despite the dreams of objective value, there is no such thing. Hence, we needed to take the next step to say, "the act of pollination could serve as a metaphor, yet we invite you to become part of a movement where it goes from metaphor to literal value". It is an invitation to a slow movement of internalisation—and this slow movement is one in which the map and the territory change alongside one another. There is no such thing as an objective value; it is always an interpretation and always at stake between interests. Thus, we are looking not for externalisation but the opposite. At this point, we might go from talking about the translation of the non-monetary value into one that is monetary (Dasgupta, 2021) and start to talk about multiplying the curved lines of value visible to the mapping of pollinations and the pollination of mapping.

Our experiment approaches the serendipity of any relationship that the map shows while discovering how the relationship might be both classifiable and singularly valuable. We seek to map value in a search for (re)generation. Including the lichen in the act welcomes the language of lichens as a possible definition of meaning. It is poetry—and since it is poetry, it contains within it a poetics. Following Stengers, we might say that the 'pollination learning' pathway is a path that holds a bag of science in one hand and a bag of poetry in the other, never being allowed to give more weight to one or the other. Through this act, we could move from economics into a much more composed form of understanding value. This chapter will only hint at such understanding. It is an act of hijacking, hacking, and kidnapping economics and insisting that it goes the same way as ecology: out of the house and into the world. The 'world' is not a whole (Stengers, 2020), or if it is, it is an implosive one (Morton & Boyer, 2021); it is not a landscape (Jullien, 2014); it is a myriad of houses and worlds that appear and disappear, it is highly unstable. Sheldrake notes that "studies of symbiotic interactions must reach across disciplinary boundaries" (Sheldrake, 2020, p. 239). First of all, they need to reach across specist boundaries. There is a need to understand not only the pragmatic or even economic value of beings, but their intrinsic being in a symbiotic or sympoïetic (Margulis in Haraway, 2016) web.

Concluding Remarks

There are strings that we have not tied at this point. The bigger picture, urban sustainability, leadership, and strategy are waiting for us to take more steps. We do not know yet how to overcome what emerged as epistemic or even ontological barriers. We invited residents to a first presentation of the idea of mapping lichens as part of a bigger learning journey. We were met with curiosity and a big gap. What is a lichen? That is where we start, once again (Fig. 17.7).

What we know at this point is already deeply intertwined with the movement from invisibility to visibility or from less than affect to mutually growing learning pathways. To write the next chapter of our journey, we will have to wait and see how the others respond. The lichens. The soil. The bees. The chemicals underground and the oddity of humans trying not to land. We consider digging a hole in the crust (again?) with the board at ØK. Baring the soil under the tarmac, an almost situationist idea—only this time, we would bring in the soil people and the mycelia. We have experienced that we need to find the spots where the gentle learning might take place. We might combine something like critical zone

Fig. 17.7 Nearby development zone, possible scene of common experiment for the Pollination Academy from 2022

observatories (Latour & Weibel, 2020) with something like 'gentle activism' (Positive News, 2021), with small interventions making the unnoticed noticeable.

Certainly, there is a gap between the micro level that we work and the level of "strategic leadership" of cities. We have conversations with directors, mayors, heads of departments in municipalities and other landowners. They hesitate to take in the perspective of the critical zone, of the mycelic layer, of the potential of *growing* a neighbourhood rather than *building* it (the instrumentalist distinction between the grown and the built is expressed clearly in the work of one of the leading landscape architects in Southern Scandinavia, Stig L Andersson) (e.g., Andersson, 2014, 2021).

In an otherwise deeply critical piece on the modern inability to grasp the collapse of its own system, Glen Kuecker claimed in 2011 that "the release phase of the modern world-system is the space, place, and time for transmodernity. It is the time of the global majority, when the meek will inherit the earth. In the release phase, the modern epistemic will be a hindrance and its bag of tricks will not be able to counter the transmodern insurgency" (Kuecker, 2014, p. 166). Ten years later, this analysis stands as naïve. It forgets to include all the repercussions and feedback loops instored when the "system" reacts to its own crises. Pandemics were predicted in the 2000s. A global economic collapse threatened in 2008–2009. Huge research projects have shown the impact of the climate crisis, the biodiversity crisis, the energy crisis. This does not pass unnoticed. The growth of cities in the past ten years is a reaction; it is an investment in the force of the vortex.

Hence, our experiences must be read in that context. When we are reduced to whispers of creative sensitivity, inspired by strange concepts like the 'zoöp' (The New Instituut, 2021), and forced to move in the cracks between the vortex centres, maybe that is the right place to start. However, our experience is also that the people we seek to connect to are under pressure. They express curiosity but also a need to be assured that this makes sense. So to paraphrase the title of a recent conference on decentralised regenerative platforms (Platform Design Toolkit, 2021), we are in a phase of sense-making. We are in the need to form ontologies with new centres, if any. A necessary step in that process is the creation of local, but also of translocal learning platforms, communities, ecologies, where an important step is to make ecologies sensible and intelligible to their human inhabitants again (cf. Johar, 2021).

We have tried to initiate a learning process. Next, we might learn to balance the continuous uncovering of the critical zone with the caring for bodies fumbling in hesitation (including our own). We stand on the shoulders of 150 years of trying to destroy the critical zone. Now, we have to learn how to embrace its response.

References

Amhøj, C. B. (2019). Sustainability as an ecology of learning, thinking and acting: Using the World Health Organization's six P's as an action-research intervention to create public value with multiple bottom lines. *Local Economy, 34*(5), 439–455. https://doi.org/10.1177/0269094219847124

Andersson, S. L. (2014). Catalogue for the Danish pavilion at the 14th International Architecture Exhibition 2014. Danish Architects Association.

Barad, K. (2007). *Meeting the universe halfway. Quantum physics and the entanglement of matter and meaning*. Duke University Press.

Barrios-Negron, L. (2020). Breathtaking greenhouse parastructures: A supplement to the arcades project from a caribbean perspective [and a call for a careful practice of epistemológica. KTH, Skolan för arkitektur och samhällsbyggnad (ABE), Arkitektur. Konstfack.(KTD) ORCID-id: 0000-0002-3422-8463

Beinhofer, E. (2007). *The origin of wealth: Evolution, complexity, and the radical remaking of economics*. Random House Business.

Benjamin, W. (1989). *Theses on the philosophy of history*. Routledge.

Bennett, J. (2009). *Vibrant matter. Toward a political ecology of things*. Duke University Press.

Brenner, N., & Schmid, C. (2013). The 'Urban Age' in question. *International Journal of Urban and Regional Research*. https://doi.org/10.1111/1468-2427.12115

Burø, T. (2020). *Land of light: Assembling the ecology of culture in Odsherred 2000–2018*. Copenhagen Business School.

Burø, T., & Koefoed, O. (2021a). Organising spirit. *Culture and Organization, 27*(2), 171–190. https://doi.org/10.1080/14759551.2020.1821201

Burø, T., & Koefoed, O. (2021b). The name of the world is chaos. Learning in the Anthropocene. In Bjerg Petersen, von Brömssen, Garsdal, Jacobsen, Paulsen, & Koefoed (Eds.), *Rethinking education in the light of global challenges*. Routledge.

C40. (2016). Copenhagen. https://www.c40.org/case_studies/c40-good-practice-guides-copenhagen-public-private-finance-scheme. Visited May 14, 2021.

Conti M. E., & Cecchetti, G. (2001). Biological monitoring: Lichens as bioindicators of air pollution assessment—a review. *Environmental Pollution, 114*(3), 471–492. https://doi.org/10.1016/s0269-7491(00)00224-4. PMID: 11584645.

Dasgupta, P. (2021). *The economics of biodiversity: The Dasgupta review*. London: HM Treasury.

Dickinson, C. (2019). Bruno Latour and the myth of autonomous academic discipline: Rethinking education in the light of various modes of existence. *Transformation in Higher Education, 4*(0), a75. https://doi.org/10.4102/the.v4i0.75

Duxbury, N., Garrett-Petts, W. F., & MacLennan, D. (Eds.). (2015). *Cultural mapping as cultural inquiry*. Routledge.

Ekhtiari, H., Nasseri, P., Yavari, F., Mokri, A., & Monterosso, J. (2016). Neuroscience of drug craving for addiction medicine: From circuits to therapies. In Ekhtiari, Hamed, & M. Paulus (Eds.), *Progress in brain research* (Vol. 223, pp. 115–141). Elsevier. ISSN 0079-6123, ISBN 9780444635457. https://doi.org/10.1016/bs.pbr.2015.10.002

Falanga, R. (2020). *Democracy and human rights citizen participation during the COVID-19 pandemic. Insights from local practices in European cities*. Friedrich Ebert Stiftung. https://library.fes.de/pdf-files/bueros/lissabon/17148.pdf

Freund, B., & Padayachee, V. (Eds.). (2002). *(D)urban vortex: South African city in transition*. University of Natal Press.

Fullerton, J. (2015). Regenerative capitalism. How universal principles and patterns will shape our new economy. Capital Institute. http://capitalinstitute.org/wp-content/uploads/2015/04/2015-Regenerative-Capitalism-4-20-15-final.pdf

Gaillardet, J. (2020). The critical zone, a buffer zone, the human habitat. In Latour & Weibel (Eds.), *Critical zones—The science and politics of landing on earth*. ZKM Books.

Hall, S. M., & Savage, M. (2015). Animating the urban vortex: New sociological urgencies. *International Journal of Urban and Regional Research, 40*(1), 82–95.

Haraway, D. (2016). *Staying with the trouble. Making Kin in the Chthulucene*. Duke University Press.

Haraway, D. (2020). Carrier bags for critical zones. In Latour & Weibel (Eds.), *Critical zones—The science and politics of landing on earth*. ZKM Books.

Harders, A. K. (2014). *Stædige infrastrukturer og genstridige praksisser: Et praksisteoretisk studie af byudviklingsprojekter mellem vision og realitet*. PhD-afhandling, Aalborg Universitet.

Holt, E. A., & Miller, S. W. (2010). Bioindicators: Using organisms to measure environmental impacts. Nature Education Knowledge 3(10): 8. https://www.nature.com/scitable/knowledge/library/bioindicators-using-organisms-to-measure-environmental-impacts-16821310/#:~:text=Lichens%20and%20bryophytes%20serve%20as,of%20contaminants%20from%20the%20air. Visited May 14, 2021.

Jullien, F. (2014). *Vivre du paysage ou l'impensé de la raison*. Nrf Gallimard.

Katz, B., & Nowak, J. (2018). How a dying city transformed into one of the wealthiest in the world. *LinkedIn Newsletter*. https://www.linkedin.com/pulse/how-dying-city-transformed-one-wealthiest-world-bruce-katz/

Kindon, S., Pan, R., & Kesby, M. (2007). Participatory action research: Origins, approaches and methods. In S. Kindon, R. Pain, & M. Kesby (Eds.), *Participatory action research approaches and methods: Connecting people, participation and place* (pp. 9–18). Routledge.

Koefoed, O. (2013). Tingenes Teater. Om Frivillig Suspension of Vilje og Magt. In K. Gørtz & M. Mejlhede Frivlliig Ledelse. Erfaringsbaserede refleksioner, psykologiske forhold og filosofiske synsvinkler. DJØFs forlag.

Koefoed, O. (2019). Urban nature as transformed practice—A case of multi-dimensional processing to increase public value in Copenhagen. *Journal of Local Economy, 34*(6), 525–544. Article first published online: November 26, 2019; Issue published: September 1, 2019. SAGE.

Koefoed, O., & Burø, T. (2022). The name of the world is chaos: Learning in the Anthropocene. In K. B. Petersen, K. V. Brömssen, G. H. Jacobsen, J. Garsdal, M. Paulsen, & O. Koefoed (Eds.), Rethinking education in light of global challenges: Scandinavian perspectives on culture, society, and the anthropocene (pp. 193–203). Routledge. Routledge Research in Education, Society and the Anthropocene. https://doi.org/10.4324/9781003217213-15

Kuecker, G. (2014, May). From the alienation of neoliberal globalization to trans-modern ways of being: Epistemic change and the collapse of the modern world-system. *Journal of Globalization Studies, 5*(1), 154–170.

Latour, B. (1999). Politiques de la nature. Éditions La Découverte.

Latour, B. (2017). *Facing Gaia: Eight lectures on the new climatic regime*. Polity Press.

Latour, B. (2018). *Down to earth: Politics in the new climatic regime*. Polity Press.

Latour, & Weibel (Eds.). (2020). *Critical zones—The science and politics of landing on earth*. ZKM Books.

Le Guin, U. K. (2020/1986). *The carrier bag theory of fiction* (D. Haraway, Intro.). Terra Ignota.

Lovins, L. H., Wallis, S., Wijkman, A., & Fullerton, J. (2020). *A finer future. Creating an economy in service to life*. New Society Publishers.

Malone, K., Dymont, J., Hill, A., & Cutter-Mackenzie, A. (2016, March). Reconsidering children's encounters with nature and place using posthumanism. *Australian Journal of Environmental Education, 32*(1), 42–56. https://doi.org/10.1017/aee.2015.48

Merrifield, A. (2014). *The new urban question*. Pluto Press.

Mogel, L., & Bhagat, A. (Eds.). (2007). *An Atlas of radical cartography*. Journal of Aesthetics and Protest Press.

Morton, T. (2009). *Ecology without nature. Rethinking environmental aesthetics*. Harvard University Press.

Morton, T., & Boyer, D. (2021). *Hyposubjects. On becoming human*. Open University Press.

Paulsen, M. (2021): From late Holocene to early Anthropocene educational thinking (humanism revisited). Bjerg Petersen, von Brömssen, Garsdal, Jacobsen, Paulsen, & Koefoed (Eds.), *Rethinking education in the light of global challenges*. Routledge.

Puskas, N., Abunnasr, Y., & Naalbandian, S. (2021). Assessing deeper levels of participation in nature-based solutions in urban landscapes—A literature review of real-world cases. *Landscape and Urban Planning, 210*, 104065. https://doi.org/10.1016/j.landurbplan.2021.104065

Raworth, K. (2018). *Doughnut economics—Seven ways to think like a 21st-century economist*. Cornerstone.

Sassen, S. (2014). *Expulsions: Brutality and complexity in the global economy*. Harvard University Press.

Sheldrake, M. (2020). *Entangled life. How fungi make our worlds, change our minds, and shape our futures.* The Bodley Head.
Stamets, P. (2011). *Mushroom running. How Mushrooms can help save the world.* Ten Speed Press.
Stapper, N. J., & John, V. (2015, January). Monitoring climate change with lichens as bioindicators. *Pollution Atmosphérique.* https://doi.org/10.4267/pollution-atmospherique.4936
Stengers, I. (2020). The earth won't let itself be watched. In *Latour & Weibel: Critical zones—The science and politics of landing on earth* (pp. 122–129). ZKM Books.
Stiegler, B. (2013). *What makes life worth living: On pharmacology.* Polity Press.
Storm, L., & Hutchins, G. (2019). *Regenerative leadership: The DNA of life-affirming 21st century organizations.* Wordzworth Publishing.
Tsing, A. L. (2015). *The Mushroom at the end of the world.* Princeton University Press.
Tsing, A. L., Swanson, H. A., Gan, E., & Bubandt, N. (Eds.). (2017). *Arts of living on a damaged planet. Ghosts and monsters of the Anthropocene.* Minnesota University Press.
UN. (2018). https://www.un.org/development/desa/en/news/population/2018-revision-of-world-urbanization-prospects.html
WEF (World Economic Forum). (2021). Four infographics to help you understand the state of the planet. https://www.weforum.org/agenda/2021/05/infographics-environment-impact-planet/. Visited May 14, 2021.
WHO (Copenhagen Consensus of). (2018): *Healthier and happier cities for all. A transformative approach for safe, inclusive, sustainable and resilient cities for all.* World Health Organization.
Winther, R. (2017). *When maps become the world.* Chicago University Press.

Websites and Blogs

Andersson, S. L. (2021). Bynatur er ikke natur. https://politikenbyrum.dk/Debat/art8247302/Bynatur-er-ikke-natur.-Bynatur-er-noget-vi-mennesker-designer-for-at-l%C3%B8se-de-urbane-problemer-vi-selv-har-skabtutm_campaign=byrum&utm_content=22-06-2021&utm_medium=newsletter&utm_source=byrum
Barefoot College. (2021). https://www.barefootcollege.org/
Ipsos. (2019). https://www.ipsos.com/da-dk/ny-undersogelse-danskerne-og-klimaforandringerne
Johar, I. (2021). https://provocations.darkmatterlabs.org/
KL. (2019). https://www.kl.dk/nyheder/momentum/2019/nr-19/danskere-og-svenskere-er-de-mest-klimabekymrede/

Maphub. (2021). https://maphub.net/
Miljøpunkt. (2021). http://miljopunkt-amager.dk/
Positive News. (2021). https://www.positive.news/society/the-art-of-gentle-protest/
Selenadec. (2019). Isolation. In *Young People Anthropocene*. https://youngpeopleanthropocene.org/2020/06/30/isolation/
Slow Factory. (2021). https://slowfactory.foundation/
The New Instituut. (2021). https://hetnieuweinstituut.nl/en
Winge Laura. (2021). https://www.laurawinge.dk/

CHAPTER 18

Outro

jan jagodzinski, Michael Paulsen, and Shé M. Hawke

We end this book by presenting three different perspectives on how we see the future of the Earth and the future of Anthropocene pedagogies, what is likely to happen, what is needed and what can we hope for?

j. jagodzinski
University of Alberta, Edmonton, AB, Canada
e-mail: jj3@ualberta.ca; jan.jagodzinski@ualberta.ca

M. Paulsen (✉)
Department of the Studies of Culture, University of Southern Denmark, Odense, Denmark
e-mail: mpaulsen@sdu.dk

S. M. Hawke
Mediterranean Institute for Environmental Studies, Science and Research Centre of Koper, Koper, Slovenia
e-mail: she.m.hawke@zrs-kp.si

M. Paulsen et al. (eds.), *Pedagogy in the Anthropocene*, Palgrave Studies in Educational Futures,
https://doi.org/10.1007/978-3-030-90980-2_18

Outro 1: Somebody Save Us!?

jan jagodzinski

Pedagogy in such a time of the Anthropocene seems impotent, and it is perhaps this very impotency that is also its strength. If we look around us today an 'apocalypse' has already happened as there is a never-ending parade of natural and human disasters which are linked in ways that are unforeseeable given that the 'forces' at work are themselves invisible yet measurable. The populace, however, has lost 'faith' in numbers, as it is too easy to manipulate them; the cynics know the game well. Fascism, corruption and a wave of new religiosity have spread to believe that some higher power will surely save 'them'. As the string of climate change conferences have shown, the bottom line is always an economic one by countries whose standard of living is at the expense of planetary fairness and justice. To believe that some form of planetary consciousness will eventually emerge is a fool's dream. We have seen how 'hospitable' countries have been to refugees, diaspora, immigrants and asylum seekers. The COVID pandemic has shown globally how 'cheap' life is, and how money is to be made on people's suffering: an old story.

Pedagogy is hand-tied, especially when it is in the hands of the state, preventing evolution to be taught in some countries, or the whitewashing of history to stop the teaching of critical race theory and the 1619 project in the US, or initiatives to ban the teaching of sexual diversity in the name of some over-arching religious belief. This turn to conservative values is fuelled by the fear that the comfort of life as we know it in the past is fading. The 'barbarians' are coming. The COVID pandemic has brought it to 'home'. Against this bleak backdrop, the narratives available are few: the redemptive narrative is the most common. It comes in two varieties. The first is one that we will 'save' the earth—for us. This is pervaded by new vitalist animisms wherein we become attuned and sensitive to the more-than-human other, the 'things of the Earth' are personified and given protective status' sustainability is forwarded, a 'good' Anthropocene is imagined as we wake-up to the pollution that we are causing. The colonialist legacy will be overcome as indigenous peoples rise and 'show us the way' for pluriversal worlds. This environmental ethics will surely save us. What is needed is a good stewardly 'clean-up' to get the oikos-house in order. If we 'all' just pull together, we can do it.

The second narrative, closely wedded to the first only in sense of its repositioning in its priority: We can do it! Save ourselves and the earth through the new green technologies that we will invent to stave off the worse disasters. After all, we have developed vaccines to stop COVID: be damn that not everyone is willing to be inoculated—a portion of the population will always embrace its own death drive, its own belief in the necessity of war and so on. But, a peaceful species we are not. A death drive runs through us. There are just not enough 'guns' in the world, not enough nuclear bombs, not enough territory to go around, not enough—not enough. The embracing of this death drive may be, uncomfortably, more as an escape towards death on a dying planet. Starvation, the loss of all one's belongings via natural and human disasters, living (forever) in refugee camps, watching your children starve, having your homes constantly bombed, enduring endless strife and struggle. What future? Should 'death' be any worse? Such a thought!

The last narrative is not so comforting if the earth is not ours to save? The earth is undergoing its own phase change with us giving it a 'little' boost with our carbons and plastics. What to do now? Can pedagogy 'save us' here? The rich think it's best to terraform Mars and re-settle. Why not? There are lots of sci-fi scenarios that play with this projected narrative of one sort or other. There is always the transhuman to turn to, to the wonders of science—to the moment in the near future when fusion energy is made tenable: we will then be saved. With the continuous integration of nano-bio-info-cogno (NBIC) sciences, what can we not accomplish? Or, in the wizard-like projections of Michio Kaku, we may well indeed reach the status of planetary 1 consciousness where all forms of energy have been harnessed—with the proviso that we don't blow ourselves up first.

In the meantime, those of us as educators who toil with this problematic might take some heart in Foucault's reflection when he said that he was not a pessimist, but just considered failure to be a real possibility. Hope for him was not some 'beyond' but was immanent to what is possible in the here and now; there is nothing 'beyond' that must be conquered, but to work with the cards that have been given to us. If failure is not involved, then it is not even worth trying.

Outro 2: Live Life!

Michael Paulsen

The capacity of life, as well as non-life, to produce life is unknown. We live, but do not know why we exist, why there is life, how it arises. We know perhaps how we can survive; we know poorly enough what a good life is. The science that we have created, and which can uncover the anthropogenic effects (that are reported and can be read about every day when we open a newspaper, here in 2021), only knows about actualities and possibilities, not about the virtual capacities—the cosmic forces from which everything springs, and which are also transformed by what is produced, and through which life and transformation arise. The exact science can only reveal the current effects and extrapolate opportunities and risks, but not tell what constitutes a good life. Thus, exact science has specific narrative limitations when it comes to all life-essentials. It does not take much more than a child with a few curious questions to discover this. Where does life come from? What is life? What can life be? What is a good life? How to create a good life?

The complex problem we face today is, in part, because of our own way of life. It comes to the fore in the form of global warming, global environmental disaster, biodiversity crisis and the sixth mass extinction; it is in some ways virtual, and thus difficult to recognize as life itself. The problem we thus face, and which is identified with the notion that we now live in an Anthropocene age, is not identical with the many effects (including future forecast effects) that natural science can uncover; these are merely results of the problem.

What is the problem then? The problem is the whole way of life, which configured itself in the Holocene epoch, starting approx. 11,500 years ago, especially in the Late Holocene. Because of the way humans evolved in relation to everything else on earth during this period, it is fair to say that we today have entered the Anthropocene. A key factor has been that humans, increasingly in the unique stable climate of the Holocene epoch, developed the solution of becoming sedentary settlers, creating agriculture, cities, property, stocks and written language and organizing large hierarchical communities. The abundance and new inequality this brought about, as an answer to the problem of how to live well in the environment that constituted the Holocene period, created the idea that the horizontal expansion possibilities seemed endless.

Both the capacity of life and the complex problem we are confronted with today are thus virtual and consist of cosmic forces that make it difficult to distinguish solutions from problems. Any genuine solution means a transformation of the problem, and thus the creation of a new problem. Problem and solutions are two sides of the same chaosmos. Thus, 'man' himself is both a problem and part of a possible solution to the reduction of the capacity of life in the Anthropocene. As human beings, we have a certain influence on ourselves, and should consider what we can do with ourselves and our way of life. Here, pedagogy and the idea of education is one significant force field, which can be assumed—and required—to co-generate self-transformation that adequately responds to the basic problem of the Anthropocene age. However, current pedagogy is conceived within the circle of Late Holocene logics, like most of our other societal institutions What is therefore required is a transformation of itself as well, to be able to respond adequately to the problem that has been produced through Holocene logics. Is this possible? And if so: Into what? And how?

In the film *Snowpiercer* (2013) by the acclaimed South Korean director Bong Joo Ho, humanity finds itself in 2031 in a train circulating around the globe. Outside, the earth is covered in a barren ice age, which has arisen after a failed attempt to stop global warming using climate technology. Inside the train, people live in classes. Thus, the elite are in the front compartments, while a poor underclass is kept as resources in the rear carriages. The entire train's class division, mechanics and way of life are permeated by resource optimization, which constitutes the highest context. In the film, we follow a revolt that is apparently being waged from the poorest rear wagons, against and with the direction of occupying the driver's wagon. However, the film unmasks that this revolt in no way overthrows the polar basic structure, but instead supports the resource optimization logic. Apparently, there are no life-possibilities 'outside' this structure—also because it seems to be impossible to live outside the train. Thus, all possibilities lie in the polar axis, between resources and resource users/managers. Without revealing too much of the film's plot, however, it is necessary here to reveal that it is possible to get off the train and move towards 'the great unknown', a third and uncertain possibility of life.

I think the film *Snowpiercer* captures the situation we—and pedagogy—face today: We can continue along a Holocene vector, where all self-transformation is tied to the task of resource optimization, where we maintain and try to minimize the problem, or resign to it. For those who give up, suffer from depression or are drained from life, pedagogy is given a task of psychological, physical and possibly religious character: to save the fallen. For those who have faith in the resource machinery, pedagogy is given the task of further developing the technical (hierarchical) school that can provide the necessary human resources to keep the machine running, ensure that there are enough resources, and ensure that the self-initiated 'natural disasters' are kept at distance, postponed or technically countered.

All manner of variants of alliances between the technical school and the school for the apostate and depressed can be merged, as the film *Snowpiercer* shows! However, it is surprisingly easy, but without reliable results, to deviate from the Holocene vector, thereby creating a more genuine Anthropocene response to the Anthropocene problem complex. This response, which is in fact an infinite series of brave responses and thus deviating vectors, consists in freeing the self-transformation (from the binding to the requirement of resource optimization as the highest context), and resorting to a richer life of alternative life forms, which are excluded by the late Holocene logics.

Here I am thinking in particular of the possibilities of creating interaction, togetherness, community, dialogue, and life with more-than-human beings. This constitutes an inexhaustible source for creating a different and far richer life than that which is centred around optimization of available resources for purely human-human relationships (where everything else is perceived as a scene, background or resource—including that large parts of the human and many people (as shown in *Snowpiercer*) also are perceived as resources). Such a deviation could, in my view, create zoëlogical pedagogical policies. This could be understood as a swarm of experiments that create alternative ways of life that could possibly respond better, or at least differently to the basic problems of the Anthropocene era, than responses that continue the Late Holocene logic's reduction of life to bios—and politics/pedagogy to bio/resource politics/pedagogy.

The great advantage of the zoëlogical alternative (as many of the chapters in this book have narrated, notably Chap. 11, where the concept of the zoëlogical is explained) is that it catalyses a gigantic surplus of passion for life: no matter how inefficient and poor the results are that must come out of this movement, it is my experience that by opening up to attempts at new dialogue and interaction with more-than-human beings (plants, animals, waterscapes) in all possible forms—also as pure thoughts in form of, for example, art, stories, literature and installations—a more enriching and vital life is reached. This can include fantastic and life-affirming meetings with other people, who are also in the process of passionately discovering and (re) creating these new life opportunities. In contrast to the Holocene axis, between depression and technical domination, in the diverse variants of the zoëlogical alternative, an unknown, new, enriching, exciting, liveable and extremely life-affirming myriad of possibilities arise, which may co-restore and transform the vital cosmic life energies anew, forced by the Anthropocene problematic and the transition period by such a new openness, which we are in right now. Conversely, it must be considered that this salutation of life can be closed by technical ingenuity (if it succeeds in gaining control of the situation) or a total ecological collapse if we are too late in our response. However, never underestimate life, life itself seems to tell us. Live life!

Outro 3: Nothing Changes If Nothing Changes

Shé M. Hawke

Our collection has highlighted the re-thinking, re-doing and re-activating of pedagogical practice in the Anthropocene Epoch, a term which in itself is variously understood, and as we have shown contested. Our intention has not been to invert any perceived or existing binaries, but instead, to open up the multi-focal, multi-species and multi-agency spectrum to a more resilient sense of inclusivity and diversity in which more-than-human species are 'a part of'. Yet, as many chapters have demonstrated, this noble approach and intention is far from straight-forward in its design, delivery and evaluation. The confluence in the diversity of approach and intention, however, provides a rich braiding of life possibilities.

We have raised questions about education and pedagogy, such as who is educated by whom, for whom and for what purpose. Embedded in this discussion is an understanding about differences in place, and place-based education from rural and remote communities to metropolitan assemblages of learning spaces and their entanglements. In this, a politics of complexity and a broad spectrum of ideology materialize; this collection hopes to be a springboard for further intervention, collision and pedagogical co-creation.

Some past examples of pedagogical theory and practice regarding sustainability are obsolete because of their subscription to binary methods as our chapters have deftly noted and critiqued. But we have not sought to throw everything out with the tide, rather to shift with the tide, and see what it washes in on different shores. This can enable us to re-craft an architecture of being (as much as doing) that includes re-thinking and re-winding the past to elicit fragments that are of incalculable value to our present living situation, and, towards a more coherent and in-tact future on the planet known as earth. This means 're-doing' ourselves conscientiously, and re-wilding not only damaged environments and teaching practices, but importantly our own consciousness. We will continue to be confronted by what we discover. Confrontation is part of any revolution, and education and pedagogy for a sustainable planet and inter-species coevolvement are no exception, because every body of life is a part of both climate change and planetary existence.

Yet the reparative action rests with the human species, and for us in this collection through educational and pedagogical renovation, that has in many parts of the world been dominated by an 'educational violence' against nature and indeed each other, as Kerrie Willis and Nick Kleese make clear in Chap. 15 in this book. The increasing divide in access and equity plays out globally, despite sustained and g/local efforts at inclusivity, as well as the covert conscription into bad environmental teaching practices and/or complacency.

Be-friending each other in unified response (without compromising the specificity of our own contexts) is as important as be-friending the earth. Paulsen and Nørreklit [in Chap. 11] radically explore through their treatise on love and dialogical relationships with our 'irreplaceable more-than-human co-fellows', that Earth-forgetfulness of the late Holocene can be over-turned. But it is mindfulness that is necessary to enact, and indeed, embody the earth through environmental respect, awareness and sustainability that are absolutely tied to culture, politics, class and gender. This collection clearly and intentionally articulates the need to remain 'awake', if the age of a new earth—predicated on love—is to become possible, sustainable and ultimately knowable: temporally and spatially, sensorially and elementally, as well as personally and politically.

My hope for a new earth remains buoyant through the recent flourishing and respect for First Nations pedagogy that has embodied and shared environmental knowledge despite the effects of colonization. Other transformative stories from dominant culture, such as that of Aldo Leopold (*A Sand County Almanac*, [1949] 2020, Oxford), also inspire hope. His hunting passion dissolved the moment he saw the green fire die in the eyes of the mother wolf he had just killed, as narrated in the section 'Thinking Like a Mountain':

> We reached the old wolf in time to watch a fierce green fire dying in her eyes. I realized then, and have known ever since, that there was something new to me in those eyes—something known only to her and to the mountain. I was young then, and full of trigger-itch; I thought that because fewer wolves meant more deer, that no wolves would mean hunters' paradise. But after seeing the green fire die, I sensed that neither the wolf nor the mountain agreed with such a view.

In his chapter called 'The Land Ethic' he describes his journey to love of the land, and appeals to our moral responsibility to care unequivocally for all life around us and to sustain balance: 'to enlarge the boundaries of the community' (192), in other words, the inter-species community. He wrote over sixty years ago, and still we are in infancy it seems with the realization of the consequences of our actions. In the twenty-first century our moral obligation is also to those who look to teachers, elders and communities for inspiration, motivation and intelligent and purposeful direction. In her book, *Despair and Personal Empowerment in the Nuclear Age* (1983, New Society Pub.), Joanna Macey explains that it is normal and healthy to experience feelings of pain for our world. But she adds that this pain can become morbid in individuals if denied, or if the person is left unsupported (22). Education, through its access to young people, has the opportunity to support pain and grief over the Anthropocene challenges that we all face, as many of our chapters have indicated via new and different methods and pedagogical reformation—to keep hope alive amid the despair.

So, the final words in this book address the immediacy of our shared catastrophe born of our over sights, minimisations and exclusionary practices. The imperative now is to act consciously, to teach innovatively and ethically and think like a mountain, a fish, a stream, a wolf, a tree, a fire, a snowfall, a bird or a child, and enter into a slower more intimate environmental pedagogy, in which there is space for cherishing all life.

Index[1]

[1] Note: Page numbers followed by 'n' refer to notes.

M. Paulsen et al. (eds.), *Pedagogy in the Anthropocene*, Palgrave Studies in Educational Futures,
https://doi.org/10.1007/978-3-030-90980-2

D

GPSR Compliance
The European Union's (EU) General Product Safety Regulation (GPSR) is a set of rules that requires consumer products to be safe and our obligations to ensure this.

If you have any concerns about our products, you can contact us on

ProductSafety@springernature.com

In case Publisher is established outside the EU, the EU authorized representative is:

Springer Nature Customer Service Center GmbH
Europaplatz 3
69115 Heidelberg, Germany

www.ingramcontent.com/pod-product-compliance
Ingram Content Group UK Ltd.
Pitfield, Milton Keynes, MK11 3LW, UK
UKHW020833290926
15470UKWH00004B/1330
* 9 7 8 3 0 3 0 9 0 9 7 9 6 *